Malignancies of the Groin

Keith A. Delman • Viraj A. Master
Editors

Malignancies of the Groin

Surgical and Anatomic Considerations

Springer

Editors
Keith A. Delman
Division of Surgical Oncology
Department of Surgery
Emory University School of Medicine
Atlanta, GA
USA

Viraj A. Master
Department of Urology
Emory University School of Medicine
Atlanta, GA
USA

ISBN 978-3-319-60857-0 ISBN 978-3-319-60858-7 (eBook)
https://doi.org/10.1007/978-3-319-60858-7

Library of Congress Control Number: 2017955813

Printed on acid-free paper

This Springer imprint is published by Springer Nature
The registered company is Springer International Publishing AG
The registered company address is: Gewerbestrasse 11, 6330 Cham, Switzerland

Contents

Contributors

Alexander C.J. van Akkooi, M.D., Ph.D. Department of Surgical Oncology, Netherlands Cancer Institute - Antoni van Leeuwenhoek, Amsterdam, The Netherlands

Angela Cheng, M.D. Division of Plastic Surgery, Department of Surgery, Winship Cancer Institute of Emory University, Atlanta, GA, USA

Avinash Chenam, M.D. Division of Urology and Urologic Oncology, Department of Surgery, City of Hope National Medical Center, Duarte, CA, USA

Chiara Colombo, M.D. Sarcoma Service, Department of Surgery, Fondazione IRCCS Istituto Nazionale dei Tumori, Milan, Italy

Juanita Crook, M.D., F.R.C.P.C. Division of Radiotherapy and Developmental Radiotherapeutics, Department of Radiation Oncology, University of British Columbia, BC Cancer Agency, Kelowna, British Columbia, Canada

Keith A. Delman, M.D., F.A.C.S. Division of Surgical Oncology, Department of Surgery, Emory University School of Medicine, Atlanta, GA, USA

Karan A. Desai, M.D. Division of Plastic Surgery, Department of Surgery, Emory University Hospital Midtown, Emory University School of Medicine, Atlanta, GA, USA

Clara Farley, M.D. Department of Surgery, Emory University, Atlanta, GA, USA

Marco Fiore, M.D. Sarcoma Service, Department of Surgery, Fondazione IRCCS Istituto Nazionale dei Tumori, Milan, Italy

Andre Granger, M.D. Department of Anatomical Sciences, St. George's University School of Medicine, Grenada, W.I.

Department of Neurology, NYU Langone Hospital - Brooklyn, Brooklyn, NY, USA

Alessandro Gronchi, M.D. Sarcoma Service, Department of Surgery, Fondazione IRCCS Istituto Nazionale dei Tumori, Milan, Italy

Christine Ibilibor, M.D. Department of Urology, Texas Tech University Health Sciences Center, Lubbock, TX, USA

S. Mohammad A. Jafri, M.D. Department of Urology, Beaumont Health System, Oakland University William Beaumont School of Medicine, Royal Oak, MI, USA

Stephanie Kirkpatrick, P.T., D.P.T., C.L.T. Department of Radiation Oncology, Winship Cancer Institute of Emory University, Lymphedema Clinic, Atlanta, GA, USA

Theofanis Kollias, M.D. Department of Anatomical Sciences, St. George's University School of Medicine, Grenada, W.I.

Martin Korzeniowski, M.D., F.R.C.P.C. Division of Radiotherapy and Developmental Radiotherapeutics, Department of Radiation Oncology, University of British Columbia, BC Cancer Agency, Kelowna, British Columbia, Canada

Andrew J. Lambour, M.D. Department of Surgery, Dartmouth-Hitchcock Medical Center, Geisel School of Medicine at Dartmouth, Lebanon, NH, USA

Albert Liao, M.D. Division of Plastic Surgery, Department of Surgery, Emory University Hospital Midtown, Emory University School of Medicine, Atlanta, GA, USA

Albert Losken, M.D., F.A.C.S. Division of Plastic Surgery, Department of Surgery, Emory University Hospital Midtown, Emory University School of Medicine, Atlanta, GA, USA

Marios Loukas, M.D., Ph.D. Department of Anatomical Sciences, St. George's University School of Medicine, Grenada, W.I.

Michael Lowe, M.D. Division of Surgical Oncology, Department of Surgery, Emory University School of Medicine, Atlanta, GA, USA

Carrie Luu, M.D. Department of Cutaneous Oncology, Moffitt Cancer Center, Tampa, Florida, USA

Viraj A. Master, M.D., Ph.D., F.A.C.S. Department of Urology, Emory University School of Medicine, Atlanta, GA, USA

Marcio Covas Moschovas, M.D. Section of Urologic Oncology, Department of Urology, ABC Medical School, São Paulo, Brazil

Antonio Augusto Ornellas, M.D., Ph.D. Section of Urologic Oncology, National Institute of Cancer, Rio de Janeiro, Brazil

Matthew C. Perez, M.D. Department of Surgery, Emory University School of Medicine, Atlanta, GA, USA

Stefano Radaelli, M.D. Sarcoma Service, Department of Surgery, Fondazione IRCCS Istituto Nazionale dei Tumori, Milan, Italy

Seyed Amirhossein Razavi, M.D. Division of Plastic Surgery, Department of Surgery, Emory University Hospital Midtown, Emory University School of Medicine, Atlanta, GA, USA

Amod A. Sarnaik, M.D., F.A.C.S. Department of Cutaneous Oncology, Moffitt Cancer Center, Tampa, Florida, USA

Pranav Sharma, M.D. Department of Urology, Texas Tech University Health Sciences Center, Lubbock, TX, USA

Vasileios D. Sioulas, M.D., Ph.D. Gynecology Service, Department of Surgery, Memorial Sloan Kettering Cancer Center, New York, NY, USA

Antonio Sommariva, M.D. Surgical Oncology Unit, Veneto Institute of Oncology, IOV-IRCCS, Padova, Italy

Yukio Sonoda, M.D. Gynecology Service, Department of Surgery, Memorial Sloan Kettering Cancer Center, New York, NY, USA

Philippe E. Spiess, M.D., M.S., F.R.C.S., F.A.C.S. Department of Genitourinary Oncology, Moffitt Cancer Center, Tampa, FL, USA

Andrew J. Spillane, M.D., F.R.A.C.S. Melanoma Institute Australia, The University of Sydney, Sydney, NSW, Australia

John F. Thompson, M.D., F.R.A.C.S., F.A.C.S. Melanoma Institute Australia, The University of Sydney, Sydney, NSW, Australia

Marcos Tobias-Machado, M.D., Ph.D. Section of Urologic Oncology, Department of Urology, ABC Medical School, São Paulo, Brazil

Sandra L. Wong, M.D., M.S. Department of Surgery, Dartmouth-Hitchcock Medical Center, Geisel School of Medicine at Dartmouth, Lebanon, NH, USA

Jonathan S. Zager, M.D., F.A.C.S. Department of Cutaneous Oncology, Moffitt Cancer Center, Tampa, Florida, USA

Chapter 1
History of Minimally Invasive Inguinal Lymphadenectomy

Marcos Tobias-Machado, Marcio Covas Moschovas, and Antonio Augusto Ornellas

Twentieth Century: Traditional Surgery Era

Cancer of the penis has been recognized for centuries, and its management was first described by Celsus [1] in the seventeenth century when he recommended an amputation for a penile carcinoma with cauterization of the raw stump to control bleeding. However, only in the 1800s was inguinal lymphadenectomy recognized as a routine to treat inguinal metastases [1, 2].

Almost 200 years from the first procedure to treat penile cancer, Young, in 1907, recommended bilateral inguinal lymphadenectomy simultaneously with the penectomy for penile carcinoma [3].

Daseler and associates, in 1948, after the dissection of 450 cadavers, found the precise inguinal anatomic parameters, and Baronofsky, in the same year, advocated one technique that is still widely employed, the transposition of sartorius muscle over the femoral vessels [4].

The first technique described for inguinal lymph node dissection (ILND) for penile cancer found in Medline resources was attributed to Zenker and Pichlmaier in 1966 [5]. Since then, hundreds of modifications and new approaches have been published.

Fegen and Persky first associated the success of a complete inguinal dissection with an increase in penile cancer survival rates [6].

M. Tobias-Machado (✉) • M.C. Moschovas
Section of Urologic Oncology, Department of Urology,
ABC Medical School, São Paulo, Brazil
e-mail: tobias-machado@uol.com.br; marcio.doc@hotmail.com

A.A. Ornellas
Section of Urologic Oncology, National Institute of Cancer, Rio de Janeiro, Brazil
e-mail: ornellasa@hotmail.com

K.A. Delman, V.A. Master (eds.), *Malignancies of the Groin*,
https://doi.org/10.1007/978-3-319-60858-7_1

Johnson and Lo, in 1984, reported that the most important factor determining survival in patients with penile cancer is the extent of lymph node metastases [7], and an aggressive lymphadenectomy is associated with improved long-term survival and potential cure [8].

All those early procedures enumerated many complications, leading the surgeons to have negative perspectives related to the inguinal lymphadenectomy. Since then, two concepts have been explored by investigators to reduce morbidity related to ILND.

The first one supports the template dissection reduction. Small incisions associated with limited lymphatic resection area could result in lower morbidity.

In 1977, Cabanas and colleagues, awaiting the reduction of the templates and morbidity of the procedure, proposed the anatomic lymph node sentinel theory. The lymph node sentinel theory advocates that penile cancer metastasizes first to a sentinel lymph node in the groin. If the biopsy of that node proved negative, the ilioinguinal dissection would be obviated. Some reports showed that this procedure ensures a high rate of recurrence and disease progression. This procedure is no longer routine [9, 10].

A refining approach based on Cabanas' idea was the functional sentinel lymph node. In this case, the anatomical region of dissemination in the lymphatic system is individualized for each patient and considered the fifth place of radioisotope capitation on the lymph node.

In 1994, the group at the Netherlands Cancer Institute (NKI) pioneered dynamic sentinel lymph node biopsy (DSNB) for staging in penile cancer. This procedure was included in the 2009 European Association of Urology (EAU) guidelines on penile cancer [10].The technique for DSNB was described by Hadway et al., Leijte et al., and Lam et al. [11–13]. DSNB is an acceptable staging procedure at experienced centers, remains a diagnostic procedure for clinically negative lymph nodes, and is helpful in avoiding unnecessary IFLND.

Another option described to achieve morbidity reduction was the simplified inguinal dissection. Catalona et al. described the technique and preliminary results of simplified inguinal lymphadenectomy in the beginning of the 1980s [14] in a series of six patients, of whom three had clinical stage 2 disease and three had clinical stage 3 disease with limited positive nodes. The rationale for this modification is that all the positive nodes were in the medial Daseler's quadrant. In that technique the resection lymph nodes area is restricted to the medial area of the saphenous vein.

In later reports, reduced approaches by Cabanas and Catalona were considered not completely accurate with a significant number of patients with inguinal recurrence leading to death [9, 15–17].

In 1995, Pompeo et al. and Puras-Baez et al. described that a positive deep fascia lata node would only occur if the superficial nodes were also positive [18, 19]. Spies and colleagues confirmed those results 12 years later in their study with DSNB [20]. For some authors, superficial ILND reduces morbidity of radical dissection and could be considered the standard template of dissection. The deep dissection will be indicated only in patients with positive nodes in the superficial area.

All of those surgical technique options illustrated a significant decrease in the procedure morbidity but with a long-term follow-up that has demonstrated nonnegligi-

Table 1.1 Surgical morbidity series of conventional ILND

Author (year)	Patients skin	Necrosis skin (%)	Infection (%)	Seroma (%)	Lymphocele (%)	Lymphedema (%)
Ravi (1962–1990)	112	62	17	7	–	27
Ornellas et col. (1972–1987)	200	45	15	6	–	23
Ayyappan et col.	78	36	70	–	87	57
Lopes et col. (1953–1985)	145	15	22	60	–	30
Bevan-Thomaz et col.	53	8	10	10	–	23
Bouchot et col. (1989–2000)	88	12	7	19	–	22
Kroon et col. (1994–2003)	129	15	27	9	12	31
Pandey et col. (1987–1998)	128	20	17	16	–	19
Pompeo (1984–1997)	50	6	12	6	–	18
Spiess et col. (2008)	43	11	9	–	2	17

ble recurrence rates, ranging from 5–15%. DNSB proves to be a good option in reference centers but had low *reproductivity* worldwide, resulting in the need for a high level of standardization to achieve acceptable results [21–23].

The second strategy for the morbidity reduction is selected ILND only in high-risk patients for developing inguinal metastasis. Independent prognostic factors related to inguinal nodes are the disease stage, grade, and lymph vascular invasion.

The most important classification of risk stratifications is validated based on these factors (Solsona from EAU and Ornellas from National Institute of Cancer, Brazil) [10].

Long-term studies proved that prophylactic lymphadenectomy for high-risk patients promotes better survival than rescue lymphadenectomy for patients in "wait and see protocol" [8, 22]. These data stress the importance of ILND in patients with occult microscopic metastasis favoring an early approach.

Considering important advances in perioperative care over time as intermittent leg compression, prophylactic anticoagulation, and antibiotic prophylaxis, the surgical morbidity of traditional surgery is still relatively high. Reports from the beginning of the 2000s indicate that the complication rate was at least 50% (Table 1.1). Bevan-Thomas et al. reported 106 lymphadenectomy procedures in 53 patients with complications (major or minor) in 58% of them [24].

Two years later, Nelson et al. reported a retrospective analysis of 40 inguinal lymphadenectomies and demonstrated lymphedema in 4 of 40 cases (10%), minor wound infection in 3 (7.5%), and minor wound separation in 3 (7.5%), and 5 of 40 patients (12.5%) had a lymphocele, which was spontaneously resolved. Late complications were lymphedema in 2 of 40 patients (5%), flap necrosis in 1 (2.5%), and lymphocele in 1 (2.5%), requiring percutaneous drainage [24].

Other authors reported complications such as seroma or lymphocele in 0–26%, lymphorrhea in 9–10%, and wound infections or skin necrosis in 0–15% [21, 27–30].

Twenty-First Century: Minimally Invasive Surgical Era

Minimally invasive surgery, including the endoscopic and laparoscopic techniques, is widely accepted and performed in urology with proven benefits for morbidity reduction. Consequently, the concept of minimally invasive surgery is supported for ILND.

Bishoff et al. in 2003 reported the first endoscopic inguinal node dissection in two cadavers and one patient. The dissection was possible in the human cadavers, but it was not possible in the patient due to the adherence of the enlarged lymph nodes to the femoral vessels [29].

Tobias-Machado et al. in 2006 reported the first successful video-endoscopic inguinal lymphadenectomy (VEIL) in humans [30]. The idealized technique allows a superficial and deep excision of the inguinal lymph nodes, analogous to the radical conventional surgery utilizing laparoscopic instruments.

In a comparative study published in 2007 in the *Journal of Urology*, they performed, in the same patient, a standard open lymphadenectomy on one side and endoscopic on the other side. The initial impression, obtained in ten patients undergoing bilateral lymphadenectomy for non-palpable lymph nodes, was lower postoperative morbidity with no skin complications when compared to the conventional technique [31, 32]. No disease progression was described at 25 months follow-up.

In the same year, Sotelo et al. [33] reported in the *Journal of Endourology* the outcomes after 14 inguinal endoscopic lymphadenectomy (IEL) in eight patients with clinical stage T2 squamous cell penis carcinoma. Median operative time was 91 min, and the average node yield was nine. No wound-related groin complications occurred.

In 2009, Master and colleagues reported 25 endoscopic inguinal lymphadenectomies (LEG procedures) in 16 patients and 5% morbidity [24].

In 2011, Tobias-Machado et al. reported the feasibility of less applied to ILND [34].

In 2013, Pompeo et al. reported a bilateral simultaneous veil as an alternative to reduce operative time [35].

The complication rate of video-endoscopic surgery performed in series with more than ten groins is at least half the rate of conventional surgery (Table 1.2).

Romanelli et al. advocated that long-term oncological results were exactly the same as those reported by open series [36].

With the advance of the robot-assisted surgery field in the twenty-first century, surgeons worldwide are using the laparoscopic technique to perform robotic inguinal lymphadenectomy for penile cancer.

Table 1.2 Reported series performing video-endoscopic inguinal lymphadenectomy

Author	Cutaneous event (%)	Lymphatic event (%)	Morbidity (%)
Sotelo et al. (2007) 8 cases/14 VEIL	0	23	23
Thyavihaly et al. (2008) 10 cases/10 sides	0	15	15
Tobias-Machado et al. (2009) 20 cases/30 VEIL	5	10	15
Master et al. (2009) 16 cases / 25 sides	0	5	5
Rawal Sudhir et al. (2012) 22 cases/ 39 VEIL	2.5	10	12.5
Romanelli and Tobias (2013) 20 cases/33 VEIL	0	27.2	27.2

In 2009, Josephson et al. [37] reported the first staged bilateral endoscopic operation performed robotically (RAVEIL) with no metastatic involvement in six superficial and four deep lymph nodes.

Robot applicability is new in this surgical field. Only a small series have been presented [38], and a prospective evaluation is needed to compare with standard laparoscopic endoscopic procedures. The results resemble the laparoscopic procedure; however, the robot has its ergonomic advantages. In fact, more than 30 centers routinely perform minimally invasive ILND worldwide with acceptable outcomes.

The standardized robotic technique was recently described in the eleventh edition of *Campbell-Walsh Urology*.

Future possibilities in this field include better preoperative definitions of patients with positive nodes on PET scans, intraoperative fluorescence [39], and lymphatic morbidity reduction with better surgical techniques.

References

1. Murphy L. The history of urology. Springfield, IL: Charles C Thomas; 1972.
2. Ficarra V, Akduman B, Bouchot O, Palou J, Tobias-Machado M. Prognostic factors in penile cancer. Urology. 2010;76(2 Suppl 1):S66–73.
3. Young HH. A radical operation for the cure of cancer of the penis. J Urol. 1931;26:285.
4. Daseler EH, Anson BJ, Remaimann AF. Radical excision of the inguinal and iliac lymph glands: a study based upon 450 anatomical dissections and upon supportive clinical observations. Surg Gynecol Obstet. 1948;87:679.
5. Zenker R, Pichlmaier H. Extended radical lymph node resection in the inguinal region. Chirurg. 1966;37(4):162–4.
6. Fegen P, Persky L. Squamous cell carcinoma of the penis: its treatment, with special reference to radical node dissection. Arch Surg. 1969;99(1):117–20.

7. Johnson DE, Lo RK. Complications of groin dissection in penile cancer. Experience with 101 lymphadenectomies. Urology. 1984;24:312–4.
8. McDougal WS, Kirchner FK, Edwards RH, et al. Treatment of carcinoma of the penis: the case for primary lymphadenectomy. J Urol. 1986;136:38–41.
9. Cabanas RM. An approach for the treatment of penile carcinoma. Cancer. 1977;39(2):456–66.
10. Pizzocaro G, Algaba F, Horenblas S, et al. EAU penile cancer guidelines 2009. Eur Urol. 2010;57:1002–12.
11. Hadway P, Smith Y, Carbishley C, et al. Evaluation of dynamic lymphoscin- tigraphy and sentinel lymph-node biopsy for detecting occult metastases in patients with penile squamous cell carcinoma. BJU Int. 2007;100:561–5.
12. Leijte JAP, Kroon BK, Valdes Olmos RA, et al. Reliability and safety of current dynamic sentinel node biopsy for penile carcinoma. Eur Urol. 2007;52:170–7.
13. Lam W, Alnajjar HM, La-Touche S, et al. Dynamic sentinel lymph node biopsy in patients with invasive squamous cell carcinoma of the penis: a prospective study of the long-term outcome of 500 inguinal basins assessed at a single institution. Eur Urol. 2013;63:657–63.
14. WJ C. Modified inguinal lymphadenectomy for carcinoma of penis with preservation of saphenous vein. J Urol. 1988;140:306.
15. Lopes A, Rossi BM, Fonseca FP, Morini S. Unreliability of modified inguinal lymphadenectomy for clinical staging of penile carcinoma. Cancer. 1996;77:2099–102.
16. Perinetti E, Crane DB, Catalona WJ. Unreliability of sentinel lymph node biopsy for staging penile carcinoma. J Urol. 1980;124:734.
17. Wespes E, Simon J, Schulman CC. Cabanas approach: is sentinel node biopsy reliable for staging penile carcinoma? Urology. 1986;28:278.
18. Pompeo AC, Mesquita JL, Junior WA, et al. Staged inguinal lymphadenec- tomy (SIL) for carcinoma of the penis (CP). J Urol. 1995;153:246A.
19. Puras-Baez A, Rivera-Herrera J, Miranda G, et al. Role of superficial inguinal lymphadenectomy in carcinoma of the penis. J Urol. 1995;153:246A.
20. Spiess PE, Izawa JI, Bassett R, et al. Preoperative lymphoscintigraphy and dynamic sentinel node biopsy for staging penile cancer: results with pathological correlations. J Urol. 2007;177:2157–61.
21. d'Ancona CA, de Lucena RG, Querne FA, et al. Long-term followup of penile carcinoma treated with penectomy and bilateral modified inguinal lymphadenectomy. J Urol. 2004;172:498–501.
22. Kroon BK, Horenblas S, Lont AP, Tanis PJ, Gallee MP, Nieweg OE. Patients with penile carcinoma benefit from immediate resection of clinically occult lymph node metastases. J Urol. 2005;173(3):816–9.
23. Neto AS, Tobias-Machado M, Ficarra V, Wroclawski ML, Amarante RD, Pompeo AC, Del Giglio A. Dynamic sentinel node biopsy for inguinal lymph node staging in patients with penile cancer: a systematic review and cumulative analysis of the literature. Ann Surg Oncol. 20111;18(7):2026–34.
24. Bevan-Thomas R, Slaton JW, Pettaway CA. Contemporary morbidity from lymphadenectomy for penile squamous cell carcinoma: the M.D. Anderson cancer center experience. J Urol. 2002;167(4):1638–42.
25. Parra RO. Accurate staging of carcinoma of the penis in men with nonpalpable inguinal lymph nodes by modified inguinal lymphadenectomy. J Urol 1996;155:560–3. noma of the penis. J Urol. 2003;169:78A.
26. Bouchot O, Rigaud J, Maillet F, et al. Morbidity of inguinal lymphadenec- tomy for invasive penile carcinoma. Eur Urol. 2004;45:761–5.
27. Spiess PE, Hernandez MS, Pettaway CA. Contemporary inguinal lymph node dissection: minimizing complications. World J Urol. 2009;27:205–12.
28. Coblentz TR, Theodorescu D. Morbidity of modified prophylactic inguinal lymphadenectomy for squamous cell carcinoma of the penis. J Urol. 2002;168:1386–9.
29. Bishoff JT, Basler JW, Teichman JM, et al. Endoscopic subcutaneous modi- fied inguinal lymph node dissection (ESMIL) for squamous cell carcinoma of the penis. J Urol. 2003;169:78A.

30. Tobias-Machado M, Tavares A, Molina WR Jr, Forseto PH Jr, Juliano RV, Wroclawski ER. Video endoscopic inguinal lymphadenectomy (VEIL): minimally invasive resection of inguinal lymph nodes. Int Braz J Urol. 2006;32(3):316–21.
31. Tobias-Machado M, Tavares A, Ornellas AA, Molina WR Jr, Juliano RV, Wroclawski ER. Video endoscopic inguinal lymphadenectomy: a new minimally invasive procedure for radical management of inguinal nodes in patients with penile squamous cell carcinoma. J Urol. 2007 Mar;177(3):953–7.
32. Tobias-Machado M, Tavares A, Silva MN, Molina WR Jr, Forseto PH, Juliano RV, Wroclawski ER. Can video endoscopic inguinal lymphadenectomy achieve a lower morbidity than open lymph node dissection in penile cancer patients? J Endourol. 2008;22(8):1687–91.
33. Sotelo R, Sánchez-Salas R, Carmona O, Garcia A, Mariano M, Neiva G, et al. Endoscopic lymphadenectomy for penile carcinoma. J Endourol. 2007;21:364–7.
34. Tobias-Machado M, Correa WF, Reis LO, Starling ES, de Castro NO, Juliano RV, Pompeo AC. Single-site video endoscopic inguinal lymphadenectomy: initial report. J Endourol. 2011;25(4):607–10.
35. Pompeo A, Tobias-Machado M, Molina WR, Lucio J II, Sehrt D, Pompeo AC, Kim FJ. Extending boundaries in minimally invasive procedures with simultaneous bilateral video endoscopic inguinal lymphadenectomy (veil) for penile cancer: initial Denver health medical center and ABC school of medicine experience and surgical considerations. Int Braz J Urol. 2013;39(4):587–92.
36. Romanelli P, Nishimoto R, Suarez R, Decia R, Abreu D, Machado M, et al. Video endoscopic inguinal lymphadenectomy: surgical and oncological results. Actas Urol Esp. 2013;37(5):305–10.
37. Josephson DY, et al. Robotic-assisted endoscopic inguinal lymphadenectomy. Urology. 2009;73(1):167–70.
38. Matin SF, et al. Phase 1 prospective evaluation of the oncological adequacy of robotic assisted video-endoscopic inguinal lymphadenectomy in patients with penile carcinoma. BJU Int. 2013;111(7):1068–74.
39. Frimberger D, Linke R, Meissner H, Stepp H, Zaak D, Hungerhuber E, Waidelich R, Schmeller N, Hofstetter A, Schneede P. Fluorescence diagnosis: a novel method to guide radical inguinal lymph node dissection in penile cancer. World J Urol. 2014;22(2):150–4.

Chapter 2
Epidemiology of Diseases of the Groin

Michael Lowe

Introduction

The epidemiology of diseases of the groin varies rather markedly by the type of disease. Patients with melanoma tend to be younger than patients with nonmelanoma skin cancers and than those with squamous cell carcinomas of the penis and vulva. UV exposure predisposes to melanoma and nonmelanoma skin cancers, while HPV infection and tobacco use are associated with squamous cell carcinomas of the vulva and penis. Management of cancers that affect the groin depends on the type of cancer, but surgery remains the best chance for cure in almost cases, with anal squamous cell carcinoma being the exception. This chapter introduces the comparative epidemiology of diseases of the groin in an effort to provide context for treatment algorithms for each of the specific disease processes.

Melanoma

Melanoma accounts for less than 5% of all skin cancers but has the highest mortality. It is estimated to be the fifth most common cancer in males and seventh most common cancer in females. A total of 87,110 new cases of melanoma and 9730 deaths from melanoma are estimated in the United States in 2017 [1]. The incidence of melanoma has been steadily increasing over the past several decades with an annual increase varying between 3 and 7% [2]. This is thought to be related to

M. Lowe, M.D.
Division of Surgical Oncology, Department of Surgery,
Emory University School of Medicine,
1365 Clifton Rd NE, Atlanta, GA 30322, USA
e-mail: mlowe3@emory.edu

K.A. Delman, V.A. Master (eds.), *Malignancies of the Groin*,
https://doi.org/10.1007/978-3-319-60858-7_2

changing behaviors regarding sun exposure and increased detection of early lesions [3]. While the incidence of melanoma has consistently increased, in general mortality rates have stabilized given the trend toward more frequent diagnosis of early-stage melanomas that are more likely to be curable.

There is however marked variability in mortality rates based on age. Older patients tend to have more aggressive primary tumors that are more often on the head and neck; these patients have higher mortality rates compared with other age groups [4]. In addition, according to SEER data, mortality rates continue to decrease for patients 20–44 years old but continue to increase for patients older than 65 [5].

Overall incidence of and mortality from melanoma are higher in men than in women. Of the estimated melanoma cases in 2017, 59.9% are expected to be diagnosed in males, and 65.5% of patients that die from melanoma will be males [1]. In patients aged 18–39 years, women have had a significantly higher increase in incidence of melanoma compared to males of the same age. Males more commonly develop melanomas on the head and neck whereas females are more likely to have extremity or trunk melanomas.

Melanoma is essentially a disease of Caucasians, who account for 95% of new melanoma diagnoses. African Americans are the next most commonly affected group, accounting for approximately 0.5% of cases. Non-Caucasians are more likely to present with advanced-stage disease and tend to have worse overall survival compared to Caucasians [6]. The lifetime risk of developing melanoma is 2.5% for Caucasians, 0.1% for Blacks, and 0.5% for Hispanics.

Risk factors contributing to the development of melanoma include ultraviolet light exposure, personal or family history of melanoma, fair complexion, immunocompromised states, advanced age, male sex, congenital melanocytic nevus, and familial melanoma syndromes such as dysplastic nevus syndrome. People with any of these risk factors are advised to undergo routine dermatologic exams and self-skin exams and practice safe sun exposure habits.

Nonmelanoma Skin Cancer

Squamous Cell Carcinoma

While it is difficult to assess the total number of cases of nonmelanoma skin cancers given the lack of population-based registries, it is estimated that over five million basal and squamous cell carcinomas are diagnosed each year. Of these, approximately two in ten are squamous cell carcinomas, which we will focus on here given their greater propensity to be more aggressive than basal cell carcinomas. The overwhelming majority of squamous cell carcinomas (approximately 80%) develop in the head and neck. This is likely related to UV light exposure, which is one of the strongest risk factors for the development of nonmelanoma skin cancers. Other risk factors include previous radiation exposure, actinic keratoses, immunosuppression, previous scars or burns, and disorders such as xeroderma pigmentosum, epidermolysis bullosa, and pansclerotic morphea of childhood.

Given that most patients with cutaneous squamous cell carcinoma present with primaries on the head and neck, there is little data about the frequency and management of inguinal lymph node involvement from perineal or lower extremity primaries. Lymph node metastasis is often associated with adverse pathologic findings such as lymphovascular invasion, poor differentiation, and perineural invasion. High-risk lesions, defined as primaries greater than 2 cm on the extremities with adverse pathologic findings or in the setting of immunosuppression, tend to have higher rates of lymph node and distant metastases. Patients with nodal metastases have expected 10-year survival of less than 20% [7]. In general, rates of nodal involvement are low, but early detection and treatment may significantly alter the prognosis of patients with lymph node metastases.

Merkel Cell Carcinoma

Merkel cell carcinoma (MCC) is a rare neuroendocrine cancer of the skin with a historically poor prognosis. It is one of the least common types of skin cancer, with an estimated 1500 case diagnoses per year. MCC is essentially a disease of older whites, with two-thirds of patients over the age of 70 and nine out of ten patients being white. Risk factors include extensive sun exposure, older age, and immunosuppression. Immunosuppressed patients that develop MCC tend to be much younger (approximately 50% younger than 50 years old) than immunocompetent patients with MCC. A novel polyomavirus has been identified in a majority of cases of MCC, but a causal link between this virus and the development of MCC has not been established [8].

Approximately 15% of patients with MCC present with primary lesions on the lower extremity and 27% present with lymph node involvement [9]. Nodal involvement is associated with a decrease in 5-year survival from 64% with local disease to approximately 39%. Nodal disease is detected in the sentinel lymph node in approximately one-third of patients without clinically detectable lymph node metastasis. Patients with pathologically negative sentinel lymph nodes appear to have a survival advantage compared to patients that undergo only clinical nodal evaluation, which confirms the utility of sentinel lymph node biopsy.

Penile Cancer

Squamous cell carcinoma (SCC) represents the most common type of cancer of the penis but is rare, representing only 0.5% of malignancies in men. There will be an estimated 2120 new cases of and 360 deaths from penile SCC in 2017 in the United States [1]. The incidence is significantly higher in men in developing countries, particularly Asia and Africa.

Median age at diagnosis in the United States is 68 years. Risk factors for the development of penile SCC are phimosis, balanitis, penile trauma, tobacco use, lichen sclerosus, poor hygiene, and a history of sexually transmitted disease, particularly human papillomavirus (HPV) [10]. Up to 80% of penile SCC is related to HPV; HPV types 16 and 18 are strongly correlated with the development of penile SCC [11]. Patients with HIV are also at significantly increased risk of developing penile cancer, although this increased risk may be related to increased incidence of HPV among males with HIV. The protective effects of neonatal circumcision against penile SCC are thought to be lost in adults that undergo circumcision. This is likely related to the elimination of phimosis and lower incidence of HPV infections in neonates undergoing circumcision compared to adults [12].

Involvement of inguinal lymph nodes in penile SCC is one of the most important prognostic factors affecting survival of patients with invasive disease. Five-year survival of patients with inguinal lymph node involvement is approximately 40%, compared to over 85% for patients without inguinal lymph node involvement. Essentially no patients survive to 5 years when pelvic lymph nodes or distant sites are involved. A thorough physical exam is essential to the detection of inguinal lymph node disease; however, detection of a palpable inguinal lymph node may not represent nodal metastasis given that up to 50% of palpable adenopathy will be secondary to inflammation. Of patients without palpable adenopathy, up to 25% will have micrometastatic disease in the inguinal lymph nodes. Advanced T stage, poor differentiation, and lymphovascular invasion are independent risk factors for inguinal lymph node involvement. Of patients with a single palpable lymph node, approximately a third will harbor contralateral inguinal lymph node metastasis that is not palpable.

Comparative Epidemiology: Melanoma and Penile SCC

Given that videoscopic inguinal lymphadenectomy was originally developed for penile cancer and further refined in melanoma, a comparison between patients undergoing this procedure for these indications is warranted. In unpublished data from Emory University, patients undergoing videoscopic inguinal lymphadenectomy for melanoma were younger (50.1 vs. 61.4 years), had lower BMI (27.6 vs. 31.0), were less likely to have a smoking history (19% vs. 56%), and were less likely to be diagnosed with diabetes mellitus (11% vs. 33%). Although the technical aspects of and complications resulting from this procedure will be discussed later in this book, it should be noted that lymph node retrieval, operative time, conversions to open surgery, and postoperative complications did not differ between patients undergoing videoscopic inguinal lymphadenectomy for melanoma or penile cancer.

Vulvar Cancer

Numerous histologic subtypes of cancer can occur on the vulva, which includes the vaginal opening, the labia majora, the labia minora, and the clitoris. The most common type of cancer of the vulva is squamous cell carcinoma (SCC), which accounts for nine in ten vulvar cancers. The next most common cancers are adenocarcinoma, or extramammary Paget's disease, and melanoma. Rare cancers include Bartholin gland adenocarcinoma, verrucous carcinoma, basal cell carcinoma, and sarcoma. It is estimated that 6020 women will be diagnosed with vulvar cancer and 1150 will die in 2017 [1].

Risk factors for vulvar SCC include increasing age, human papillomavirus (HPV) infection, tobacco use, and states of immunodeficiency. Up to 70% of cases are related to HPV infection, most frequently by the HPV-16 and HPV-18 strains. The availability of a vaccine to these strains may significantly alter the incidence of this disease as vaccination becomes more widespread.

Like penile SCC, inguinal lymph node involvement in vulvar cancer is the most important prognostic factor affecting survival. Approximately one-third of patients with vulvar cancer have micrometastatic disease in the sentinel lymph node, although this number may underrepresent the true number of patients with micrometastatic disease since sentinel lymph node biopsy is only recommended for patients with clinically negative nodes, a unifocal primary less than 4 cm, and no previous history of vulvar surgery. Patients that do not fulfill these criteria should undergo elective lymphadenectomy. Patients with sentinel lymph node involvement have a 10-year disease-specific survival of 65%, which compares unfavorably to patients without sentinel node involvement, whose 10-year disease-specific survival is 91% [13].

Extremity Soft Tissue Sarcoma

Soft tissue sarcomas represent a heterogeneous group of tumors originating from the fat, muscle, nerve and nerve sheath, blood vessels, bone, and other connective tissues. An estimated 12,390 people will be diagnosed with and 4990 people will die from soft tissue sarcoma in the United States in 2017, with slightly more males being affected (55.6%). Risk factors include prior radiation and genetic predispositions such as Li-Fraumeni syndrome, Gardner's syndrome, hereditary retinoblastoma, and neurofibromatosis. Sarcomas are typically further classified into soft tissue sarcomas of the extremity and trunk, retroperitoneal sarcomas, gastrointestinal stromal tumors, desmoid tumors, and pediatric sarcomas, most commonly rhabdomyosarcoma. We will focus here on lower extremity sarcomas since the majority of patients with inguinal involvement have lower extremity sarcomas.

In the largest series in the literature from a single institution, 28% of patients with soft tissue sarcoma presented with sarcoma of the lower extremity. The most common histologies affecting the lower extremity are liposarcoma, undifferentiated pleomorphic sarcoma, synovial sarcoma, myxofibrosarcoma, and leiomyosarcoma. Local recurrence rates for extremity sarcoma approach 25% at 10 years, and disease-specific survival is approximately 60% at 10 years. Predictors of recurrence and survival include size, grade, and depth of invasion. Most distant metastases from lower extremity sarcomas occur in the lung [14].

Lymph node involvement in sarcoma is rare, with series quoting rates of lymph node metastases between 2 and 4%. Patients with lymph node involvement without systemic disease have significantly better overall survival than patients with synchronous nodal and systemic disease (71% vs. 21%, respectively, in one study) [15]. The lower extremity is the most common site for lymph node metastases in sarcoma. In one series, inguinal lymph node involvement accounted for 41.3% of all nodal metastases [16]. The histologic tissue types with the highest prevalence of lymph node involvement are angiosarcoma, embryonal rhabdomyosarcoma, and epithelioid sarcoma. It is recommended that patients with lymph node involvement undergo radical lymphadenectomy, which has been shown to extend survival from 4 to 16 months.

Anal Cancer

Anal cancer is much less common than cancers of the colon and rectum, with an estimated 8200 new cases and 1100 deaths in the United States in 2017. Anal cancer is more common in women; approximately 64% of the estimated new cases will be diagnosed in women [1]. The incidence of anal cancer has been steadily increasing in both men and women over the past several decades. Risk factors for anal cancer include human papillomavirus (HPV) infection, history of receptive anal intercourse or sexually transmitted disease, immunosuppression, hematologic malignancies, and tobacco use. HPV DNA can be found in up to three-quarters of anal cancer tumors, with HPV-16 and HPV-18 strains seen most commonly. Most primary cancers of the anal canal are squamous cell carcinomas, although adenocarcinoma of the anal glands, small cell carcinoma, undifferentiated cancers, and melanomas can occur in the anal canal.

Lymphatic drainage of anal cancers depends on the location of the tumor. Cancers arising from the perianal skin and the anal canal distal to the dentate line drain almost uniformly to the inguinal lymph nodes. Cancers more proximally drain to the perirectal, internal iliac, or even mesenteric nodal basins. As with most cancers, prognosis is adversely related to the presence of lymph node metastases. In a review of the SEER database, approximately 29% of anal cancer patients presented with regional nodal metastases, and these patients had a 5-year overall survival rate of 60%. This compares to 5-year survival rates of 80% for patients with localized disease and 30% for patients with distant metastases. Evaluation of the clinically negative inguinal nodal basin is not routinely recommended, but sentinel lymph node biopsy has been shown to be effective. Rates of sentinel lymph node involvement vary significantly in the literature, with estimates as high as 44% [17]. Management of the involved inguinal nodal basin is usually with radiation along with the primary anal cancer.

Conclusions

The epidemiology of the diseases of the groin varies markedly. Patients with melanoma tend to be younger, while nonmelanoma skin cancers and primary cancers of the penis and vulva tend to occur in older patients. Risks factors for each cancer type vary as well, but exposure to HPV and tobacco use are important risk factors for squamous cell carcinomas that may affect the groin. Exposure to UV light and age are risk factors for melanoma and other skin cancers. Management of cancers that affect the groin depends on the type of cancer, but surgery remains the best chance for cure in almost cases with anal squamous cell carcinoma being the exception.

References

1. Siegel RL, Miller KD, Jemal A. Cancer statistics, 2017. CA Cancer J Clin. 2016;67:7–30.
2. Garbe C, Leiter U. Melanoma epidemiology and trends. Clin Dermatol. 2009;27(1):3–9.
3. Linos E, Swetter SM, Cockburn MG, Colditz GA, Clarke CA. Increasing burden of melanoma in the United States. J Invest Dermatol. 2009;129(7):1666–74.
4. Balch CM, Soong SJ, Gershenwald JE, Thompson JF, Coit DG, Atkins MB, et al. Age as a prognostic factor in patients with localized melanoma and regional metastases. Ann Surg Oncol. 2013;20(12):3961–8.
5. Howlander NA, Krapcho M, Miller D, Bishop K, Altekruse SF, Kosary CL, Yu M, Ruhl J, Tatalovich Z, Mariotto A, Lewis DR, Chen HS, Feuer EJ, Cronin KA, editors. SEER Cancer Statistics Review, 1975–2013. Bethesda, MD: National Cancer Institute; 2016. [Based on November 2015 SEER data submission, posted to the SEER web site, April 6.] http://seer.cancer.gov/csr/1975_2013/
6. Wu XC, Eide MJ, King J, Saraiya M, Huang Y, Wiggins C, et al. Racial and ethnic variations in incidence and survival of cutaneous melanoma in the United States, 1999–2006. J Am Acad Dermatol. 2011;65(5 Suppl 1):S26–37.
7. Alam M, Ratner D. Cutaneous squamous-cell carcinoma. N Engl J Med. 2001;344(13):975–83.
8. Amber K, McLeod MP, Nouri K. The Merkel cell polyomavirus and its involvement in Merkel cell carcinoma. Dermatol Surg. 2013;39(2):232–8.
9. Lemos BD, Storer BE, Iyer JG, Phillips JL, Bichakjian CK, Fang LC, et al. Pathologic nodal evaluation improves prognostic accuracy in Merkel cell carcinoma: analysis of 5823 cases as the basis of the first consensus staging system. J Am Acad Dermatol. 2010;63(5):751–61.
10. Pow-Sang MR, Ferreira U, Pow-Sang JM, Nardi AC, Destefano V. Epidemiology and natural history of penile cancer. Urology. 2010;76(2 Suppl 1):S2–6.
11. Daling JR, Madeleine MM, Johnson LG, Schwartz SM, Shera KA, Wurscher MA, et al. Penile cancer: importance of circumcision, human papillomavirus and smoking in in situ and invasive disease. Int J Cancer. 2005;116(4):606–16.
12. Morris BJ, Gray RH, Castellsague X, Bosch FX, Halperin DT, Waskett JH, et al. The strong protective effect of circumcision against cancer of the penis. Adv Urol. 2011;2011:812368.
13. Te Grootenhuis NC, van der Zee AG, van Doorn HC, van der Velden J, Vergote I, Zanagnolo V, et al. Sentinel nodes in vulvar cancer: long-term follow-up of the GROningen INternational Study on Sentinel nodes in Vulvar cancer (GROINSS-V) I. Gynecol Oncol. 2016;140(1):8–14.

14. Brennan MF, Antonescu CR, Moraco N, Singer S. Lessons learned from the study of 10,000 patients with soft tissue sarcoma. Ann Surg. 2014;260(3):416–21. discussion 21–2
15. Riad S, Griffin AM, Liberman B, Blackstein ME, Catton CN, Kandel RA, et al. Lymph node metastasis in soft tissue sarcoma in an extremity. Clin Orthop Relat Res. 2004;426:129–34.
16. Fong Y, Coit DG, Woodruff JM, Brennan MF. Lymph node metastasis from soft tissue sarcoma in adults. Analysis of data from a prospective database of 1772 sarcoma patients. Ann Surg. 1993;217(1):72–7.
17. Noorani A, Rabey N, Durrani A, Walsh SR, Davies RJ. Systematic review of sentinel lymph node biopsy in anal squamous cell carcinoma. Int J Surg. 2013;11(9):762–6.

Chapter 3
Anatomy of the Groin

Andre Granger, Theofanis Kollias, and Marios Loukas

Arteries and Veins of the Groin

Arteries

The structures of the groin are in close proximity to major vessels—most notably the femoral and external iliac arteries (Fig. 3.1). Thus, its highly vascular nature comes to no surprise. To clearly describe the arterial blood supply to the inguinal area, it can be divided into two parts: superficial and deep. The superficial vessels supply the skin and subcutaneous tissue, while the deep vessels supply structures such as the muscles and fascia. One must note, however, that anastomoses exist between the superficial and deep vessels.

Superficial Arterial Supply

Proximal superficial branches of the femoral artery supply this part of the groin. The specific branches, from lateral to medial, are the superficial circumflex iliac artery, the superficial epigastric artery, and the superficial external pudendal artery (Fig. 3.2).

Illustrations for this chapter were done by Jessica Holland M.S., C.M.I.; Angelica Ortiz M.Sc., B.M.C.; Xochitl Vinaja M.S.; Charles Wesley Price M.S., C.M.I.

A. Granger, M.D.
Department of Anatomical Sciences, St. George's University School of Medicine, Grenada, W.I.

Department of Neurology, NYU Langone Hospital - Brooklyn, Brooklyn, NY, USA
e-mail: agrange1@sgu.edu

T. Kollias, M.D. • M. Loukas, M.D., Ph.D. (✉)
Department of Anatomical Sciences, St. George's University School of Medicine, Grenada, W.I.
e-mail: tkollia1@sgu.edu; mloukas@sgu.edu

K.A. Delman, V.A. Master (eds.), *Malignancies of the Groin*,
https://doi.org/10.1007/978-3-319-60858-7_3

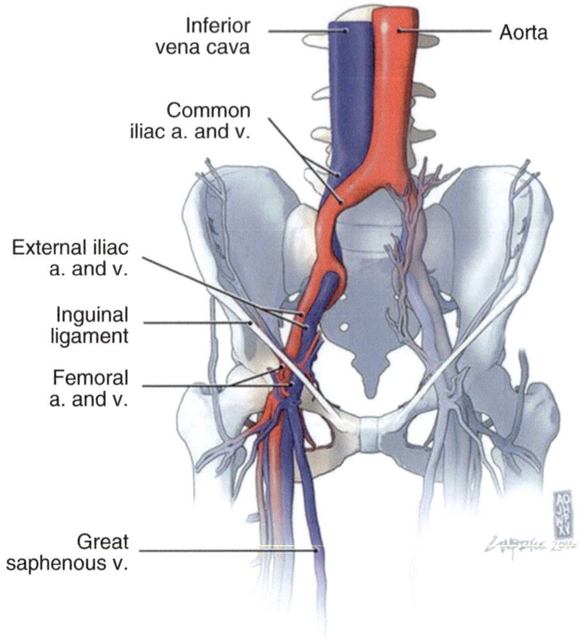

Fig. 3.1 An anterior view of the pelvis illustrating the large vessels traversing the inguinal region

Superficial Circumflex Iliac Artery

The superficial circumflex iliac artery is the smallest of the three main superficial branches. It pierces through the fascia lata as its origin is usually lateral to the saphenous opening. It travels superolaterally, in the direction of the anterior superior iliac spine, to become highly branching and supplies the region of the skin over the lateral third of the inguinal ligament and the iliac crest. This artery also supplies the superficial fascia and superficial inguinal nodes along its course. It anastomoses with branches of the deep circumflex iliac, superior gluteal, and lateral circumflex femoral artery.

The superficial circumflex artery can be divided into three subtypes: type 1, type 2, and type 3. In type 1 or archetype, it originates below the inguinal ligament and may have branches. Type 2 is seen when it arises from the deep circumflex iliac artery. In type 3, the artery is absent. Type 3 has a prevalence of about 17%.

Superficial Epigastric Artery

Medial to the superficial circumflex iliac artery, and following a more vertical course, is the superficial epigastric artery. Typically, it originates from the anterior aspect of the femoral artery about 2–5 cm distal to the inguinal ligament. It ascends anterior to the inguinal ligament up to the region just below the umbilicus. It supplies the skin, superficial fascia, and inguinal nodes in midinguinal area. Branches of the superficial epigastric artery anastomose with those of the contralateral artery. It also communicates with the inferior epigastric artery.

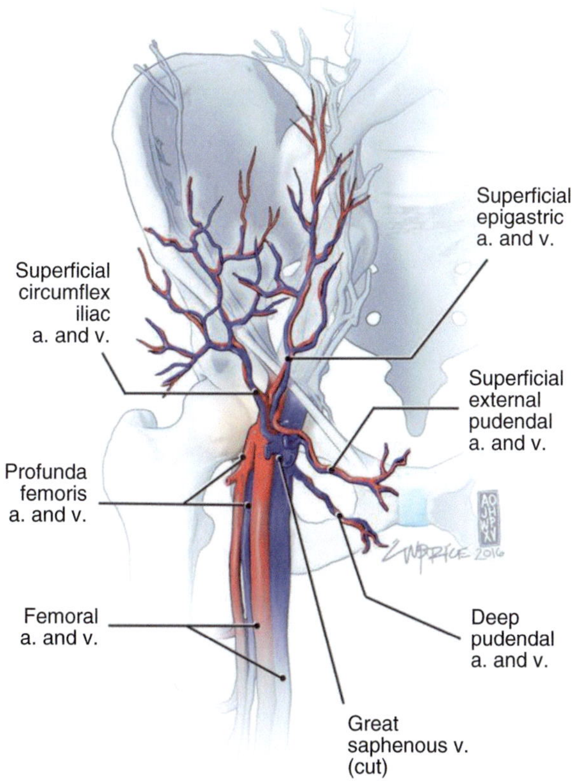

Fig. 3.2 Illustration of the superficial branches that supply the inguinal area

Variations in the origin of the superficial epigastric artery have been previously reported. It often originates from a trunk that is shared with the superficial circumflex iliac artery. It may also branch from the pudendal artery or the profunda femoris artery. Results from studies on the prevalence of this vessel vary widely from 58 to 90%.

Superficial External Pudendal Artery

The superficial external pudendal artery has a medial origin on the femoral artery. It has a medial route, coursing in the direction of the pubic symphysis where it traverses the spermatic cord in males and the round ligament in females. It supplies cutaneous blood flow to the inferior abdomen, the penis, and the scrotum in males and the labia majora in females. It anastomoses with branches of the internal pudendal artery.

The source of the superficial external pudendal artery is almost always the femoral artery, but rarely it may originate from the profunda femoris artery. It has also been observed to share a common trunk with the superficial epigastric artery.

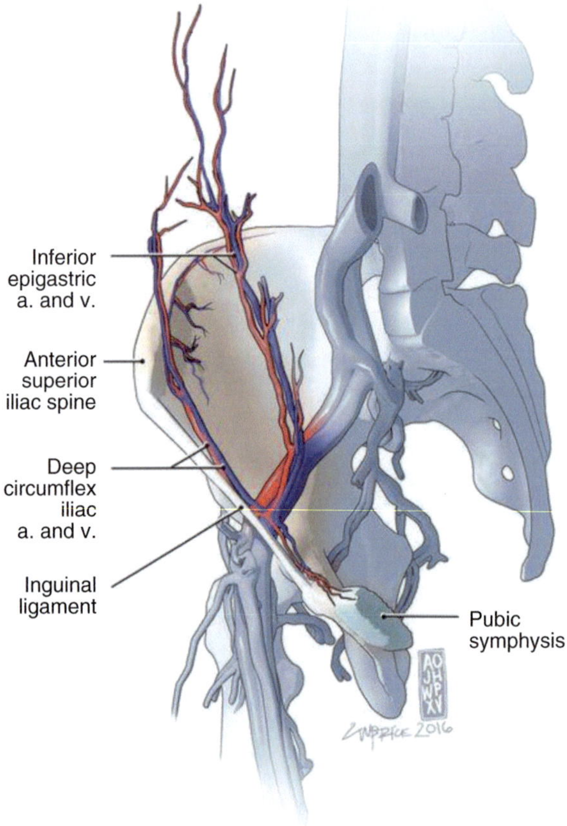

Fig. 3.3 Illustration of the deep vessels that supply the inguinal area

Deep Arterial Supply

The deep layer of the groin obtains its blood supply mainly from two arteries: the deep circumflex iliac and the inferior epigastric (Fig. 3.3). More superiorly, some of its supply may be derived from the anterior branches of the subcostal and lumbar arteries. The ascending branch of the deep circumflex iliac, along with the anterior branches of the subcostal and first four lumbar arteries, can be found between the internal oblique and transversus abdominis muscles. Here, their vascular networks supply the muscles that they come into contact with.

Deep Circumflex Iliac Artery

The deep circumflex iliac artery branches off of the lateral aspect of the external iliac artery and supplies the deep lateral groin. It's a laterally running nerve that forms many anastomoses. Its initial direction is toward the anterior superior iliac spine. Up to this point, it remains enclosed in a sheath of connective tissue formed from fibers of the transversalis fascia and the iliac fascia. Near the anterior superior iliac spine, three important things occur: (1) it anastomoses with the ascending branch of the lateral circumflex femoral artery, (2) the artery pierces the transversalis fascia and continues laterally along the inner lip of the ilium, and (3) a large ascending branch is given off. The deep circumflex iliac artery continues laterally and posteriorly to anastomose with the iliolumbar and superior gluteal arteries. The ascending branch passes through the transversus abdominis muscle just superior to its origin and continues in a cephalad direction. Here, it runs between, and supplies, the transversus abdominis and inner oblique muscles. It continues in this intermuscular plane to form anastomoses with the lumbar and inferior epigastric arteries. Small proximal branches may also anastomose with the superficial circumflex iliac artery.

Inferior Epigastric Artery

The inferior epigastric artery, a branch of the external iliac artery, has its origin just medial to that of the deep circumflex iliac artery. The inferior epigastric gives off two branches: the pubic and the external spermatic (or cremasteric). The pubic branch crosses the conjoint tendon to travel inferiorly toward the obturator artery. It forms an anastomosis with the obturator artery. When this pubic branch is large (20–30% of cases), it takes the place of the obturator to become the aberrant obturator artery. Notably, this pubic branch may also enter the inguinal (Hesselbach's) triangle. The external spermatic branch joins with the contents of the spermatic cord in the male. Here it supplies the cremasteric muscle and other fascial layers within the cord. It also anastomoses with the testicular artery. In females, the artery is relatively smaller and supplies the round ligament.

The inferior epigastric artery, together with its accompanying vein, forms the lateral border of the inguinal (Hesselbach's) triangle. The artery courses superomedially toward the rectus abdominis muscle, passing near the medial border of the deep inguinal ring. Thus it lies deep to the origin of the spermatic cord. While traveling anterior to the parietal peritoneum, the artery causes an observable elevated fold that is evident from an intraabdominal view of the anterior abdominal wall. This fold is called the lateral umbilical fold. Near the lateral border of the rectus abdominis muscle, in the region of the apex of the inguinal triangle, the artery pierces the transversalis fascia. It continues superiorly, just lateral to the midline, along the posterior aspect of the rectus abdominis where it passes anterior to the arcuate line. It remains between the rectus abdominis and the posterior lamina of the rectus sheath to anastomose with the superior epigastric artery (above the umbilicus) and the lower posterior intercostal arteries. During its ascension, the inferior epigastric artery also gives off several branches that anastomose with other arteries. Its branches join with branches of the superficial epigastric, circumflex iliac, and lumbar arteries. In the end, the inferior epigastric artery supplies muscles of the abdominal wall, peritoneum, and even some areas of skin over the lower abdomen via its cutaneous branches.

A few variations in the origin of the inferior epigastric artery have been noted. It may sometimes branch off of the femoral artery, in which case it ascends to enter the abdominal cavity. It is also not uncommon to have it arising from the external iliac artery via a common trunk with the obturator artery. Rarely, it may originate from the obturator artery itself from the internal iliac artery.

Veins

The veins of the inguinal region generally tend to accompany their similarly named arteries. Thus, the superficial groin will be drained by the superficial circumflex iliac, the superficial epigastric, and the superficial external pudendal veins. These three superficial veins transport their deoxygenated blood to the saphenous vein. The deeper structures of the groin will be drained mainly by the deep circumflex iliac veins and the inferior epigastric veins. These veins usually occur in pairs, or venae comitantes, for each artery, eventually combining to form one common vein. These deep veins drain into the external iliac vein—about 1 cm above the inguinal ligament for the inferior epigastric vein and about 2 cm above the inguinal ligament for the deep circumflex iliac vein.

The great saphenous vein usually joins with the femoral vein at the saphenous opening. However, it may also pierce the fascia lata prior to reaching the saphenous ring, it may be duplicated, or a venous network may replace it. Variations of the venous drainage in the vicinity of the saphenous opening are numerous. Most commonly, the superficial circumflex iliac and the superficial epigastric veins combine before joining the saphenous vein. In one variant, all three superficial veins drain directly into the femoral vein.

An inconstant vein, the thoracoepigastric vein, may be observed on the anterior abdominal wall and connects the inferior epigastric vein, or the femoral vein, with the lateral thoracic veins. This essentially communicates the inferior vena cava drainage area to that of the superior vena cava.

Variation of the external iliac vein, which drains the inferior epigastric vein and the deep circumflex iliac vein, is uncommon. Unilateral aplasia, bilateral aplasia, and duplication have all been reported.

Nerves of the Groin

The lumbar plexus originates from the ventral rami of lumbar nerves one to four. The inguinal region receives its somatic motor and sensory innervation from the terminal branches of the lumbar plexus (Fig. 3.4). The groin receives its innervation mainly from three nerves: the iliohypogastric, the ilioinguinal, and the genitofemoral. The femoral and obturator nerves, though they don't innervate the groin, are at

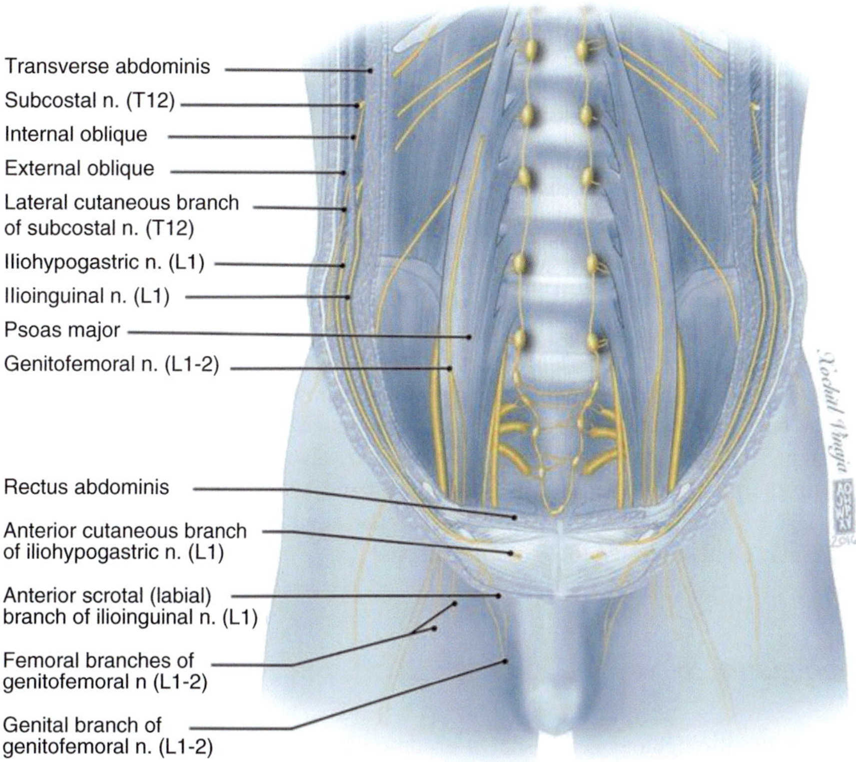

Fig. 3.4 This illustration highlights the innervation of the inguinal region by nerves originating from the lumbar plexus

risk of damage due to pathology or intervention in this region. Thus, they will also be briefly mentioned as knowledge of their course is a necessity prior to manipulation of this area.

Iliohypogastric Nerve

The iliohypogastric nerve commonly shares its origin with the ilioinguinal nerve. Its fibers originate mainly from L1, but it may also have some contribution from T12. It emerges anterior to the quadratus lumborum and posterolateral to the origin of the psoas muscles. It continues to course on the anterior belly of the quadratus lumborum muscle until it crosses its lateral margin. At this point, it pierces the transversus abdominis muscle and continues in the intermuscular plane between the transversus abdominis and the internal oblique muscles. It continues superior and parallel to the iliac crest and gives off a lateral cutaneous branch (iliac branch) near to the anterior superior iliac spine. The lateral cutaneous branch sits between the internal and external oblique above the iliac crest. It innervates the skin in the posterolateral gluteal region. An anterior cutaneous branch (hypogastric branch) continues between the internal oblique and the transversus abdominis. As it continues on its medial course, it penetrates the internal oblique and the external oblique aponeurosis to enter the subcutaneous area about 3 cm above the superficial inguinal ring. It innervates the skin just superior to the pubic symphysis. Apart from this sensory function, it also innervates the abdominal muscles that it comes into contact with. The iliohypogastric nerve communicates with neighboring nerves, namely, the subcostal and ilioinguinal nerves.

The lateral cutaneous branch of the inguinal nerve may sometimes be replaced by the lateral thoracic branch of T12. Another possible variation occurs when the anterior cutaneous branch supplies the pyramidalis muscle. At times the hypogastric branch may be replaced by the ilioinguinal nerve in the region of the external inguinal ring. The fibers of the hypogastric branch may also combine with those of the twelfth thoracic nerve. Instead of originating from L1, the iliohypogastric nerve may also originate from T12 and may even obtain some of its fibers from T11. The nerve may be absent in up to 20.6% of persons.

Ilioinguinal Nerve

Though the ilioinguinal nerve shares a common origin with the iliohypogastric nerve, its nerve fibers are usually solely from the L1 nerve root. It takes a similar but more inferior course to the iliohypogastric nerve. It travels above the iliac crest, piercing the transversus abdominis near the anterior superior iliac spine. Further medially it penetrates the inner oblique. It provides motor innervation to these muscles that it comes into contact with. It exits medially through the superficial inguinal ring and branches into an anterior scrotal (labial) branch, a small pubic branch, and crural branches. The anterior scrotal (or labial) branches conduct sensory stimuli from the anterior scrotum or labia majora. The small pubic branch innervates a small area at the base of the penis or clitoris and mons pubis. The crural branches innervate the upper inner thigh and inguinal crease.

Several deviations from the textbook norm have been observed with the ilioinguinal nerve. One such example is seen when it originates from L2 instead of L1. Additionally, in about 5% of cases, it may be formed from two spinal nerve roots. Previous studies have shown that the nerve originates from the lumbar plexus in about 72.5%. In 25% of cases, it arises from a common trunk with the iliohypogastric nerve.

It is surgically important to note that the ilioinguinal nerve may completely bypass the inguinal ring. At times it may pass deep to the inguinal ligament. Also, the ilioinguinal nerve may join with the iliohypogastric in cases where the former is very small. In such cases, it is replaced by a branch of the iliohypogastric nerve. In cases where the ilioinguinal nerve is absent, the iliohypogastric nerve (most commonly), the genital branch of the genitofemoral, or the femoral branch of the genitofemoral innervates its region. On the other hand, the ilioinguinal nerve may innervate nearby areas if other nerves are absent. It may partially or totally replace the lateral femoral cutaneous nerve or the genital branch of the genitofemoral nerve.

Genitofemoral Nerve

The first and second lumbar nerves both contribute to the genitofemoral nerve. As the nerve emerges, it pierces through the cephalad portion of the psoas major muscle. It continues caudally, anterior to the belly of the psoas major muscle and deep to the psoas fascia before it divides into a lateral femoral branch and a medial genital branch. The femoral branch travels lateral to the external iliac artery and continues deep to the inguinal ligament into the femoral sheath. It then pierces the anterior lamina of the femoral sheath and the fascia lata superficial and lateral to the origin of the femoral artery. The femoral branch provides sensation to an area of skin on the anterior central thigh, just inferior to the inguinal ligament. It also provides sensory innervation to the femoral artery through its connections with the femoral intermediate cutaneous nerve. The genital branch enters the deep ring of the inguinal canal and innervates the cremasteric muscle within the wall of the spermatic cord in men. Alternatively, it may pass superficial to the deep inguinal ring, traveling in the aponeurosis of the external abdominal oblique. This branch also provides sensory innervation to the anterior scrotum in men and the mons pubis and anterior labia majora in women.

The genital and femoral branches can arise from different locations in the lumbar plexus, either from L1, L2, or occasionally L3. The genital branch may also contain fibers from the T12 ventral ramus. The nerve may divide prior to exiting the psoas muscle in about 20% of cases. When the genitofemoral nerve is absent, the distribution for the genital branch is covered by the ilioinguinal nerve, while the anterior and lateral cutaneous femoral nerves innervate the territory of the femoral branch. Similarly, when the ilioinguinal nerve is absent, branches of the genitofemoral nerve may replace it.

Femoral Nerve and Obturator Nerve

The femoral nerve contains fibers originating from the L2–L4 spinal nerve roots. Initially, it travels between the lateral aspect of psoas major and the iliacus muscle. On its way to the inguinal region, it innervates the iliacus muscle. It then passes deep to the inguinal ligament and into the femoral sheath that is formed, in part, by the transversalis fascia. Also within the femoral sheath are the femoral artery and vein and the node of Cloquet. Within the femoral canal, it lies lateral to the femoral artery, though it may at times be found between the artery and the vein. The femoral nerve innervates the flexors of the hip and the extensors of the knee. It provides sensorineural innervation to the anterior thigh, anteromedial knee, medial leg, and medial foot.

Similar to the femoral nerve, the obturator nerve has contributions from L2–L4 lumbar nerves. It emerges medial to the psoas major muscle and enters the lesser pelvis dorsal to the common iliac vessels then lateral to the internal iliac vessels.

It exits the pelvis through the obturator foramen after dividing into anterior and posterior branches. The anterior branch innervates the medial thigh, the hip joint, and the adductors. The posterior branch passes posterior to the adductor brevis to also supply innervation to the adductors of the hip.

Lymphatics of the Groin

The lymphatic system consists of an interconnected network of channels that begin as close-ended, porous tubes that converge to drain into large veins in the subclavian region. They function to return both nutrients lost during cell-to-cell exchange and interstitial fluid back into circulation. Also contained in the lymphatic fluid are pathogens, immune cells, metabolic products of cells, and cellular debris. Along the course of these lymphatic channels, collections of small encapsulated structures, or lymph nodes, appear in fairly predictable locations. Two such groups, located in the groin, are the superficial inguinal nodes and the deep inguinal nodes as depicted in Fig. 3.5.

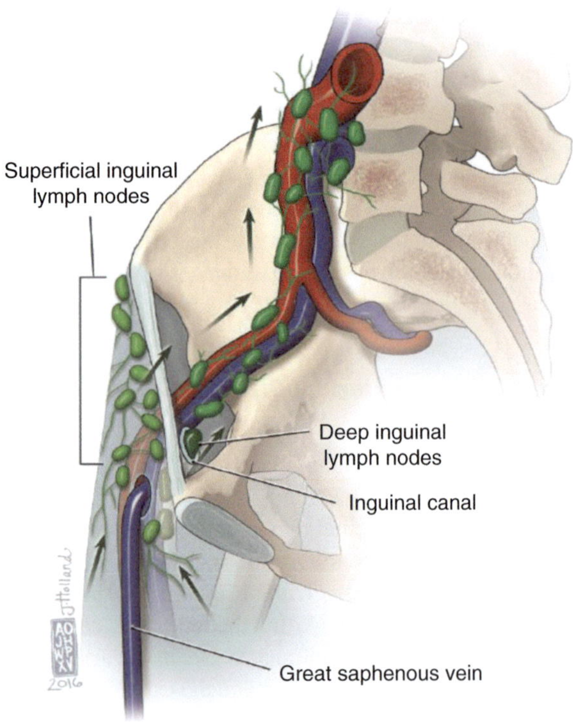

Fig. 3.5 In this medial-to-lateral view of the right hemipelvis, the relationship between the inguinal and pelvis nodes can be seen

Superficial Inguinal Nodes

The superficial inguinal lymph node group is located in the most proximal region of the lower limb. The lymph nodes are arranged parallel to the inguinal ligament (Fig. 3.6). An important differentiating factor between the deep and superficial groups is their relationship to fascias of the inguinal region. The superficial group is located deep to Camper's fascia (also referred to as the superficial fascia) and superficial to the fascia lata (or deep fascia). On the other hand, the deep group lies deep to the fascia lata and is arranged parallel to the femoral vein. The drainage channels that initiate the collection of lymph in the superficial group are located in the dermis. The channels carrying lymph toward the superficial group

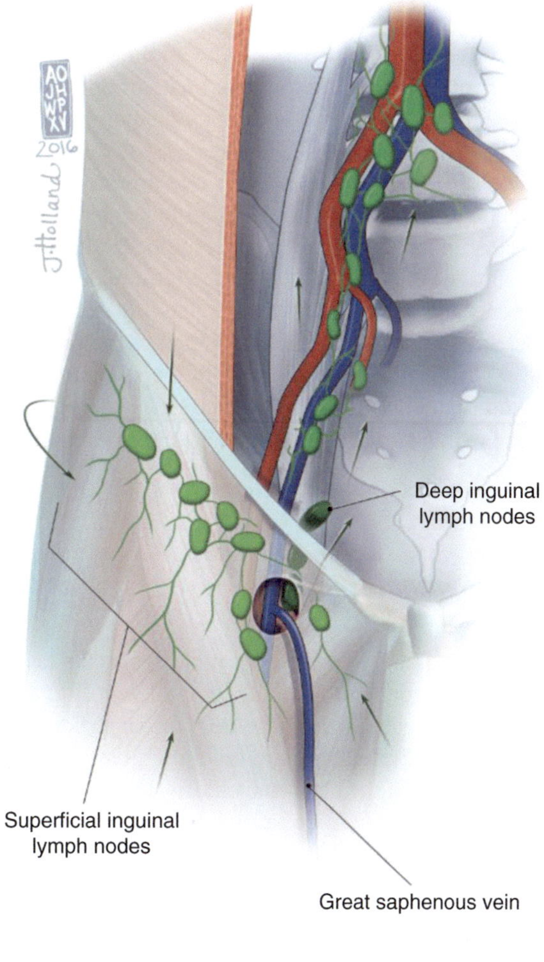

Fig. 3.6 In this figure, the superficial inguinal nodes can be seen on the surface of the fascia lata, inferior to the inguinal ligament

of nodes do not follow the course of the vessels even though they share a similar distribution with the vessels. For the deep inguinal system, their initial vessels are located in the fascia, muscles, periosteum, perichondrium, aponeurosis, and near joints. The channels comprising the deep system follow the course of the blood vessels in the area.

There have been a few proposed methods of dividing the superficial inguinal lymph node group into smaller subgroups. Rouviere divided this group of approximately ten lymph nodes (range of about 4–25) into five groups. In this subclassification system, the junction of the great saphenous vein and the femoral vein was used as the point of reference. Thus, the five groups were named superolateral, superomedial, central, inferolateral, and inferomedial. In Romanes' version of subgroups, the point where the great saphenous vein is terminated was also used as the reference point. However, in this case, only two categories were named. The proximal group was defined as the group that was located in close proximity to the inguinal ligament (about 1 cm distal to its inferior margin), whereas the distal group was that group of nodes closely associated with the termination of the great saphenous vein. Further subdivisions of the proximal group have been described. They can be divided into superolateral and superomedial groups in a similar fashion to that proposed by Rouviere.

The superolateral nodes drain two areas in particular. Their anterior source of lymph originates from the infraumbilical anterior abdominal wall, and the posterior source originates from the gluteal region. The superomedial nodes drain some of the structures of the genitourinary system. In males, these structures are the prepuce of the penis and the scrotum. Also drained by this group of nodes are the inferior anal canal and the perianal region. In females, the prepuce of the clitoris, the superolateral aspect of the uterus along with the structures of the external female genitalia, and the vaginal orifice are drained by the superomedial nodes. The lower limb, with the exception of the posterolateral thigh, is drained by the distal group of lymph nodes.

Deep Inguinal Nodes

The deep inguinal nodes are located deep to the fascia lata of the thigh and tend to be in close association with the femoral vein (Fig. 3.7). This cluster of one to three lymph nodes receives drainage from the superficial group of lymph nodes as well as deep lymphatics that run with the femoral artery. They also receive direct drainage from the glans of the penis and clitoris. In cases where three deep nodes are present, their locations seem to be fairly constant. One is located just lateral to the femoral ring, one is found within the femoral canal, and one distal to the termination of the great saphenous vein, but still deep to the fascia lata. Though the presence of the node lateral to the femoral ring is variable, the node within the femoral canal is

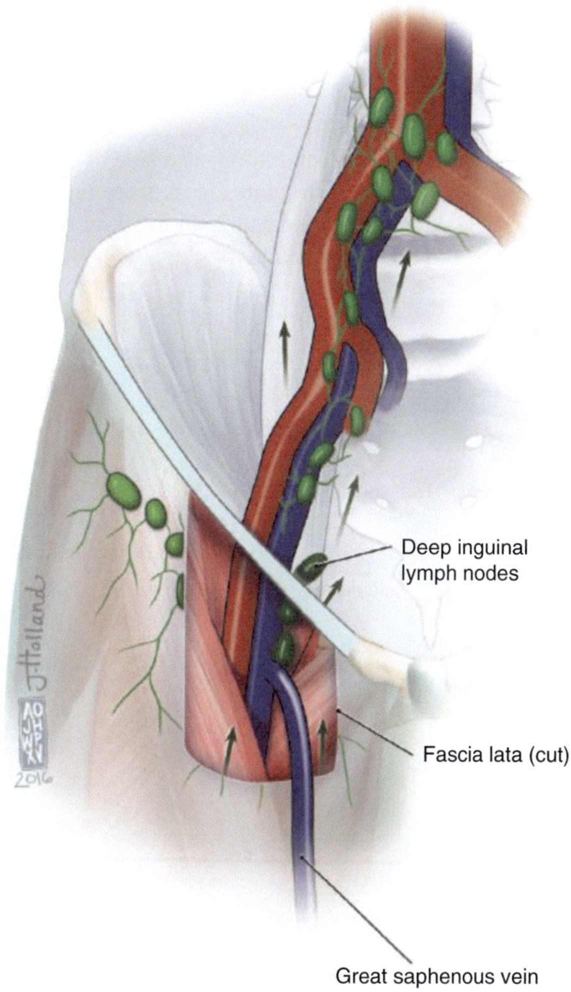

Fig. 3.7 With a segment of the fascia lata removed, the deep inguinal nodes can be easily visualized

almost always present. This frequently present node is also known as the node of Cloquet. Cloquet's node drains the deep thigh and communicates directly with the iliac and obturator nodes.

Aberrant Nodes

Aberrant nodes are small nodes that may be located at the base of the penis, anterior to the pubic symphysis, or within the inguinal canal.

Anatomical Spaces of the Groin

Inguinal Canal

The inguinal canal is a conduit on each side of the midline of the lower abdominal wall that allows certain structures to pass through (Fig. 3.8). In adults, its length is approximately 4 cm and located 2–4 cm above the inguinal ligament. The canal is defined by the following boundaries:

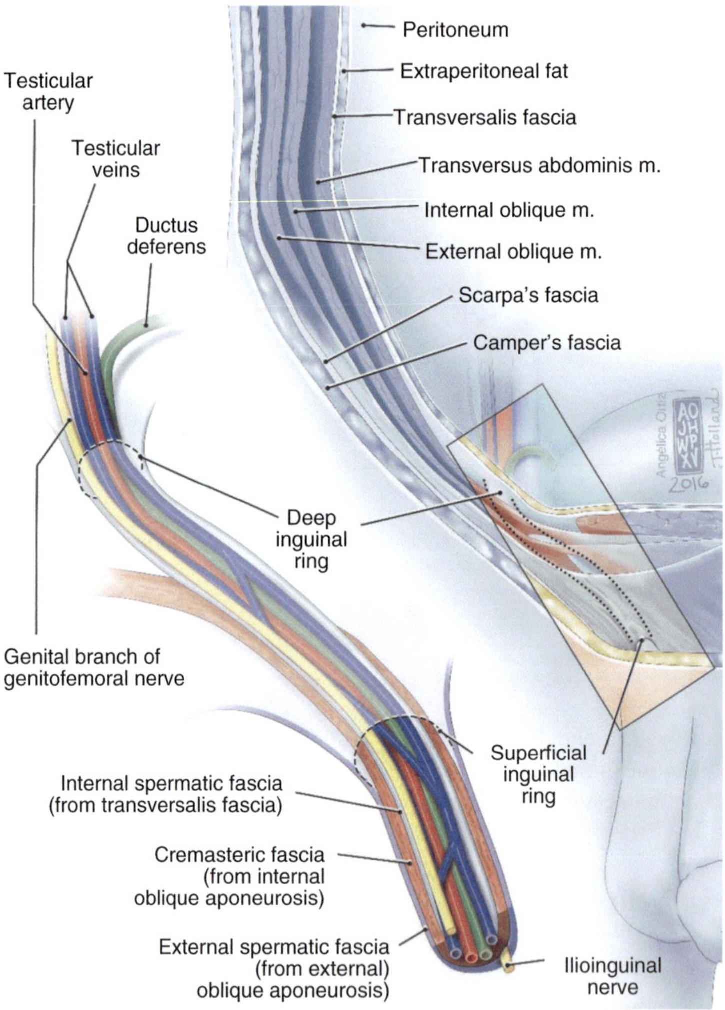

Fig. 3.8 This illustration highlights the course of the inguinal canal—from the deep inguinal ring to the superficial inguinal ring. Also seen here are the contents of the inguinal canal

Anterior: The external oblique aponeurosis forms the anterior wall. On the lateral part, it is also reinforced by the aponeurosis of internal oblique muscle.

Posterior (floor): In most of the population, the lateral side of the posterior wall is formed by the aponeurosis of transversus abdominis muscle and the transversalis fascia. The medial side of the posterior wall is reinforced by the internal oblique aponeurosis.

Superior (roof): The superior wall of the inguinal canal is composed of the arched fibers of the lower edge of the internal oblique muscle and by the transverse abdominis muscle and aponeurosis.

Inferior: The inferior wall of the canal is formed by the inguinal ligament (*Poupart's ligament*) and the lacunar ligament (*Gimbernat's ligament*).

Variations

In about 25% of individuals, the lateral side of the posterior wall is formed only by the transversalis fascia.

Contents of the Inguinal Canal

The contents of inguinal canal will enter through the internal (deep) inguinal ring, pass through the canal, and exit from the external (superficial) inguinal ring.

Internal (deep) inguinal ring: It is a normal defect, where an opening/evagination of the transversalis fascia and transversus aponeurosis forms a shape that appears like an inverted "V" or "U." The internal (deep) inguinal ring's location corresponds to the midpoint of the inguinal ligament, lying superiorly to it. The ring's anterior and posterior arms are thickened parts of the transversalis fascia. The inferior border is the iliopubic tract.

External (superficial) inguinal ring: This opening is formed by the external oblique aponeurosis. It is shaped like a triangle with its base being a part of the pubic crest and its two borders formed by two crura, the superior (medial) crus and the inferior (lateral) crus. The superior crus is formed by the external oblique aponeurosis and the inferior crus by the inguinal ligament.

Variations of the Inguinal Canal Contents

The contents of the inguinal canal differ between men and women. In men, the canal contains the spermatic cord and the ilioinguinal nerve and in females the round ligament of the uterus and the ilioinguinal nerve. Even though the ilioinguinal nerve exits from the superficial inguinal ring, it does not enter from the deep ring; therefore it only travels through a part of the canal. In addition, the ilioinguinal nerve in males is not a component of the spermatic cord, rather it is located on the outside traveling next to it.

Spermatic Cord

The components of the spermatic cord include:

Three veins: testicular vein, cremasteric vein, and deferential vein
Three arteries: testicular artery, cremasteric artery, and deferential artery
Two nerves: genital branch of the genitofemoral nerve and the testicular plexus
The pampiniform plexus, the ductus deferens, and **lymphatics**

The pampiniform plexus lies anteriorly to the cord. Posteriorly to the cord lies the ductus deferens and the remnant of processus vaginalis or hernial sac.

The contents of the spermatic cord are covered by three layers of fascia: the external spermatic fascia (from *external oblique fascia*), the cremasteric fascia (from *internal oblique muscle and fascia*), and the internal spermatic fascia (from *transversalis fascia*).

Femoral Triangle

The femoral triangle is a structural landmark that can be delineated in a standing position where the muscles of the anterior compartment of the thigh are flexed with the hip externally rotated and slightly abducted. Its triangular shape has the apex pointing inferomedially, and it is defined by the following boundaries (Fig. 3.9):

Superiorly (Base): The base of the femoral triangle is the inguinal ligament.
Medially: It is formed by the lateral border of the adductor longus muscle.
Laterally: It is formed by the medial border of the sartorius muscle.

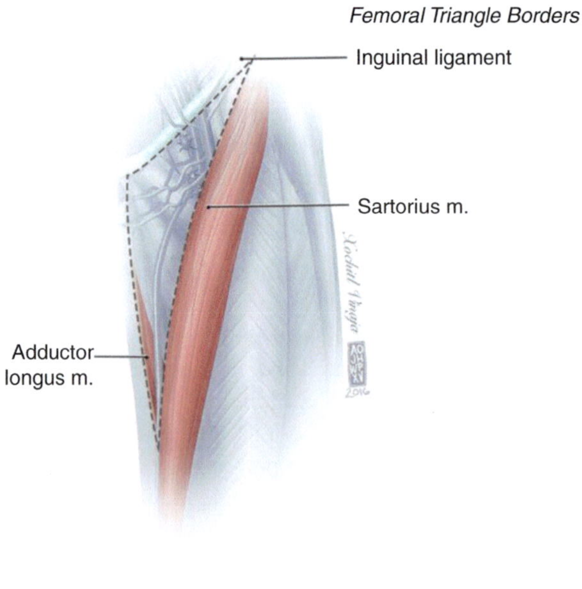

Fig. 3.9 The borders of the femoral triangle: the base is formed by the inguinal ligament; the medial border is formed from the lateral border of the adductor longus muscle; and the lateral border is by the medial edge of the sartorius muscle

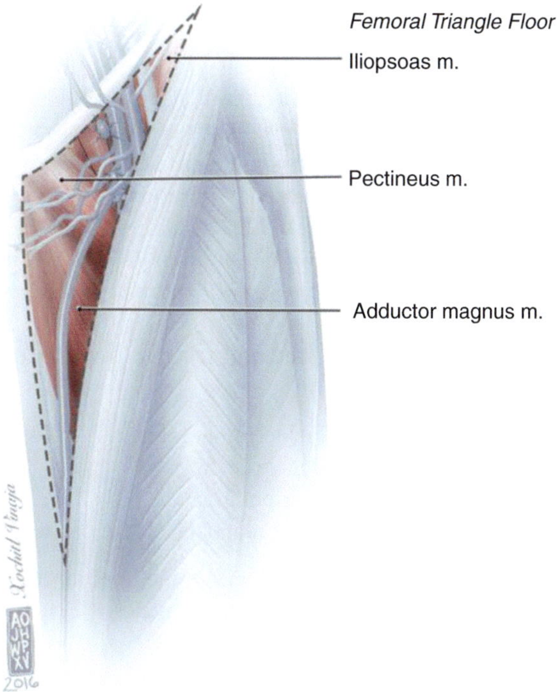

Fig. 3.10 The floor of the femoral triangle is formed from the iliopsoas, the pectineus, and the adductor magnus muscles

Floor: It is formed medially by the adductor longus and pectineus muscles and laterally by the iliacus and psoas major muscles (Fig. 3.10).

Roof: It is formed by the fascia lata.

Apex: It is formed by the intersection of the medial border of the sartorius muscle and the lateral boarder of the adductor longus muscle.

Landmarks

The *inguinal ligament* can be felt running from the anterior superior iliac spine to the pubic tubercle when the thigh is externally rotated and abducted.

The *sartorius* muscle can be best outlined, in a sitting position, as a "strap-like" muscle when the hip is flexed, the knee extended, and the thigh slightly abducted and externally rotated.

While tracing the sartorius: **Proximally**, its course will be from the anterior superior iliac spine, running inferomedially, half way to the thigh. **Distally**, it may appear as a soft longitudinal ridge passing toward the posterior part of the medial femoral condyle.

The medial part of *adductor longus* can be felt as a crease when the thigh is adducted against resistance.

Contents of the Femoral Triangle

The femoral triangle is a passageway where structures will pass from the abdominal/pelvic cavity to the lower limb. As the inguinal ligament crosses from the anterior superior iliac spine to the pubic tubercle, it produces a space called **retroinguinal** space, which is further separated into two compartments (lateral and medial) by the **iliopectineal arch** (a thickened portion of the iliopsoas fascia). The **lateral compartment**, also called *muscular compartment of the retroinguinal space*, will contain the iliopsoas muscle and the femoral nerve. The **medial compartment**, also called *vascular compartment of the retroinguinal space*, will contain arteries, veins, and lymphatics.

The contents of the femoral triangle from lateral to medial are the following (Fig. 3.11):

- **Femoral nerve** and its terminal branches
- **Femoral artery** and its branches
- **Femoral vein** and its tributaries
- **Femoral canal** which contains fat, loose connective tissue, and lymphatics

Cloquet's node is located within the femoral canal and is thought to be the link between inguinal and iliac/obturator nodes.

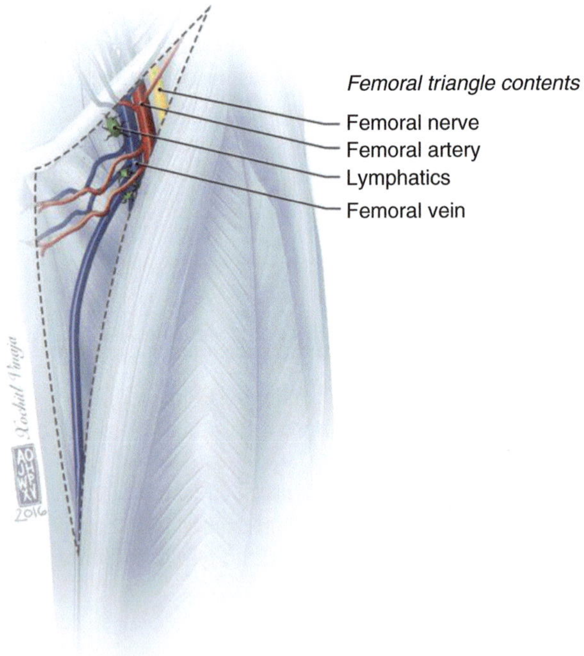

Fig. 3.11 The contents of the femoral triangle are highlighted here: the femoral nerve, femoral artery, femoral vein, and lymphatics

Variations

It is important to note that, occasionally in the femoral triangle, the femoral nerve can appear as a collection of nerve branches. There is a degree of variability of the location where the femoral nerve divides as described in the literature. Sometimes, it can branch after entering the thigh, or below the inguinal ligament, or upon its course in the femoral triangle.

A conventional way to remember the contents of the femoral triangle is by applying the mnemonic *NAVEL* (nerve, artery, vein, empty space, and lymphatics). However, there have been some instances where the femoral artery lies medially to the femoral vein. Also, a case report described the femoral artery crossing the femoral vein deep into the inguinal ligament and overlying it at the femoral triangle base.

The Femoral Sheath

The femoral vessels and the lymphatics are the deepest structures, passing from midbase to apex of the femoral triangle, and are enclosed by the **femoral sheath** (Fig. 3.12). The femoral nerve with its terminal branches runs lateral and is located outside the femoral sheath.

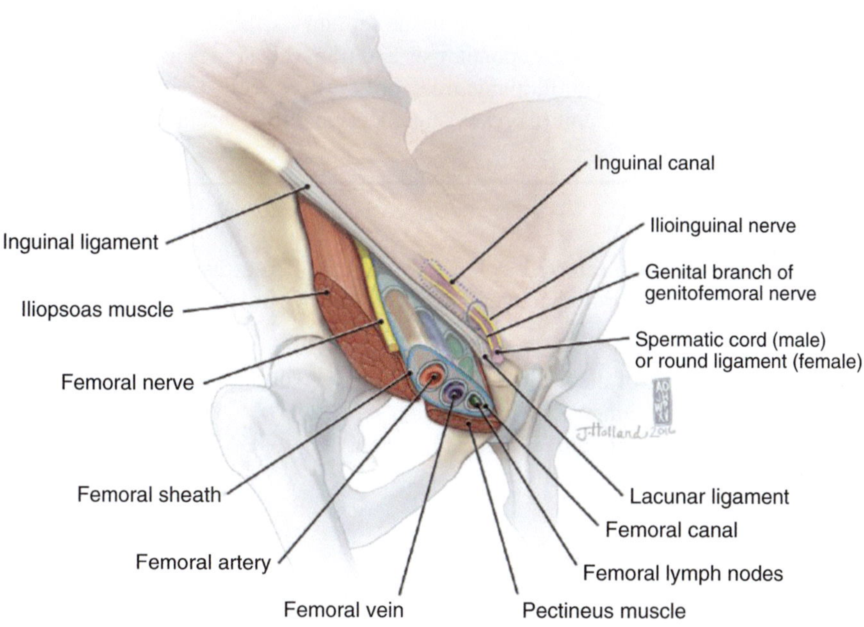

Fig. 3.12 This anterior view of the inguinal region shows the contents of the femoral sheath. The femoral artery, femoral vein, and lymph nodes and vessels are found within the sheath. The femoral nerve lies outside the sheath. The femoral canal represents the most medial region enclosed by the femoral sheath and contains lymph nodes and lymphatic vessels

The femoral sheath is a distal prolongation of extraperitoneal fascia that encloses the vascular compartment. It is formed anteriorly by the transversalis fascia and posteriorly by the iliac fascia. It has the shape of a cone, with its proximal end being wider as it narrows distally and fusing with the vascular adventitia, approximately 3–4 cm from the inguinal ligament. Its role is to provide protection to the vessels, especially during hip joint movements.

The femoral sheath encases the vascular compartment, which is further subdivided by connective tissue into septa, thereby forming the three compartments:

Lateral compartment: It contains the femoral artery.
Intermediate compartment: It contains the femoral vein.
Medial compartment: It contains the femoral canal, which contains lymph vessels and occasionally a deep inguinal node.

The Femoral Canal

The femoral canal (Fig. 3.12) apart from containing lymphatic vessels and lymph nodes, and allowing the femoral vein to distend, is also of high importance in clinical medicine, while it can serve as a conduit for femoral hernias. The canal has the shape of a cone and approximately measures 1.25 cm in length. It is bounded by the following structures:

Lateral: Femoral vein and a connective tissue septum
Medial: Transversus abdominis aponeurosis and transversalis fascia
Anterior: Inguinal ligament, iliopubic tract, or both
Posterior: Pectineal ligament (*Cooper's* ligament) and iliac fascia

The Proximal Ring

The proximal ring, also referred as *the femoral ring*, serves as an entrance to the femoral canal. It is wider in diameter, and its boundaries are as follows:

Lateral: Femoral vein
Medial: Lateral edge of the lacunar ligament, transversus aponeurosis, or both
Anterior: Inguinal ligament
Posterior: Pectineal ligament (*Cooper's* ligament)

The Distal Ring

The saphenous hiatus will serve as an exit of the femoral canal. It is secured by the distal ring whose boundaries are as follows:

Lateral: Femoral sheath
Medial: Lacunar ligament or iliopubic tract
Anterior: Fascia lata and cribriform fascia
Posterior: Pectineal fascia

Variations

In women, the femoral canal is larger than it is in men. This is due to the fact that women have a wider pelvis and smaller femoral vessels. In addition, the round ligament of the uterus is located above the anterior margin in women, where in men, it is the spermatic cord.

Bibliography

1. Agur AM, Dalley AF II. Grant's atlas of anatomy. 11th ed. Philadelphia, PA: Lippincott, Williams and Wilkins-Wolters Kluwer; 2005. p. 364.
2. Bandyopadhyay M, Biswas S, Roy R. Vessels in femoral triangle in a rare relationship. Singap Med J. 2010;51(1):e3–5.
3. Bannister LH. Functional neuroanatomy of man; being the neurology section. In: Williams PL, Warwick R, Dyson M, editors. Gray's anatomy (vol. 378). Edinburgh: Churchill Livingstone; 1989.
4. Bardeen CR. A statistical study of the abdominal and border nerves in man. Am J Anat. 1902;1(2):203–28.
5. Bergman RA, Thompson SA, Afifi AK. Catalog of human variation. Berlin: Urban & Schwarzenberg; 1985.
6. Bergman RA, Thompson SA, Afifi AK, Saadeh FA. Compendium of human anatomic variation. Baltimore, MD: Urban & Schwarzenberg; 1988.
7. Cesmebasi A, Baker A, Du Plessis M, Matusz P, Shane Tubbs R, Loukas M. The surgical anatomy of the inguinal lymphatics. Am Surg. 2015;81(4):365–9.
8. Chen H, Sonneday C, Lillemoe K. Arterial and venous access. Manual of common bedside surgical procedures. 2nd ed. Philadelphia, PA: Lippincott Williams & Wilkins; 2000. p. 77.
9. Dwight T, McMurrich JP, Hamann CA, White JW. Human anatomy: including structure and development and practical considerations, 1st edn, vol. 2. Philadelphia, PA: JB Lippincott Company; 1913.
10. Enneking FK, Chan V, Greger J, Hadzic A, Lang SA, Horlocker TT. Lower extremity peripheral nerve blockade: essentials of our understanding. Reg Anesth Pain Med. 2005;30:4–35.
11. Falce L, Lanzoni O, Ambrosio JD, Souza RRD. The anatomy of the superficial external pudendal artery: a quantitative study. Clinics. 2006;61(5):441–4.
12. Fathi M, Hatamipour E, Fathi HR, Abbasi A. The anatomy of superficial inferior epigastric artery flap. Acta Cirur Brasil. 2008;23(5):429–34.
13. Fischer JE. Mastery of surgery. 6th ed. Philadelphia, PA: Wolters Kluwer Health/Lippincott Williams & Wilkins; 2012. Chapters 195, 226
14. Ghassemi A, Furkert R, Prescher A, Riediger D, Knobe M, O'dey D, Gerressen M. Variants of the supplying vessels of the vascularized iliac bone graft and their relationship to important surgical landmarks. Clin Anat. 2013;26(4):509–21.
15. Griffin M. Some varieties of the last dorsal and first lumbar nerves. J Anat Physiol. 1891;26(Pt 1):48.
16. Hansen, J. T., Benninger, B., Brueckner-Collins, J., Hoagland, T., Tubbs, R. S. (eds). (2014). Atlas of human anatomy. 6th edn. Philadelphia, PA, Elsevier Health Sciences.
17. Hester TR Jr, Nahai F, Beegle PE, Bostwick J. Blood supply of the abdomen revisited, with emphasis on the superficial inferior epigastric artery. Plastic Reconstruct Surg. 1984;74(5):657–66.
18. Hollinshead WH. Anatomy for surgeons. 2. The thorax, abdomen, and pelvis. New York, NY: Hoeber-Harper; 1961.
19. Kawai K, Honma S, Koizumi M, Kodama K. Inferior epigastric artery arising from the obturator artery as a terminal branch of the internal iliac artery and consideration of its rare occurrence. Ann Anat. 2008;190(6):541–8.

20. Koh YX, Chok AY, Zheng H, Xu S, Teo MC. Cloquet's node trumps imaging modalities in the prediction of pelvic nodal involvement in patients with lower limb melanomas in Asian patients with palpable groin nodes. Eur J Surg Oncol. 2014;40(10):1263–70.
21. Kutsuna M. The anatomy of the lymphatic system in the Japanese. Tokyo: Kanehara Shuppan; 1968. p. 219–25.
22. Lonchena TK, McFadden K, Orebaugh SL. Correlation of ultrasound appearance, gross anatomy, and histology of the femoral nerve at the femoral triangle. Surg Radiol Anat. 2016;38(1):115–22.
23. Moore KL, Dalley AF, Agur AM. Clinically oriented anatomy. Lippincott Williams & Wilkins: Baltimore, MD; 2013.
24. Moosman DA, Oelrich TM. Prevention of accidental trauma to the ilioinguinal nerve during inguinal herniorrhaphy. Am J Surg. 1977;133(2):146–8.
25. Onkar D, Onkar P, Mitra K. Isolated bilateral external iliac vein aplasia. Surg Radiol Anat. 2013;35(1):85–7.
26. Penteado CV. Venous drainage of the groin flap. Plast Reconstr Surg. 1983;71(5):678–82.
27. Putz R, Pabst R. Sobotta-atlas of human anatomy: head, neck, upper limb, thorax, abdomen, pelvis, lower limb. Two- volume set. Starkville, MS: Campus Book Mart; 2006.
28. Reardon CM, O'Ceallaigh S, O'Sullivan ST. An anatomical study of the superficial inferior epigastric vessels in humans. Br J Plast Surg. 2004;57(6):515–9.
29. Sinna R, Hajji H, Qassemyar Q, Perignon D, Benhaim T, Havet E. Anatomical background of the perforator flap based on the deep branch of the superficial circumflex iliac artery (SCIP flap): a cadaveric study. Eplasty. 2010;18(10):e11.
30. Skandalakis JE. Skandalakis' surgical anatomy: the embryologic and anatomic basis of modern surgery. 1st ed. Athens: Paschalidis Medical Publications; 2004.
31. Sperling M, Winter G, Weiland W. Duplication of the external and common iliac veins with a venous trunk anterior to the artery (author's transl). Munch Med Wochenschr. 1975;117(39):1547–50.
32. Standring S. Gray's anatomy: the anatomical basis of clinical practice. 40th ed. Madrid: Churchill Livingstone; 2008.
33. Stevenson TR, Hester TR, Duus EC, Dingman RO. The superficial inferior epigastric artery flap for coverage of hand and forearm defects. Ann Plast Surg. 1984;12(4):333–9.
34. Taylor GI, Daniel RK. The anatomy of several free flap donor sites. Plast Reconstr Surg. 1975;56(3):243–53.
35. Thein T, Kreidler J, Stocker E, Herrmann M. Morphology and blood supply of the iliac crest applied to jaw reconstruction. Surg Radiol Anat. 1997;19(4):217–25.
36. Tubbs RS, Shoja MM, Loukas M. Bergman's comprehensive encyclopedia of human anatomic variation. Hoboken, NJ: John Wiley & Sons Inc.; 2016.
37. Williams A, Newell RL, Davies MS, Collins P. Pelvic girdle and lower limb. In: Standring S, editor. Gray's anatomy. 39th ed. Philadelphia, PA: Churchill Livingstone; 2005. p. 1503–5.
38. Zaluska S. Internal structure of the ilioinguinal nerve in postfetal life in man. Folia Morphol (Warsz). 1976;35(1):41.

Chapter 4
Complications of Inguinal Lymphadenectomy

Avinash Chenam and S. Mohammad A. Jafri

Abbreviations

DVT	Deep vein thrombosis
ILND	Inguinal lymph node dissection
LND	Lymph node dissection
PE	Pulmonary embolism, endoscopic inguinal lymphadenectomy

Introduction

Penile squamous cell carcinoma is a rare entity with an incidence of less than 1 per 100,000 males [1]. At initial presentation, 50% of patients with penile squamous cell carcinoma have inguinal lymphadenopathy, but only half of them have metastatic lymph node involvement [2]. It is one of the few urologic malignancies potentially curable by regional lymphadenectomy. The presence and severity of these nodal metastases have been shown to be the single most important predictor of cancer-specific survival [3]. In addition to refining pathologic staging, inguinal

A. Chenam, M.D.
Division of Urology and Urologic Oncology,
Department of Surgery, City of Hope National Medical Center,
1500 E. Duarte Rd, MOB L002H, Duarte, CA 91010, USA
e-mail: achenam@coh.org

S.M.A. Jafri, M.D. (✉)
Department of Urology, Beaumont Health System,
Oakland University William Beaumont School of Medicine,
3535 W. 13 Mile Road, Suite 438, Royal Oak, MI 48073, USA
e-mail: s.jafri@beaumont.org

K.A. Delman, V.A. Master (eds.), *Malignancies of the Groin*,
https://doi.org/10.1007/978-3-319-60858-7_4

lymph node dissection (ILND) remains the most effective means of eradicating minimal metastatic disease in invasive penile cancer patients. Over time, the strategies concerning the indication of ILND as well as surgical technique have changed dramatically. Thus, different template extensions and perioperative management—as well as inconsistent methodology of complication definition, grading, and reporting—have contributed to a great variability of ILND complication rates reported in the literature.

Traditional groin dissection has been associated with a high complication rate [4–7] (Table 4.1). Radical ILND involves a 10-cm length skin incision with extensive dissection field involving the superficial and deep inguinal nodes with complete exposition of the femoral vessels, division of the great saphenous vein, and transposition of the sartorius muscle. The boundaries of the dissection are as follows: proximally, the inguinal ligament; distally, the entrance of Hunter's canal where the femoral vessels go under the muscles of the leg, medially is the adductor muscle, and laterally is the sartorius muscle. The floor of the dissection consists of the fascia lata, the femoral vessels, and the pectineus muscle. Morbidity with this template includes phlebitis, wound infection, pulmonary embolism, wound dehiscence, flap necrosis, and lymphedema. Due to the frequency and severity of complications, many physicians have been reluctant to offer ILND to patients with penile cancer particularly in the absence of palpable inguinal lymphadenopathy.

Over time, significant advances in the surgical approach to penile cancer have been made. It is now realized that not all patients require radical surgery to stage or even treat the inguinal region. The improved pre- and postoperative care, modification of the extent of the dissection, advances in surgical technique, plastic surgery consultation for myocutaneous flap coverage, and preservation of the dermis, Scarpa's fascia, and saphenous vein have decreased the incidence of ILND complications [8, 10–12]. Contemporary surgical series report decreased ILND complication rates during the last two decades and have been lower for patients undergoing modified ILND [16–18, 20, 23–26].

In this chapter, we review common complications of ILND, modifications of surgical technique and its effect on surgical morbidity, and potential preventative and management strategies of these problems.

Modified Dissections

Both superficial inguinal and modified complete dissections have been proposed as staging tools for the patient without palpable inguinal lymphadenopathy. Superficial node dissection involves removal of those nodes superficial to the fascia lata. The rationale for the superficial dissection is based on some series, which haven't shown positive nodes deep to the fascia lata unless superficial nodes were also positive [21, 27].

Coblentz et al. proposed a modified ILND to reduce the morbidity and to preserve oncologic control [25]. It involves a smaller skin incision (6–7 cm), preservation of the saphenous vein, and thicker skin flaps. It also narrows the field of inguinal

Table 4.1 Incidence of common complications after inguinal lymph node dissection for penile cancer

Series	Year	Patients (n)	Overall complication rate (%)	Wound infection (%)	Wound dehiscence + necrosis (%)	Lymphocele (%)	Lymphedema (%)
Johnston et al. [8]	1984	67	82	14	50	9	50
Darai et al. [9]	1988	85	–	12	14	3	32
Ornellas et al. [10]	1991	200	–	15	45	–	23
Ravi et al. [11]	1993	112	–	–	25	9	16
Kamat et al. [12]	1993	31	87	–	–	–	–
Bouchot et al. [13]	1993	32	–	29	44	–	44
Ayyappan et al. [14]	1994	78	–	70	36	87	57
Lopes et al. [15]	1996	145	–	–	18	–	30
Bevan-Thomas et al. [16]	2002	53	57	10	8	–	23
Coblentz et al. [17]	2002	22	45	9	9	27.2	0
Bouchot et al. [18]	2004	88	42	3	12	–	22
Nelson et al. [19]	2004	22	–	8	10	15	15
Spiess et al. [20]	2009	43	49	9	11	2	17
Gopman et al. [21]	2015	327	55	32	–	8	22
Koifman et al. [22]	2013	170	10	1	2	2	4

dissection excluding the area lateral to the femoral artery and caudal to the fossa ovalis [23, 25]. This technique also avoids transposition of the sartorius muscle to cover exposed femoral vessels. Unlike in superficial dissection, deep nodes within the fossa ovalis are also removed in a modified ILND. These maneuvers result in less severe disruption of the lymphatic collaterals and less vascular damage compared to a radical ILND. A long-term follow-up in two series showed that this method was reliable with less morbidity than standard ILND [16, 18]. The incidence of flap necrosis (2.5%), lymphedema (3.4%), and deep venous thrombosis (none) in a group of patients with modified lymphadenectomy was remarkably decreased in comparison with a historical control group of radical lymphadenectomy (skin necrosis 8.6%, lymphedema 22.4%, and deep vein thrombosis (DVT) 12%) [18]. Of note, cases with greater metastatic disease are more likely to be associated with increased morbidity. Although increasing the number of lymph nodes removed increases the likelihood of complications, surgical excision of suspected lymph nodes is necessary for staging as well as therapeutic treatment of the disease [8, 19, 22, 28].

Wound Infection

Wound infection after ILND tends to be one of the most prevalent complications. Historical series [8, 10, 11, 29–31] have reported wound infection rates following ILND between 12 and 29% with one series showing a 70% wound infection rate [32]. The skin is a dynamic home to a large number of bacteria. Microorganisms isolated from groin wounds have included gram-negative rods, *Staphylococcus* species, diphtheroids, and *Peptostreptococcus* [32]. With improved operative technique, timely administration of preoperative antibiotics, and a variety of measures aimed at neutralizing the threat of contamination, infection rates have decreased in contemporary series [20, 27].

Preoperative skin sterilization with an antiseptic is important to remove transient organisms from the skin and decrease wound colonization prior to proceeding with surgical intervention. Additionally, patients should undergo clipping of the surgical site as needed as studies have shown that shaving the skin as compared with clipping results in a statistically significant increase in the rate of surgical site infection [32, 33]. Shaving results in microscopic cuts and abrasions, thus acting as a disruption of the skin's barrier, whereas clippers should not cut into the patient's skin potentially explaining the differences in infection rates.

Even though no comparative studies have been done on the use of prophylactic antibiotics, the potential benefit of decreased wound infection from antimicrobial prophylaxis (broad-spectrum antibiotics, e.g., ampicillin/gentamicin or ampicillin/ciprofloxacin) prior to skin incision is advisable. This type of surgery should be considered a contaminated procedure because of the often coexisting inflammatory reactions in the lymph nodes. Furthermore, in patients with active infection of the groin, bacterial cultures should be obtained and culture-specific antibiotics should be given preoperatively. If the primary tumor is infected, a staged procedure is recommended. Antibiotic therapy for 4–6 weeks has been advised after treatment of the primary penile tumor, to remove the infected source and allow resolution of septic lymphadenitis before ILND [6, 13, 15]. After the groin dissection, it has been suggested that antibiotics should be continued for 1 week or until the wound drains have been removed as migration of bacteria along the drain can increase the risk of infection [6, 9].

Patient characteristics may also guide duration of antibiotic therapy post ILND. Diabetes, cigarette smoking, obesity, and coincident remote site infections or colonization have each shown significant independent association for surgical site infection prediction [14]. It is hypothesized that increased susceptibility to surgical site infections in obese patients results from tissue hypoperfusion, which in turn may lead to greater risk of ischemia or necrosis and suboptimal neutrophil-oxidative killing [34]. Nonetheless, no clear guidelines exist for duration of antibiotics after ILND. Additionally, postoperatively, it is imperative to keep the wound site clean and dry, especially in obese patients as the groin provides a moist environment that may predispose to fungal overgrowth [20].

Intraoperatively, meticulous atraumatic tissue handling should be performed to reduce the risk of wound-related problems such as a lymphocele or hematoma, which could potentially become infected [20]. Excess skin should be excised as well in order to reduce dead space and prevent fluid collections, which similarly could get infected. Additionally, the subcutaneous tissue superficial to the fibrous layer of Camper's fascia should be preserved as devitalized skin flaps are at an increased risk of ischemia, necrosis, infection, and wound dehiscence [4, 18, 20, 23].

Wound Dehiscence and Skin Necrosis

Historically, ILND has been associated with a high rate (25–50%) of wound dehiscence and skin necrosis [8, 10, 11, 13, 14]. A decreased wound complication rate depends on preservation of the blood supply to the skin along with maintenance of collateral lymphatics, which is why knowledge of the vascular surgical anatomy of the groin is imperative. The blood vessels supplying the skin of the inguinal region arise from the superficial branches of the inferior epigastric, external pudendal, and circumflex iliac arteries. All three of these vessels are transected and ligated during the course of an ILND, and the flaps must rely on anastomotic branches and microcirculation for viability. These vessels run parallel to the inguinal ligament and lie in the fat of the superficial layer of the superficial fascia (Camper's fascia) [35, 36]. Consequently, the most physiological incision is parallel to the natural skin folds transecting as few anastomotic vessels in Camper's fascia as possible and maximizing the likelihood of primary wound healing without flap necrosis.

A variety of incisions have been described in the literature including but not limited to horizontal, vertical, T-shaped, S-shaped, and Gibson. Incisions that interrupt the anastomotic vessels in Camper's fascia are vertical incisions, S-shaped incisions, or T-shaped incisions. Postoperative edema leads to excessive traction and tension along the line of the incision resulting in increased incidence of skin necrosis. Ravi and colleagues reported the incidence of flap necrosis was greatest using a T-shaped incision compared with a horizontal or vertical incision [11]. Tonouchi and colleagues compared the operative morbidity of an S-shaped incision versus a straight incision, and the authors noted the incidence of wound infections was significantly higher after S-shaped incisions [37]. Ornellas and colleagues found skin-edge necrosis in 82% of patients with bi-iliac incision, 72% with an S-shaped incision, and only 5% with a Gibson incision [10]. In a 170 patients series by Koifman and colleagues, a Gibson incision was used with a 1.5% rate of wound dehiscence or skin necrosis [22].

The length of the hospital stay as a function of wound morbidity has also been correlated with incision types, and the highest likelihood of primary wound healing occurred with oblique straight-line incisions [38, 39]. If enlarged nodes are present extending superficially toward the skin and subcutaneous tissues, an oblique skin incision can easily be modified to circumscribe and excise the skin en bloc with the nodal packet [40]. Additionally, the oblique incision allows access for simultaneous pelvic lymph node dissection if warranted [40]. For the most part, para-inguinal horizontal incisions that avoid the groin crease have been preferred due to their preservation of the blood supply [6, 20, 22, 41].

The key to minimal morbidity after lymphadenectomy is proper skin handling and meticulous dissection of the skin flaps [42]. A 2 mm thickness of fat is recommended to be left on the undersurface of the skin to accommodate the microcirculation of the skin flaps. Thin skin flaps are at an increased risk of ischemia, skin necrosis, and subsequent wound dehiscence. After dissection, the wound edges should be inspected, and any areas with doubtful vascularization should be removed. Some have suggested use of intravenous fluorescein to better detect the viability of the skin edges [8, 43, 44]. However, extensive experience has not been reported, and this surgical adjunct has not gained wide acceptance. In order to eliminate dead space and prevent fluid collection, the subcutaneous tissue should be anchored to the underlying muscles with interrupted absorbable sutures [6, 37]. Sartorius muscle transposition, previously recommended to protect the femoral vessels during ILND, has recently shown to increase the risk of complications postoperatively [21, 22, 45, 46]. A prospective randomized controlled trial examining the effect of transposition of the sartorius muscle on morbidity after ILND in vulvar cancer patients showed no favorable effects and a possible negative impact on seroma formation [40].

Whenever the skin has been sacrificed by the removal of a portion of the groin dissection flap, primary closure is rarely possible except under tension. Tension frequently tents the flaps up. This leads to underlying dead space permitting the formation of fluid collections, delayed healing, with the resultant increased risk of surgical site infection. Inguinal reconstruction with myocutaneous flaps can avoid wound dehiscence related to excessive tension [20]. Myocutaneous flaps used include gracilis, tensor fascia lata, rectus abdominis muscle, and internal oblique flaps [6, 47]. Ravi reported a 0% incidence of skin flap necrosis in a latter cohort of 30 patients undergoing therapeutic dissection with myocutaneous flap reconstruction compared with an earlier cohort of patients undergoing lymphadenectomy without flap reconstruction (skin-edge necrosis was 61–78%) [11]. Additionally, split-thickness skin grafts can be used to cover skin edges that cannot be reapproximated [47]. The prompt assistance of a plastic surgeon may be necessary for tissue transposition or skin grafts in anticipation of large defects. If a myocutaneous flap is used, mobilization should be avoided for 48–72 h to avoid compromising the blood supply to the flap [20].

Lymphedema and Lymphocele

After radical inguinal lymphadenectomy, lymphedema has the potential for causing difficulty ambulating and standing for prolonged periods. During a groin dissection, numerous major afferent lymphatics are transected and large segments of lymphatics are resected. Historically, lymphedema was a frequent complication following ILND. Kamat et al. described a total incidence of lymphedema of approximately 50% with a severe lymphedema occurring in 35% of dissections [8]. The rate of lymphedema has decreased in recent years due to more prophylactic dissection as well as other modifications in surgical technique [16–22]. Contemporary series have included a higher ratio of early prophylactic dissections of nonpalpable microscopic disease, which remove less lymphatic tissue. ILND in this setting may be less likely to produce complications than node dissection in the presence of bulky nodal metastases as alternative drainage of the limb is potentially maintained. For example, Bevan-Thomas and colleagues observed an incidence of scrotal and leg edema of 23% with only 13% severity. Notably, this rate increased to 33% when the authors excluded prophylactic dissections in clinically node-negative patients [16].

With preservation of the saphenous vein during a modified ILND for penile cancer, the risk of lymphedema has also shown to be reduced [20, 24, 26]. Zhang and colleagues showed rates of short-term lower extremity lymphedema occurring in 67% of patients who underwent saphenous vein excision versus 44% of patients who underwent saphenous vein sparing in vulvar cancer patients [46]. Four studies have reported results of lymphedema from saphenous vein sparing, and meta-analysis of these studies showed significant reduction in the rates of lymphedema in those who had preservation of the saphenous vein [48]. Transposition of the sartorius muscle, which is also not done during a modified ILND, has been associated with higher incidence of persisting lymphedema [49]. Some have also investigated the use of an omental flap after groin dissection to cover the defect of the dissected area of iliac lymph nodes with reduction of lymphedema [50–52]. The omentum is thought to facilitate absorption of any lymphatic fluid, provide good coverage for the femoral blood vessels, afford additional blood supply, and enhance wound healing. Another interesting concept in reducing lymphedema rates in ILND is preservation of the muscle fascia, which was reported by some centers with relatively low reported lymphedema rates of 14% in both studies [53, 54]. It is thought that fascia-preserving techniques cause less scarring and subsequently less lymphatic vessel occlusion. Orefice et al. performed lymphovenous anastomoses immediately after completion of ilioinguinal lymphadenectomy in 30 patients and noted reduced incidence of lymphedema (30% vs. 75%) [55].

Meticulous control of lymphatics throughout the surgical dissection and careful ligation using absorbable sutures or titanium clips has been suggested in preventing lymphedema [20, 50]. The use of an electrothermal bipolar tissue sealing system (LigaSure™, Minneapolis, MN: Covidien) during a groin dissection has also been shown to reduce lymphedema in addition to reducing operative time [56]. Fibrin glue has been used to seal capillaries and obliterate dead space. A randomized prospective trial using suture closure with or without the addition of fibrin sealant following groin dissection was evaluated in vulvar cancer patients [57]. Unfortunately, rates of lymphedema based on the use of fibrin sealant were not effective. Bouchot and colleagues utilized a vaporized tissue sealant when closing the groins and did not use suction drains leading to three seromas of 118 procedures [18]. A review of randomized controlled trials in breast cancer literature concluded that fibrin sealant did not reduce the rate of postoperative seroma, the drainage volume, or the length of hospital stay [58].

Before closing the wound, suction drains are recommended to prevent the initial formation of lymphocele and increase the chances of primary wound healing [46]. There are no reported guidelines for duration of drainage with most recommending removal when the drain output is less than 25–50 mL/day, which is typically 3–17 days postoperatively [6, 37, 46, 47]. Of note, a prospective randomized study, evaluating women undergoing axillary lymph node dissection (LND) for breast cancer, showed no significant benefit in using high versus low vacuum drainage and indicated drains did not prevent seroma formation [9].

Early ambulation, physical therapy, elastic stockings, and/or pneumatic stockings have been suggested to be used postoperatively to reduce the chance of a lower extremity lymphedema [19, 20, 59]. A stepwise approach to the management of chronic lymphedema was developed and advocated by the International Society of Lymphology [60]. It consists of initial skin care, light manual massage, elevation of the affected limb, range-of-motion exercises, and intermittent compression with low-stretch elastic stockings or multilayered bandage wrapping. With compression garments, gradient pressure is applied to the limb, in which the pressure exerted distally is greater than that exerted proximally allowing movement of lymphatic fluid proximally [61]. Elastic stockings are recommended to be used for at least 6 months after surgery. However, prospective randomized studies on these interventions are currently lacking in penile cancer patients who underwent ILND. Randomized clinical trials in breast cancer literature have evaluated the role of early postoperative physiotherapy to prevent the development of lymphedema [62, 63]. Box et al. randomized 65 women to a treatment group consisting of early physiotherapy versus a control without intervention, and at 24 months, the incidence of lymphedema in the control group was 30% versus 11% in the treatment group [64]. Another study randomized 116 women to early physiotherapy (manual lymph drainage, massage of scar tissue, exercise, and educational strategy) or to a control group (educational strategy alone) [65]. Of the 116 patients, 18 developed secondary lymphedema: 14 were in the control group and 4 were in the intervention group. Plastic surgery literature has also shown the benefit of elastic compression garments and lymphatic massage in minimizing the incidence, severity, and sequelae of lymphedema [66, 67].

DVT/PE

Venous thromboembolism is a serious complication that should be aggressively prevented when possible. Its incidence in series has ranged from 0 to 7% [8, 11, 16, 18, 20–22, 27, 28]. In terms of risk of a deep vein thrombosis (DVT)/pulmonary embolism (PE), ILND for penile cancer should be considered a high-risk procedure as it meets all three criteria of Virchow's triad: (1) endothelial injury during dissection of the femoral vessels, (2) venous stasis during immobilization, and (3) a hypercoagulable state secondary to malignancy [20, 68].

Early ambulation decreases the risk of deep vein thrombosis formation and also assists in moving the patient to a status that is consistent with the level of ambulation required for discharge [19]. Prior to anesthesia induction, antiembolic stockings or intermittent compression devices have also been recommended to prevent DVT [6, 20, 37]. Strict leg elevation may also be maintained in the hospital when the patient is not ambulating.

In regard to low-dose heparin, no comparative studies have been done on its use to reduce the incidence of DVT for penile cancer patients after ILND [6, 20, 46]. Most centers recommend low molecular weight heparin while on bed rest postoperatively, but some centers have indicated that the perioperative use of low-dose heparin may be associated with an increased risk of wound hematoma and lymph drainage without reducing the incidence of DVT [6, 19, 38]. However, in patients with a remote history of DVT/PE low dose, low molecular weight heparin must be administered perioperatively until postoperative day 28, in accordance with results from a meta-analysis of randomized trials [69]. With a history of a DVT/PE 6 months prior to ILND, therapeutic dose of heparin should be restarted when the risk of postoperative hemorrhage is minimal with subsequent conversion to oral warfarin [20, 70].

Vascular Injury and Hematoma

Vascular injury is a rarely reported complication after ILND. Although infrequent, vascular injury can have serious consequences including the need for emergent surgical exploration to prevent exsanguination or delayed interventions to drain an infected hematoma. In 106 dissections, Bevan-Thomas and colleagues reported a 4% incidence rate of vascular injuries or postoperative hemorrhage [16]. Similarly, Spiess and colleagues reported a 2% incidence rate of this complication [20].

A contemporary series of 340 procedures by Koifman and colleagues noted one case of an intraoperative femoral vein lesion that was promptly corrected [22]. Additionally, Gopman and colleagues noted 4 patients out of 327 required surgical re-exploration for hematomas [21]. The split-and-roll technique is commonly used in lymph node dissection for removal of the tumor and lymphatic tissues surrounding large vessels. Care must be taken in patients with bulky tumors surrounding the femoral vessels or palliative groin dissections in postchemotherapy patients that are deemed resectable.

To avoid vascular complications when performing an ILND, it is essential that the operating surgeon be familiar with the vascular anatomy. During the procedure, vessel ligation should be performed in a systematic fashion using sutures, surgical clips, or a vascular sealing device. At the completion of resection, the operative field should be aggressively irrigated with water or saline to uncover any potential unrecognized bleeding sources [20]. Postoperatively, any patients with suspected active bleeding should be re-explored to prevent exsanguination and minimize complications that potentially may lead to wound fibrosis obstructing lymph drainage [20].

Neurapraxia and Nerve Injury

Neurapraxia or nerve injury is rarely mentioned in the ILND for penile cancer literature. Spiess and colleagues reported a 2% incidence in their series [20]. During an ILND, the femoral nerve is the most significant nerve to the surgeon. The femoral nerve originates in the lumbar plexus from branches of the posterior division of the L2, L3, and L4 roots. Injury to the femoral nerve creates considerable morbidity because it innervates the quadriceps, sartorius, and pectineus muscles and supplies sensation to a large part of the skin of the anterior and medial portions of the thigh [47]. Injury to the femoral nerve usually produces weakness of knee extension secondary to quadriceps paresis.

There are three general categories of nerve injury: neurapraxia, axonotmesis, and neurotmesis [64]. Neurapraxia, which is a nerve contusion, is a functional injury that is caused by nerve compression or traction resulting in a conduction block without overt axonal degeneration. Recovery from neurapraxia is expected to occur within 6 weeks. Axonotmesis is a more severe injury caused by prolonged compression or excessive traction. The supporting neuronal structures allow for nerve regeneration, and function recovers slowly in 6 months to 1 year. The most severe nerve injury, neurotmesis, denotes complete division of the nerve. In this case, both neural elements and supporting structures are disrupted, and recovery is not expected.

To avoid nerve injury when performing an ILND, it is essential again that the operating surgeon be familiar with the anatomy of neurovascular structures in the groin. The femoral nerve lies lateral to the artery as these structures pass beneath the inguinal ligament and enter the thigh; the nerve divides into its many branches and immediately passes beneath the sartorius muscle out of the field of dissection. Most cases of femoral neuropathy following ILND result from direct compression injury from the placement of self-retaining retractors [6, 20]. The severity of the injury is usually related to the duration of retraction and positioning of the patient. If femoral nerve transection occurs, the nerve should be repaired immediately with the help of neurosurgery or plastic surgery [20].

If nerve injury is suspected postoperatively, prompt examination is required to determine the etiology. It is critical the clinician rule out nerve compression syndromes that require decompressive procedures [20]. Most femoral neuropathy is managed by physical therapy to prevent muscle wasting, and chronic neurogenic pain may be treated with nonnarcotic analgesics, carbamazepine, and amitriptyline [71]. Even though femoral neurapraxia almost invariably resolves spontaneously, the time to resolution remains variable [65].

Video Endoscopy and Robotic-Assisted Techniques

Minimally invasive approaches—video endoscopic and robotic assisted—have been undertaken, and recent series demonstrate that these approaches can limit surgical morbidity with inguinal lymphadenectomy. The Tobias-Machado group reported 20% of complications in 20 dissections [72]. Master et al. reported on their incidence of complications in a series of 41 groin dissections performed in 29 patients [73]. A total of 11 (27%) minor complications (3% superficial wound infection, 12% seroma/lymphocele, 5% mild–moderate lymphedema) and 6 (15%) major complications (3% flap necrosis, 5% secondary procedure, 0% venous thromboembolism, 0% severe lymphedema) were reported in their series [73]. These series as well as other small series have concluded that a minimally invasive approach produces fewer complications in comparison to historical open surgery [3, 71, 74, 75]. The reduced complication rate is thought to be due to less mechanical trauma produced by retraction, minimal use of electrocautery, smaller incisions that allow a better conservation of blood flow and lymphatic drainage of the skin, absence of flap rotation of the sartorius muscle, and easy identification of lymphatic vessels by optical magnification [76].

Recently, the incorporation of robotic assistance as an enabling tool for performing endoscopic ILND has been described [77]. Matin and colleagues reported a series of eight patients who underwent bilateral robotic-assisted surgery [78]. Of the eight patients, two were readmitted to the hospital for cellulitis, with one patient requiring incision and drainage of an abscess. Two additional patients were treated as outpatients, one for an area of wound breakdown and the other for an area of skin necrosis. There were no intraoperative vascular or neurological injuries. Further studies are needed to evaluate the incidence and type of complications as well as oncological efficacy of minimally invasive techniques. Based on these initial series, there appears to be a trend toward improved outcomes in regard to surgical morbidity.

Conclusion

Lymphadenectomy plays a paramount role in treating various malignancies, especially penile cancer. Historically, radical ILND is associated with high complication rate secondary to infections, wound healing, and lymphedema. The morbidity of ILND has declined over the past 20 years from a multitude of factors including surgical technique modifications (dissection templates, saphenous vein sparing, and thicker skin flaps), perioperative management strategies, patient selection, and surgical approach.

Disclosures Authors have nothing to disclose.

References

1. Barnholtz-Sloan JS, Maldonado JL, Pow-sang J, Guiliano AR. Incidence trends in primary malignant penile cancer. Urol Oncol. 2007;25:361–7.
2. Koifman L, Vides AJ, Koifman N Further oblique branches arise from the perforator which do not contribute to the fascial plexus and instead travel directly toward the skin to form part of the subdermal plexus.
3. Pizzocaro G, Algaba F, Horenblas S, Solsona E, Tana S, Van Der Pel H, et al. European association of urology guidelines group on penile C: EAU penile cancer guidelines. Eur Urol. 2010;57:1002–12.
4. Johnson DE, Lo RK. Complications of groin dissection in penile cancer. Experience with 101 lymphadenectomies. Urology. 1984;24:312–4.
5. Ornellas AA, Seixas AL, de Moraes JR. Analysis of 200 lymphadenectomies in patients with penile carcinoma. J Urol. 1991;72:941–5.
6. Ravi R. Morbidity following groin dissection for penile carcinoma. Br J Urol. 1993;72:941–5.
7. Kamat MR, Kulkarni JN, Tongaonkar HB. Carcinoma of the penis: the Indian experience. J Surg Oncol. 1993;52:50–5.
8. Colberg JW, Andriole GL, Catalona WJ. Long term follow up of men undergoing modified inguinal lymphadenectomy for carcinoma of the penis. Br J Urol. 1997;79:54–7.
9. Jain PK, Sowdi R, Anderson AD, Macfie J. Randomized clinical trial investigating the use of drains and fibrin sealant following surgery for breast cancer. Br J Surg. 2004;91:54–60.
10. Catalona WJ. Modified inguinal lymphadenectomy for carcinoma of the penis with preservation of saphenous veins: technique and preliminary results. J Urol. 1988;140:306–10.
11. Bevan-Thomas R, Slaton JW, Pettaway CA. Contemporary morbidity from lymphadenectomy for penile squamous cell carcinoma: the M.D. Anderson Cancer Center experience. J Urol. 2002;167:1638–42.
12. Bouchot O, Rigaud J, Maillet F, Hetet JF, Karam G. Morbidity of inguinal lymphadenectomy for invasive penile carcinoma. Eur Urol. 2004;45:761–6.
13. Catalona WJ. Role of lymphadenectomy in carcinoma of the penis. Urol Clin North Am. 1980;7:785–92.
14. Mangram AJ, Horan TC, Pearson ML, Silver LC, Jarvis WR. Guideline for prevention of surgical site infection, 1999. Centers for disease control and prevention (CDC) hospital infection control practices advisory committee. Am J Infect Control. 1999;27:97–132.
15. Crawford ED, Daneshgari F. Management of regional lymphatic drainage in carcinoma of the penis. Urol Clin North Am. 1992;19:305–17.
16. Gopman JM, Djajadiningrat RS, Baumgarten AS, Espiritu PN, Horenblas S, Zhu Y, et al. Predicting postoperative complications of inguinal lymph node dissection for penile cancer in an international multicenter cohort. BJUI Int. 2015;116:196–201.

17. Koifman LK, Hampl D, Koifman N, Vides AJ, Ornellas AA. Radical open inguinal lymphadenectomy for penile carcinoma: surgical technique, early complications and late outcomes. J Urol. 2013;190:2086–92.
18. Tsaur I, Biegel C, Gust K, Huesch T, Borgmann H, Brandt MP, et al. Feasibility, complications and oncologic results of a limited inguinal lymph node dissection in the management of penile cancer. Int Braz J Urol. 2015;41:486–95.
19. Pandey D, Mahajan V, Kannan RR. Prognostic factors in node-positive carcinoma of the penis. J Surg Oncol. 2006;93:133.
20. Yao K, Hua T, Li YH, Qin ZK, Liu ZW, Zhou FJ, et al. Modified technique of radical inguinal lymphadenectomy for penile carcinoma: morbidity and outcome. J Urol. 2010;184:546–52.
21. Puras-Baez A, Rivera-Herrera J, Miranda G. Role of superficial inguinal lymphadenectomy in carcinoma of the penis. J Urol. 1995;153:246A.
22. Novara G, Galfano A, De Marco V, et al. Prognostic factors in squamous cell carcinoma of the penis. Nat Clin Pract Urol. 2007;4:140.
23. Coblentz TR, Theodorescu D. Morbidity of modified prophylactic inguinal lymphadenectomy for squamous cell carcinoma of the penis. J Urol. 2002;168:1386–9.
24. Milathianakis C, Bogdanos J, Karamanolakis D. Morbidity of prophylactic inguinal lymphadenectomy with saphenous vein preservation for squamous cell penile carcinoma. Int J Urol. 2005;12:776–8.
25. Spiess PE, Hernandez MS, Pettaway CA. Contemporary inguinal lymph node dissection: minimizing complications. World J Urol. 2009;27:205–12.
26. Nelson BA, Cookson MS, Smith JA, Chang SS. Complications of inguinal and pelvic lymphadenectomy for squamous cell carcinoma of the penis: a contemporary series. J Urol. 2004;172:494–7.
27. Pompeo A, Mesquita J, WA J. Staged inguinal lymphadenectomy for carcinoma of the penis: 13 year prospective study in 50 patients. J Urol. 1995;153:246A.
28. Horenblas S. Lymphadenectomy in penile cancer. Urol Clin North Am. 2011;38:459–69.
29. Lopes A, Hidalgo GS, Kowalski LP, et al. Prognostic factors in carcinoma of the penis: multivariate analysis of 145 patients treated with amputation and lymphadenectomy. J Urol. 1996;156:1637–42.
30. Bouchot O, Bouvier S, Bochereau G, et al. Cancer of the penis: the value of systemic biopsy of the superficial inguinal lymph nodes in clinical N0 stage patients. Prog Urol. 1993;3:228–33.
31. Darai E, Karaitianos I, Durand JC. Treatment of inguinal lymph nodes in cancer of the penis. Apropos of 85 cases treated at the Institute Curie. Ann Chir. 1988;42:748–52.
32. Alexander JW, Fischer JE, Boyajian M, Palmquist J, Morris MJ. The influence of hair-removal methods on wound infections. Arch Surg. 1983;118:347–52.
33. Balthazar ER, Colt JD, Nichols RL. Preoperative hair removal: a random prospective study of shaving versus clipping. South Med J. 1982;75:799–801.
34. Anaya DA, Dellinger EP. The obese surgical patient: a susceptible host for infection. Surg Infect. 2006;7:473–80.
35. Baronofsky ID. Technique of inguinal node dissection. Surgery. 1948;24:555.
36. Woodhall JP. Radical groin surgery with particular reference to postoperative healing. Surgery. 1953;33:886–95.
37. Tonouchi H, Ohmori Y, Kobayashi M, Konishi N, Tanaka K, Mohri Y et al. Operative morbidity associated with groin dissections. Surg Today 34; 413–418.
38. Spratt JS, Shieber W, Dillard BM. Groin dissection. In: Anatomy and surgical technique of groin dissection. St Louis, IL: CV Mosby Company; 1965. p. 1–97.
39. Spratt J. Groin dissection. J Surg Oncol. 2000;73:243–62.
40. Judson PL, Jonson AL, Paley PJ, et al. A prospective, randomized study analyzing sartorius transposition following inguinal-femoral lymphadenectomy. Gynecol Oncol. 2004;95:226–30.
41. Horenblas S. Lymphadenectomy for squamous cell carcinoma of the penis. Part 1: diagnosis of lymph node metastasis. BJU Int. 2001;88:467–72.
42. Puras-Baez A. Indications for lymph node dissection in the patient with penile cancer. AUA News. 2000;5:1–3.

43. Smith JA, Middleton RG. The use of fluorescein in radical inguinal lymphadenectomy. J Urol. 1979;122:754–6.
44. Crawford ED. Radical ilioinguinal lymphadenectomy. Urol Clin North Am. 1984;11:543–52.
45. Stuiver MM, Djajadiningrat RS, Horenblas S, et al. Early wound complications after inguinal lymphadenectomy in penile cancer: a historical cohort study and risk factor analysis. Eur Urol. 2013;64:486–92.
46. Zhang X, Sheng X, Niu J, Li H, Li D, Tang L, et al. Sparing of saphenous vein during inguinal lymphadenectomy for vulvar malignancies. Gynecol Oncol. 2007;105:722–6.
47. Bare RL, Assimos DG, McCullough DL, Smith DP, DeFranzo AJ, Marks MW. Inguinal lymphadenectomy and primary groin reconstruction using rectus abdominis muscle flaps in patients with penile cancer. Urology. 1994;44:557–61.
48. Abbas S, Seitz M. Systematic review and meta-analysis of the used surgical techniques to reduce leg lymphedema following radical inguinal nodes dissection. Surg Oncol. 2011;17:2764–72.
49. Rouzier R, Haddad B, Dubernard G, Dubois P, Paniel B. Inguinofemoral dissection for carcinoma of the vulva: effect of modifications of extent and technique on morbidity and survival. J Am Coll Surg. 2003;196:442–50.
50. Benoit L, Boichot C, Cheynel N, Arnould L, Chauffert B, Cuisenier J, et al. Preventing lymphedema and morbidity with an omentum flap after ilioinguinal lymph node dissection. Ann Surg Oncol. 2005;12:793–9.
51. Egorov YS, Abalmasov KG, Ivanov KG, Abramov YA, Gainolin RM, Chatterjee SS, et al. Autotransplantation of the great omentum in the treatment of chronic lymphedema. Lymphology. 1994;27:137–43.
52. Longmans A, Kruyt RH, deBruin HG, Cox PH, Pillay M, Trimbos JB. Lymphedema and lymphocysts following lymphadenectomy may be prevented by omentoplasty: a pilot study. Gynecol Oncol. 1999;75:323–7.
53. Lawton G, Rasque H, Aryan S. Preservation of muscle fascia to decrease lymphedema after complete axillary and ilioinguinofemoral lymphadenectomy for melanoma. J Am Coll Surg. 2002;195:339–51.
54. Daseler EH, Anson BH, Reimann AF. Radical excision of the inguinal and iliac lymph glands: a study based upon 450 anatomical dissections and upon supportive clinical observations. Surg Gynecol Obstet. 1948;87:679.
55. Orefice S, Conti AR, Grass M, et al. The use of lympho-venous anastomoses to prevent complications from ilio-inguinal dissection. Tumori. 1988;30:347–51.
56. Gallo Rolanio FJ, Beneitez Alvarez ME, Izquierdo Garcia FM. The role of inguinal lymphadenectomy in epidermoid carcinoma of the penis: use of a ligasure and analysis of the results. Arch Esp Urol. 2002;55:535–8.
57. Carlson JW, Kauderer J, Walker JL, Gold MA, O'Malley D, Tuller E, et al. A randomized phase III trial of VH fibrin sealant to reduce lymphedema after inguinal lymph node dissection: a gynecologic oncology group study. Gynecol Oncol. 2008;110:76–82.
58. Carless PA, Henry DA. Systematic review and meta analysis of the use of fibrin sealant to prevent seroma formation after breast cancer surgery. Br J Surg. 2006;93:810–9.
59. Protzel C, Alcaraz A, Horenblas S, Pizzocaro G, Zlotta A, Hakenberg OW. Lymphadenectomy in the surgical management of penile cancer. Eur Urol. 2009;55:1075–88.
60. International Society of Lymphology. The diagnosis and treatment of peripheral lymphedema. Consensus document of the International Society of Lymphology. Lymphology. 2003;3:84–91.
61. Johnson G Jr, Kupper C, Farrar DJ, Swallow RT. Graded compression stockings: custom vs noncustom. Arch Surg. 1983;117:69–72.
62. Box RC, Reul Hirche HM, Bullock-Saxton JE, Furnival CM. Physiotherapy after breast cancer surgery: results of a randomized controlled study to minimize lymphedema. Breast Cancer Res Treat. 2002;27:2007–14.
63. Torres Lacoma M, Yuste Sanchez MJ, Zapico Goni A, Prieto Merino D, Mayoral del Moral O, Cerezo Tellez E, et al. Effectiveness of early physiotherapy to prevent lymphedema after surgery for breast cancer: randomized, single blinded, clinical trial. BMJ. 2010;340:b5396.

64. Seddon HJ. Three types of nerve injury. Brain. 1943;66:237.
65. Yossepowitch O, Bochner BH. Complications of Lymphadenectomy. In: Taneja S, editor. Complications of urologic surgery: prevention and management. Philadelphia, PA: Saunders Elservier; 2010. p. 463–77.
66. Garfein ES, Borud LJ, Warren AG, Slavin SA. Learning from a lymphedema clinic: an algorithm for the management of localized swelling. Plast Reconstr Surg. 2008;121:521–8.
67. Warren AG, Brorson H, Borud LJ, Slavin SA. Lymphedema: a comprehensive review. Ann Plast Surg. 2008;59:464–72.
68. Malone PC, Agutter PS. The etiology of deep venous thrombosis. QJM. 2006;99:581–93.
69. Ettema HB, Kollen BJ, Verheyen CC, Buller HR. Prevention of venous thromboembolism in patients with immobilization of the lower extremity: a meta-analysis of randomized controlled trials. J Thromb Haemost. 2008;6:1093.
70. Williams SK, Rabbani F. Complications of lymphadenectomy in urologic surgery. Urol Clin N Am. 2011;38:507–18.
71. Zhou XL, Zhang JF, Zhang JF, Zhou SJ, Yuan XQ. Endoscopic inguinal lymphadenectomy for penile carcinoma and genital malignancy: a preliminary report. J Endourol. 2013;27:657–61.
72. Tobias-Machado M, Tavares A, Silva MN, Molina WR, Forseto PH, Juliano RV, et al. Can video endoscopic inguinal lymphadenectomy achieve a lower morbidity than open lymph node dissection in penile cancer patients? J Endourol. 2008;8:1687–91.
73. Master VA, Jafri SM, Moses KA, Ogan K, Kooby DA, Delman KA. Minimally invasive inguinal lymphadenectomy via endoscopic groin dissection: comprehensive assessment of immediate and long term complications. J Urol. 2012;188:1176–80.
74. Master V, Ogan K, Kooby D, Hsiao W, Delman K. Leg endoscopic groin lymphadenectomy (LEG procedure): step-by-step approach to a straightforward technique. Eur Urol. 2009;56:821–8.
75. Pahwa HS, Misra S, Kumar A, Kumar V, Agarwal A, Srivastava R. Video endoscopic inguinal lymphadenectomy (VEIL) – a prospective critical perioperative assessment of feasibility and morbidity with points of technique in penile carcinoma. World J Surg Oncol. 2013;11:1–6.
76. Astigueta J, Abad-Licham M, Silva E, Yan E, Álvarez H, Agreda F, et al. Endoscopic inguinal lymphadenectomy in penile cancer: case report and literature review. E Cancer Med Sci. 2015;9:576.
77. Josephson DY, Jacobsohn KM, Link BA, Wilson TG. Robotic-assisted endoscopic inguinal lymphadenectomy. Urology. 2009;73:167–71.
78. Matin SF, Cormier JN, Ward J, Pisters JF, Wood CW, Dinney CP, Royal RE, et al. Phase 1 prospective evaluation of the oncological adequacy of robotic assisted video-endoscopic inguinal lymphadenectomy in patients with penile carcinoma. BJU Int. 2013;111:1068–74.

Chapter 5
Surgical and Nonsurgical Management of Lymphedema

Stephanie Kirkpatrick and Angela Cheng

Lymphedema

Lymphedema is swelling that is a result of the disruption of the lymphatic system's ability to adequately transport lymph fluid. This leads to the abnormal accumulation of protein-rich fluid in the interstitial tissue spaces. Lymphedema is a chronic, progressive condition for which currently there is no cure.

Lymphedema can be primary or secondary. Primary lymphedema is caused from an inherent dysfunction in the lymphatic system, which can be congenital or hereditary. Congenital lymphedema presents itself within the first 2 years of life. Some of these patients have a hereditary form termed Milroy's disease. If primary lymphedema becomes evident after birth but prior to the age of 35, it is called lymphedema praecox. This is the most common form of primary lymphedema and arises most commonly during puberty or pregnancy. Lymphedema tarda, which is less common, occurs after the age of 35. Primary lymphedema is most common in the lower extremities and in females. If bilateral lower extremities are involved, the extent of swelling is asymmetrical.

Secondary lymphedema is acquired. Worldwide, the most common cause of secondary lymphedema is filariasis, caused by a mosquito-borne nematode infection with the parasite *Wuchereria bancrofti*. Within the United States, the most common cause of secondary lymphedema is due to the treatment of breast cancer. Lymphedema can occur anywhere where there are lymphatics. It occurs most commonly though in the leg(s), arm(s), neck and facial area, and trunk/chest.

S. Kirkpatrick, P.T., D.P.T., C.L.T. (✉)
Department of Radiation Oncology, Winship Cancer Institute of Emory University, Lymphedema Clinic, Atlanta, GA, USA
e-mail: shood4@emory.edu

A. Cheng, M.D.
Division of Plastic Surgery, Department of Surgery, Winship Cancer Institute of Emory University, Atlanta, GA, USA
e-mail: angelacheng@emory.edu

K.A. Delman, V.A. Master (eds.), *Malignancies of the Groin*,
https://doi.org/10.1007/978-3-319-60858-7_5

Among the causes of secondary lymphedema are the following: surgical trauma to the lymphatics, scarring associated with wound healing, cancers compressing on lymphatic channels, and fibrosis or scarring of the channels caused by radiation therapy and infection.

Due to the high incidence in the United States of breast cancer-related lymphedema (BCRL), the most robust research is with this patient population. The majority of the research regarding secondary lymphedema of the lower extremities is associated with treatment of gynecologic cancers. The incidence reported in this population ranges from 3.6% [1] to 23% [2]. In those treated for melanoma, a 15% incidence after sentinel node biopsy has been found [3] and a 32.1% incidence after inguinal lymph node dissection [4]. Other factors affecting onset and severity of lymphedema include positive lymph nodes, obesity of the patient, presence of cellulitis, total number of lymph nodes removed, removal of circumflex iliac nodes, and ≥3 comorbidities.

Those who suffer from lymphedema demonstrate a lower quality of life than those who do not [5]. Lower extremity lymphedema also seems to have more of a negative impact on quality of life than upper extremity lymphedema [6].

Lymphedema has been classified into four stages by the International Society of Lymphology [7].

Stage 0	Latent or subclinical lymphedema No visible evidence of swelling Lymph transport has been impaired Can exist for months or years
Stage I	Visible, palpable edema that reduces with elevation High protein content allows for pitting
Stage II	Decreased reduction in limb volume with elevation Continued pitting that subsides due to excess fibrosis and deposition of fat
Stage III	Lymphostatic elephantiasis Pitting often absent due to deposition of fat and fibrosis Trophic skin changes

Pathophysiology of Secondary Lymphedema

The lymphatic system is responsible for returning proteins, water, and lipids from the interstitial tissue space to the intravascular space. The process of net filtration and reabsorption occurs within the interstitial tissue. Those molecules that are not reabsorbed into the blood capillaries are taken up by the lymphatics. However, in a lymphatic system that has been damaged, the ability of the lymphatic system to efficiently return fluid from the interstitium is compromised. This leads to the proteins and water molecules not being moved out of the tissue spaces as efficiently. According to Starling law, an increase in proteins results in an increase in colloid osmotic pressure in the tissue. The accumulation of this fluid results in distention, the development of fatty tissue, and progressive fibrosis. These processes also contribute to an increased risk of bacterial infections in those with lymphedema.

Measurement and Diagnosis

There are a variety of tools that allow limb volume to be measured. Use of circumferential measurement, water displacement, bioelectrical impedance, and perometry are the most common.

Taking circumferential measurements with a tape measure is the least expensive and most common method of tracking limb volume. There are, however, several limitations. The interrater and intrarater reliability is variable [8]. Volume calculation assumes a circular circumference which is seldom the case and also, if done correctly, is a time-consuming method.

Water displacement, while regarded as the gold standard for accuracy, is used less frequently. While it is inexpensive, it is time consuming, cumbersome, and messy. Additionally, water displacement cannot be used to localize lymphedema to a specific segment of the limb, and it cannot be used on patients with open skin lesions.

Bioelectrical impedance (bioimpedance) utilizes low-frequency electrical current that is passed through the extremity to measure the opposition to the flow of this current, or its impedance. There is some discrepancy in its ability to accurately depict limb volume changes, and it is applicable on unilateral involvement or risk only. One study compared bioimpedance with water displacement. It was found that of those with abnormal bioimpedance, few of them progressed to developing lymphedema. This demonstrated a poor correlation between the two measurement techniques [9].

The perometer is both valid and reliable, as well as time efficient [10]. It is, however, an expensive device and rather large. Many therapy clinics are challenged when it comes to clinic space and funds.

It has been determined that, while there are several means of determining limb volume, they are not accurately interchangeable nor is the same method reliable among practitioners.

There is not a consensus as to what constitutes a diagnosis of lymphedema, leading to the disparity in the literature in regard to incidence and prevalence. The use of a 200 mL or 10% difference between limbs has been made, as has a ≥2 cm difference in limb volume from the affected to nonaffected side. Subjective complaints of heaviness and achiness have also been considered. It has been cited that a 10% limb volume change from baseline is the most accurate means of diagnosing clinically evident lymphedema [8]. However, there can be up to 150 mL of fluid present prior to swelling being visible. It has been proposed that a subclinical diagnosis of a >3% increase from baseline is more proactive and allows for conservative intervention in the breast cancer population [11].

Nonsurgical Management

Complete decongestive therapy (CDT) has been the standard of care for the treatment of lymphedema. CDT includes manual lymph drainage, compression bandaging, exercises, and skin and nail care. The goals of CDT are to reduce the accumulated fluid maximally, reduce the risk of infections, and soften fibrotic tissue.

Manual lymphatic drainage is a gentle, hands-on technique that stimulates the lymphatic system to absorb fluid from the tissue spaces and move the fluid in the appropriate direction. The effects of MLD are to increase lymph production, increase the rate of contraction of the lymph angion, reverse lymph flow to reroute around the areas that have been damaged, and increase venous return. There is also a soothing and analgesic effect of MLD, as it promotes a parasympathetic response [12].

Compression for lymphedema is provided via bandages and/or garments. Short-stretch bandages are utilized during the treatment phase of CDT. They work with the muscle pump to facilitate movement of lymph fluid. The bandages exhibit a high working pressure on the tissues to promote fluid uptake. They are applied in layers, with more layers being applied distally to facilitate uptake and movement of lymph fluid back into the lymphatic system.

Once maximal reduction in limb volume has been achieved, or if the swelling was mild enough initially not to warrant CDT, compression garments are utilized to maintain the limb volume. There are a variety of daytime and nighttime garments available depending on the needs of the patient.

There is some discrepancy in the literature as to the efficacy of the individual components of CDT. A meta-analysis conducted by Huang et al. found that the addition of MLD to compression and exercise for the treatment of breast cancer-related lymphedema is unlikely to produce significant limb volume reduction. The reviewed studies were, however, poor in quality [13].

Multilayer compression bandaging has been found to be an effective means of reducing limb volume, either with or without the addition of MLD and/or exercise.

Intermittent pneumatic compression (IPC) devices have been used as an adjunct to lymphedema treatment. These devices utilize multichamber intermittent compression to facilitate the uptake and movement of lymph fluid. There has been shown to be improvement in reduction and/or maintenance of limb volume as well as increased tissue elasticity and quality of life in those who use IPC at home.

A systemic review and meta-analysis by Rogan et al. compared the effects of compression bandages, compression sleeves, IPC, and active exercise on the reduction of breast cancer-related lymphedema (BCRL). Exercise was found to contribute to volume reduction. IPC has been shown to be an effective adjunct to traditional therapy, but not as a stand-alone treatment. A compression sleeve, unless worn at a subclinical onset of lymphedema [11], is to maintain limb volume, not reduce. Compression bandaging was found to be an effective means of reducing limb volume [14].

There are a variety of other adjunct treatment options, including but not limited to low-level laser, kinesiotaping, and acupuncture. None of these have been shown to be stand-alone therapies but have been shown in some cases to enhance the effects from CDT.

Surgical Management

A variety of surgical procedures are currently offered for lymphedema patients suffering from mild to severe symptoms. Surgical procedures fall into two broad categories: ablative and physiologic. Both can offer symptomatic relief from pain, infection, and/or swelling; improve limb range of motion; facilitate hygiene; and improve cosmesis. However, patients must be optimized on nonsurgical management prior to consideration for any lymphedema surgery and to maximize success. Surgical options are not considered curative and must be customized based on each patient's etiology, comorbidities, symptoms/staging, physical examination, surgeon's preference for debulking versus physiologic procedures, and occasionally preoperative imaging (lymphoscintigraphy and/or MRI). Timing of surgical intervention may correlate to efficacy, as early and mild symptoms may respond to physiological procedures better and late-stage lymphedema may only be amenable to debulking or ablative procedures [15]. Also, patients with lymphedema secondary to obesity should first consider bariatric surgery.

Early surgical interventions consisted primarily of direct excision of the areas of enlarged skin and subcutaneous tissue. Today, two commonly used ablative procedures are direct excision and liposuction. Large pendulous masses are more easily amenable to direct excision. Often, debulking procedures were performed in series to slowly but effectively decrease the size of the limb [16]. Complications include wound healing problems, infection, and recurrence. The Charles procedure is reserved often for the most severe or advanced disease. A radical debulking is performed by removing all of the abnormal skin and subcutaneous tissue completely to the epimysium or deep fascia, and applying skin grafts from either the resected specimen or alternate donor sites. The superficial lymphatic system is completely disrupted, and any foot lymphedema may be exacerbated. The resulting limb may suffer from hyperkeratoses, papillomatosis, graft contracture, infection, and ulceration with a cosmetically poor but functionally acceptable result [17, 18].

Suction-assisted lipectomy is effective for patients complaining of diffuse circumferential lymphedema swelling with nonpitting lymphedema but risks damaging remaining lymphatics. Small incisions are used to inject tumescent solution, and large volumes of lipoaspirate can be effectively removed. A customized compression garment is applied immediately postoperatively and must be maintained. Therefore, patients must be committed to lifelong use of elastic garments. Studies have demonstrated significant volume (1000 mL) difference at 1 year with stable volume reduction of >85% at 3 and 12 months postoperatively [19, 20].

Physiological procedures to restore lymphatic drainage have gained popularity but require microsurgical expertise and are still considered experimental by some insurance providers. There is significant variability between techniques, and therefore reported results have been inconsistent. Lymphaticovenous/venular bypass or anastomoses (LVB or LVA) and vascularized lymph node transfers (vLNTx) are performed to improve lymphatic drainage in the affected limb by draining into the venous system. The LVB usually utilizes imaging to identify patent lymphatic channels in the affected region and via small incisions; the surgeon connects the distended lymphatic channel to a nearby venule to bypass the areas of blockage. The number of connections varies, and this is a technically challenging operation requiring "supermicrosurgery" skills as these vessels are commonly <1 mm. Several small studies have reported success of 35–50% volume reduction at >1 year. Most patients experience symptomatic improvement even with negligible or minimal volume reduction [21–23].

Vascularized lymph node transfers harvest several healthy lymph nodes from a donor site such as the groin, lateral chest, omentum, submental, or supraclavicular region. The blood supply to the lymph nodes is reestablished in the affected limb using standard microsurgical anastomoses, and during subsequent lymphangiogenesis, the transferred nodes develop new lymphatic connections to the surrounding tissue and function as a sump to absorb the surrounding fluid and relieve the congestion. Several studies have shown volume reduction of 30–60% [24, 25]. There is ongoing concern for donor site iatrogenic lymphedema, a dreaded and unfortunate complication [26–28], which has led to the development of reverse lymphatic mapping technique for harvest to preserve critical lymphatic drainage [29]. Both physiologic procedures have demonstrated promising results over several months in reducing symptoms and volume of the affected limb.

References

1. Mendivil AA, Rettenmaier MA, Abaid LN, Brown JV III, Micha JP, Lopez KL, Goldstein BH. Lower-extremity lymphedema following management for endometrial and cervical cancer. Surg Oncol. 2016;25(3):200–4.
2. Yost KJ, Cheville AL, Al-Hilli MM, Mariani A, Barrette BA, McGree ME, Weaver AL, Dowdy SC. Lymphedema after surgery for endometrial cancer: prevalence, risk factors, and quality of life. Obstet Gynecol. 2014;124(2 Pt 1):307–15.
3. Hyngstrom JR, Chiang YJ, Cromwell KD, Ross MI, Xing Y, Mungovan KS, Lee JE, Gershenwalk JE, Royal RE, Lucci A, Armer JM, Cormier JN. Prospective assessment of lymphedema incidence and lymphedema-associated symptoms following lymph node surgery for melanoma. Melanoma Res. 2013;23(4):290–7.
4. Friedman JF, Sunkara B, Jehnsen JS, Durham A, Johnson T, Cohen MS. Risk factors associated with lymphedema after lymph node dissection in melanoma patients. Am J Surg. 2015;210(6):1178–84.
5. Kim SI, Lim MC, Lee JS, Lee Y, Park K, Joo J, Seo SS, Kang S, Chung SH, Park SY. Impact of lower limb lymphedema on quality of life in gynecologic cancer survivors after pelvic lymph node dissection. Eur J Obstet Gynecol Reprod Biol. 2015;192:31–6.
6. Cromwell KD, Chiang YJ, Armer J, Heppner PP, Mungovan K, Ross MI, Gershenwald JE, Lee JE, Royal RE, Lucci A, Cormier JN. Is surviving enough? Coping and impact on activities of daily living among melanoma patients with lymphoedema. Eur J Cancer Care (Engl). 2015;24(5):724–33.
7. International Society of Lymphology. The diagnosis and treatment of peripheral lymphedema: 2013 Consensus Document of the International Society of Lymphology. Lymphology. 2013;46:1–11.
8. Armer JM, Stewart BR. A comparison of four diagnostic criteria for lymphedema in a post-breast cancer population. Lymphat Res Biol. 2005;3:208–17.
9. Barrio AV, Eaton A, Frazier TG. A prospective validation study of bioimpedance with volume displacement in early-stage breast cancer patients at risk for lymphedema. Ann Surg Oncol. 2015;22(Suppl 3):S370–5.
10. Stanton AW, Northfield JW, Holroyd B, Mortimer PS, Levick JR. Validation of an optoelectronic limb volumeter (perometer). Lymphology. 1997;30(2):77–97.
11. Stout Gergich NL, Pfalzer LA, McGarvey C, Springer B, Gerber LH, Soballe P. Preoperative assessment enables the early diagnosis and successful treatment of lymphedema. Cancer. 2008;112(12):2809–19.

12. Zuther JE. Lymphedema management: the comprehensive guide for practitioners. New York, NY: Thieme; 2005.
13. Huang TW, Tseng SH, Lin CC, Bai CH, Chen CS, Hung CS, Wu CH, Tam KW. Effects of manual lymphatic drainage on breast cancer-related lymphedema: a systematic review and meta-analysis of randomized controlled trials. World J Surg Oncol. 2013;11:15.
14. Rogan S, Taeymans J, Luginbuehl H, Aebi M, Mahnig S, Gebruers N. Therapy modalities to reduce lymphedema in female breast cancer patients: a systematic review and meta-analysis. Breast Cancer Res Treat. 2016;159(1):1–14.
15. Kung TA, Champaneria MC, Maki JH, Neligan PC. Current options in the surgical management of lymphedema. Plast Reconstr Surg. 2017;139(4):1003–1013e.
16. Miller TA, Wyatt LE, Rudkin GH. Stated skin and subcutaneous excision for lymphedema: a favorable report of long-term results. Plast Reconstr Surg. 1998;102:1486–98.
17. Kobayashi MR, Miller TA. Lymphedema. Clin Plast Surg. 1987;14:303–13.
18. Mavili ME, Naldoken S, Safak T. Modified Charles operation for primary fibrosclerotic lymphedema. Lymphology. 1994;27:14–20.
19. Broroson H. Liposuction in lymphedema treatment. J Reconstr Microsurg. 2016;32:56–65.
20. Boyages J, Kastanias K, Koelmeyer LA, et al. Liposuction for advanced lymphedema: a multidisciplinary approach for complete reduction of arm and leg swelling. Ann Surg Oncol. 2015;22(Suppl):S1263–70.
21. Koshima I, Inagawa K, Urushibara K, Moriguchi T. Supermicrosurgical lymphaticovenular anastomoses for the treatment of lymphedema in the upper extremities. J Reconstr Microsurg. 2000;16:437–42.
22. Chang DW. Lymphaticovenular bypass for lymphedema management in breast cancer patients: a prospective study. Plast Reconstr Surg. 2010;126:752–8.
23. Chang DW, Suami H, Skoracki R. A prospective analysis of 100 consecutive lymphovenous bypass cases for treatment of extremity lymphedema. Plast Reconstr Surg. 2013;132:1305–14.
24. Raju A, Chang DW. Vascularized lymph node transfer for treatment of lymphedema: a comprehensive literature review. Ann Surg. 2015;261:1013–23.
25. Cheng MH, Chen SC, Henry SL, Tan BK, Lin MC, Huang JJ. Vascularize groin lymph node flap transfer for postmastectomy upper limb lymphedema: flap anatomy, recipient sites and outcomes. Plast Reconstr Surg. 2013;131:1286–98.
26. Patel KM, Lin CY, Cheng MH. A prospective evaluation of lymphedema-specific quality-of-life outcomes following vascularized lymph node transfer. Ann Surg Oncol. 2015;22:2424–30.
27. Pons G, Masia J, Loschi P, Nardulli MI, Duch J. A case of donor-site lymphoedema after lymph node-superficial circumflex iliac artery perforator flap transfer. J Plast Reconstr Aesthet Surg. 2014;67:119–23.
28. Viitanen TP, Maki MT, Seppanen MP, Suominen EA, Saaristo AM. Donor-site lymphatic function after microvascular lymph node transfer. Plast Reconstr Surg. 2012;130:1246–53.
29. Dayan JH, Dayan E, Smith ML. Reverse lymphatic mapping: a new technique for maximizing safety in vascularized lymph node transfer. Plast Reconstr Surg. 2015;135:277–85.

Chapter 6
Surgical and Anatomic Considerations of Malignancies Affecting the Groin: Consideration for Melanoma

Alexander C.J. van Akkooi

Considerations of Surgical Anatomy

For surgery for melanoma of the groin, most melanoma surgeons will consider two anatomic areas. The first is the superficial groin, also called inguinal groin dissection, consisting of the femoral and inguinal lymph nodes. The second is the deep groin, also called an iliac or external iliac dissection, consisting of the external iliac and obturator lymph nodes. For the purpose of clarity, in the rest of this chapter, these will be referred to as superficial groin dissection (SGD), deep groin dissection (DGD), and combined superficial and deep groin dissection (CGD). Combined groin dissection is sometimes also referred to as ilioinguinal dissection. Another word for dissection is lymphadenectomy. While these techniques are described elsewhere in this text, it is appropriate to review them to help place the remainder of the discussion in context. The details which follow reflect the approach of this author, and there is variation among experts.

Superficial Groin Dissection (SGD)

The anatomic landmarks marking the external limits of the dissection are as follows: the sartorius muscle is the lateral limit. The medial border is the adductor magnus muscle. Be aware: this is not the same as the adductor longus muscle, which is further lateral of the adductor magnus (more medial to the adductor magnus is the gracilis muscle). Cranially, the border is formed by the inguinal (Poupart's)

A.C.J. van Akkooi, M.D., Ph.D.
Department of Surgical Oncology, Netherlands Cancer Institute - Antoni van Leeuwenhoek, Amsterdam, The Netherlands
e-mail: a.v.akkooi@nki.nl

K.A. Delman, V.A. Master (eds.), *Malignancies of the Groin*,
https://doi.org/10.1007/978-3-319-60858-7_6

ligament. The caudal limit is where the adductor magnus muscle and sartorius muscle meet. Dorsally, the femoral artery and vein are the borders of the dissection. All of the nodes inferior to the inguinal (Poupart's) ligament should be removed. While several different incisions may be used, this author considers two types of incision: a curvilinear incision starting laterally over the inguinal ligament and curving medially and inferiorly, ending over the midpoint of the adductor magnus muscle. Or others perform the dissection through a longitudinal incision in the direction of the limb up to the inguinal ligament or even above the ligament, in case of a CGD (to allow deep dissection through one single incision). The lymphatic tissue within the groin fat should be dissected carefully off of the femoral vessels and nerves all the way up to the inguinal canal and for 3 cm superior to the inguinal ligament. This last does not mean that the iliac nodes should be removed, but the subcutaneous nodes ventral from the inguinal (Poupart's) ligament, but cranially from the external entrance to the femoral canal. Some attempt to spare the saphenous vein to potentially reduce morbidity (lymph edema); however, others resect it to ensure complete excision of the lymph nodes. A sartorius transposition, as commonly used in other operations, can be performed to cover the femoral artery and vein, but this is not mandatory. Some will do this routinely, others will retain this option for redo surgery and/or in case the risk of a postoperative radiotherapy indication is considered high.

Deep Groin Dissection (DGD)

The anatomic landmarks marking the borders of the DGD are as follows: the caudal limit is the inguinal (Poupart's) ligament. The cranial border is the bifurcation of the common iliac artery to the external and internal iliac arteries. Laterally, the iliopsoas muscle is the border; medially, the bladder is the border. Dorsally, the obturator artery and nerve form the border. Again two types of incision can be used: a longitudinal incision in case of a CGD in the direction of the limb, and a separate transverse incision, approximately 3 cm superior to the inguinal (Poupart's) ligament. This incision is taken down through the external oblique, internal oblique, and transversus muscles, and the surgeon at that point stays extraperitoneally as in the approach to the iliac vessels for renal transplantation. In selected cases, surgeons might even consider an approach through a laparotomy if the disease is bulky and/or if there is extent of disease toward the aortic bifurcation.

Melanoma Treatment Considerations

Prognosis Stage III Melanoma

The American Joint Committee on Cancer (AJCC) 7th edition considers node-positive disease as stage IIIA–C. This depends on the absence (pT1a–4a) or presence (pT1b–4b) of ulceration of the primary tumor, if the metastasis is microscopic (SN, N1a–N2a) or macroscopic (N1b–N2b), the number of involved nodes (≥4 metastases; N3), and/or the presence or absence of satellite or in-transit metastases [1]. Five-year survival rates differ from 78% for stage IIIA to 40% for stage IIIC [1].

Elective Lymph Node Dissection (ELND)

Elective lymph node dissection or prophylactic lymph node dissections have now been abandoned. Originally, Herbert Snow described this approach in 1892, and it was performed routinely by some during the 1950s and 1960s. Four randomized controlled trials have prospectively analyzed the value of ELND in melanoma. Veronesi et al. only examined limb melanomas, because the draining nodal basin at risk is clear in those cases, which can sometimes be difficult to anticipate for trunk melanomas. They found no difference in outcome in 267 ELND patients versus 286 nodal observation patients [2]. Sim et al. found no differences between an immediate ELND ($n = 54$), a delayed ELND (3 months) ($n = 56$), and nodal observation ($n = 63$) [3]. Balch et al. performed the largest RCT on this topic of 383 ELND versus 356 nodal observation patients [4]. However, there were indications that some subgroups might benefit from the elective removal of lymph nodes, such as younger patients (<60 years) ($P = 0.042$), non-ulcerated melanomas ($P = 0.018$), and patients with intermediate thickness melanoma (Breslow 1–2 mm) ($P = 0.031$) [4]. Finally, a WHO study by Cascinelli et al. only analyzed trunk melanomas in 122 ELND versus 130 nodal observation patients and did not find a significant difference with survival rates of 61.7% for ELND versus 51.3% for nodal observation ($P = 0.09$) [5]. Again subgroup analyses seemed to indicate a benefit; in this case when only looking at those with nodal involvement, the survival was 48.2% in the ELND group versus 26.6% in the nodal observation group ($P = 0.04$) [5].

Since the potential benefit from an early or prophylactic lymph node dissection can only occur in patients with (microscopic) nodal involvement, it was suggested that ELND studies could not show this benefit, because it was diluted by the majority of patients with negative nodes. Moreover, only intermediate thickness melanomas were thought to potentially benefit. Therefore, Morton and coworkers developed the concept of lymphatic mapping to detect the sentinel node (SN) and potentially target node-positive patients, who might benefit from an early lymph node dissection [6].

Sentinel Node (SN)

The prognostic value of the SN in the staging of stage I/II melanoma patients to detect occult microscopic disease has been broadly demonstrated, and therefore the procedure is widely accepted as a staging tool [7–9].

The therapeutic value of the SN procedure for intermediate thickness melanoma has been analyzed in one single large prospective randomized controlled trial, the Multicenter Selective Lymphadenectomy Trial-1 (MSLT-1). Despite reporting the 10-year results, there is still an ongoing debate on the potential therapeutic value for a subgroup of patients [10–14].

This trial randomized 60:40 to wide local excision (WLE) + SN versus WLE + nodal observation. In the SN group, the patients with a positive SN underwent an early completion lymph node dissection (CLND). In the nodal observation group, patients underwent a lymph node dissection in case of clinically detected recurrence. The primary endpoint was melanoma-specific survival (MSS) and was not statistically different between the SN group (81.4 ± 1.5%) versus nodal observation group (78.3 ± 2.0%) (HR 0.84 (95% CI 0.64–1.09)) ($P = 0.18$) [11].

Despite this clear result regarding the main study endpoint, a number of alternative arguments toward a potential benefit have been posed, the first and foremost being the fact that the subgroup analysis of SN-positive patients versus node-positive patients in the nodal observation group shows a clear benefit of 62.1% (± 4.8%) versus 41.5% (± 5.6%) (HR 0.56 (95% CI 0.37–0.84)) ($P = 0.006$) [11]. This is supported by a new statistical analysis: the accelerated-failure-time latent-subgroup analysis [11, 15, 16]. Another is with respect to an improvement in terms of disease-free survival (DFS) [11].

Opposition to this interpretation for a benefit has proposed many arguments, the foremost of which being that a subgroup analysis is invalid, as, in general, subgroup analyses are post hoc, underpowered, and not preplanned. Interestingly, in the design of the MSLT-1 trial, Morton and colleagues precalculated and powered the study for an a priori subgroup analysis, which largely refutes this argument. Furthermore, as discussed subsequently, a formal statistical model has also validated this subgroup analysis. However, despite this prior planning, there remain criticism and some concern about the validity of its conclusions. This analysis does not take into account patients who were false negative and have a worse survival rate than the SN-positive or even the node-positive patients from the observation arm [14]. Excluding them from the subgroup analysis increases the survival of SN-positive patients, because not all node-positive patients are included in the subgroup analysis; especially those with a poor outcome are excluded [14]. There are also false-positive patients, which artificially improves survival in this group. There are a number of hypothetical reasons for a false-positive result, e.g., benign nevus cells, which have been incorrectly concluded to be melanoma cells [17]. By including these (actually non-metastatic) patients in the SN-positive population, the outcome of the SN-positive group may be falsely improved.

In considering the subgroup analysis, a formal statistical review using a novel technique has been applied which has not been widely utilized. In considering "accelerated-failure-time latent-subgroup analysis," it can be stated that this is a new statistical hypothesis, which has yet to be validated. It is not yet a widely accepted statistical tool. It was developed on the basis of the interim results of the MSLT-1 in 2006 [10, 15, 16] by the statistician involved in the MSLT-1, among others, and should first be validated on external studies rather than be used as proof for a survival benefit of the MSLT-1 [14].

Finally, the DFS benefit is controversial, since the group undergoing CLND obviously will recur less frequently in the regional nodes (as they are already removed) compared to the nodal observation group [18].

Thus, there is no unequivocal survival benefit for undergoing an SN in melanoma and there is ongoing debate between believers and nonbelievers and pro and con. Both parties do concur on the prognostic information gained by the SN.

Recent developments in effective systemic therapy for stage IV melanoma with targeted therapies (BRAF/MEK inhibitors) and/or immunotherapies (anti-CLTA-4/anti-PD-1) might influence the reason to perform an SN in the future [19–33]. SN-positive patients (treated by CLND) are now able to participate in adjuvant therapy trials with these new agents, which have already proven their efficacy in stage IV disease. The first results of adjuvant ipilimumab showed a significant improvement in relapse-free survival at 3 years (46.5% vs. 34.8%) [34]. Very recently, it was reported that this also turned into an MSS benefit at 5 years (65.4% vs. 54.4%, HR 0.72 (CI 0.58–0.88), $P = 0.001$) [35]. Other pivotal studies with adjuvant anti-PD-1 (nivolumab, pembrolizumab), high- versus low-dose ipilimumab, combination ipilimumab + nivolumab, and BRAF/MEK inhibitors will report first outcomes between 2017 and 2020. It is expected that this will mandate the routine use of SN staging. The completion lymph node dissection for SN+ disease might change in light of effective adjuvant therapy.

Completion Lymph Node Dissection (CLND)

A second question regarding the potential therapeutic effect of the SN procedure in melanoma is with respect to the CLND. Perhaps the SN alone is therapeutic or perhaps the CLND is the therapeutic part of the procedure. This has been the subject of the MSLT-2, which randomized SN-positive patients to CLND or nodal observation with periodic ultrasound of the (positive SN) lymph node basin [36]. The results of this MSLT-2 study are pending and expected no sooner than 2020.

In the meantime, the German DECOG-SLT study has presented its initial results. This study screened 5547 patients with melanoma of whom 1269 had a micrometastasis in the SN (23%). Four hundred and eighty-three patients agreed to be randomized between CLND and nodal observation. After a median follow-up of 35 months, the distant metastasis-free survival was 77.0% in the observation group versus 74.9% in the CLND group (HR 1.03 (90% CI 0.71–1.50)) ($P = 0.87$) [37]. Although this study has less power compared to the MSLT-2, the follow-up is not yet mature, and there were a large proportion of patients with lesser SN tumor burden; this strongly suggests that the routine use of CLND might not be of benefit for SN-positive patients.

Finally, a large number of studies have examined SN tumor burden as a potential tool to determine which patients might and might not benefit from CLND [8, 38–73]. All these (retrospective) studies have demonstrated the heterogeneous prognosis of SN-positive melanoma patients. Moreover, despite differences in ways to measure the SN tumor burden, all these different factors with different cutoffs have demonstrated their prognostic value with respect to predicting the chance of additional non-SN involvement in the CLND and/or survival. However, there are differences in the interobserver reproducibility of these respective SN tumor burden factors. Murali et al. showed that the maximum diameter and the tumor penetrative depth were the most reproducible [74]. Currently, the EORTC 1208 (Minitub) study is examining if patients with minimal SN tumor burden can safely be managed without CLND.

Palpable Node and Therapeutic Lymph Node Dissection (TLND)

Unlike in occult microscopic disease, most (if not all) surgeons will agree that a therapeutic lymph node dissection (TLND) is indicated in case of palpable (macroscopic) disease. Although these patients are at high risk for future disease relapse and also distant (visceral) metastasis, which can potentially become fatal, there is still a reasonable chance of cure of 40–59% at 5 years and 20–40% at 10 years [1].

There is no consensus between surgeons worldwide on the required extent of surgery for the groin. Some propose to always perform a combined superficial and deep groin dissection (CGD) in all cases. Others always perform a superficial groin dissection (SGD) only.

Hughes et al. reported a summary of a number of studies on elective and/or therapeutic groin lymph node dissections for melanoma [75]. There was a large spread of pelvic nodal involvement at TLND histology of 17–45% [75]. Five-year estimated survival ranged from 0 to 40% [75].

Badgwell et al. described 235 patients undergoing SGD and 97 undergoing CGD. Five-year overall survival was 42% for patients with positive deep nodes compared to 51% for those with negative deep nodes ($P = 0.11$) [76]. On multivariate analysis, positive deep nodes, male gender, and extracapsular extension were independent prognostic factors influencing survival [76]. The authors concluded that patients with involved pelvic nodes should be considered stage III and not stage IV and should be surgically treated with intended curation.

van der Ploeg et al. described 121 CGD and 48 SGD [77]. Five-year overall survival was 39.7% for patients without pelvic nodal involvement on CGD and 12.5% with pelvic nodal involvement of CGD [77]. No survival differences were seen between SGD and CGD, especially not after correcting for other prognostic factors [77].

Allan et al. demonstrated that preoperative CT scans have 60% sensitivity and an 86.2% negative predictive value, which means that a negative preoperative CT scan will be false negative in 40% of cases with respect to pelvic nodal involvement [78]. Patients with involved pelvic nodes had a higher risk of disease relapse compared to negative pelvic nodes, although this was not significant in this small series of 72 patients with 22 with pelvic nodal involvement [78].

Oude Ophuis et al. showed that 35% of CGD had involvement of the pelvic nodes [79]. Preoperative imaging, by CT scan, PET scan, or PET/CT scan, has limited sensitivity and by itself could not safely exclude pelvic nodal involvement [79].

An algorithm with negative imaging, fewer inguinal involved nodes, no extracapsular extension, and a low lymph node ratio (LNR) has low risk of pelvic nodal involvement and might safely be spared a DGD [79].

Van Wissen et al. reported on 70 stage IIIB/C melanoma patients with an indication for a groin dissection. All patients underwent preoperative PET/CT scans. The sensitivity for deep groin (iliac) involvement was 67%, specificity 91%, positive predictive value 73%, and negative predictive value of 81%. Thus the false-negative rate was 33%. Therefore the authors concluded that PET/CT alone was insufficient to safely limit the extent of surgery to the superficial groin only.

Currently, the Australia and New Zealand Trials Group is performing a prospective randomized study: Inguinal or Ilioinguinal Lymphadenectomy for Patients with Metastatic Melanoma to Groin Lymph Nodes and No Evidence of Pelvic Disease on PET/CT Scan – A Randomized Phase III Trial (EAGLE FM) (NCT02166788).

Groin-Specific Therapeutic Considerations in Melanoma

Complications can be divided into two categories: short term and long term. Short-term complications include surgical site/wound infections, seroma, skin necrosis, and fistulas. Long-term complications are mostly chronic lymph edema and rarely nerve damage.

A recent meta-analysis by Soderman et al. summarized 20 studies (including two randomized trials, two prospective cohort, and 16 retrospective series) [80]. In total the complication rate was 52% (44–60%). Infection was seen in 21% (15–27%), wound breakdown in 14% (8–21%), necrosis in 10% (6–15%), seroma in 23% (18–29%), and lymph edema in 33% (25–42%) [80].

The spread of the complication rates is most likely to be caused by the nature of the respective studies; as very few were prospective and most were retrospective, the reliability of registering all respective items is dubious. It is likely that there has been an underreporting of complications in the retrospective studies. Moreover, the methods of assessment of lymph edema differed across studies as patient reported, physician reported, and requiring treatment or measured. In general all surgeons agree that complication rates are high, and this is most frequently caused by the superficial groin dissection and not the deep groin dissection. Although the deep groin dissection part of a combined groin dissection does increase the chance of lymph edema slightly, it is usually not associated with the same problems in terms of surgical site/wound infections, seroma, skin necrosis, and fistulas as with a superficial groin dissection. A deep groin dissection (without superficial groin dissection) for an isolated iliac recurrence is usually associated with far fewer complications than a superficial or combined groin dissection.

Stuiver et al. analyzed potential covariates that might influence the chance of developing complications after groin dissections [81]. All these factors, body mass index (BMI), diabetes, other comorbidity, type of incision, use of sartorius muscle transposition, sparing of saphenous vein, palpable versus SN disease, skin excision, and bed rest did not influence the chance of complications [81]. The only significant factor was older age [81]. Adjuvant radiotherapy has been demonstrated to significantly increase the chance of chronic lymph edema [82, 83].

Attempts have been made to reduce short- and long-term complications of groin dissections. Bartlett et al. have analyzed the use of a sartorius transposition in 381 patients, but did not find any difference in surgical site/wound infections (10% vs. 14%, $P = 0.39$) [84].

Others have tried the use of fibrin sealant. Weldrick et al. have performed a systemic review of six prospective randomized controlled trials and did not find any difference in surgical site/wound infections (32% vs. 34%, $P = 0.90$) [85].

Finally, Faut et al. have analyzed different mobilization protocols (1 day of bed rest vs. 5 days vs. 10 days. vs. ≥10 days of bed rest with a Bohler-Braun splint) and did not find any differences [86].

Future Perspectives

In recent years, minimally invasive approaches to groin dissection have been developed. Sommariva et al. have analyzed ten case series of 168 patients, who have been treated by video-assisted inguinal lymphadenectomy (VEIL) for melanoma or other solid tumors [87]. They reported conversion rates between 0 and 7.7% [87]. Wound-related complications occurred in 0–13.3%; seroma was seen in 4–38.4% of cases [87]. The duration of use of the low-vacuum drain might influence these rates, since longer drain duration will reduce seroma rates but increase wound infection rates. Although these rates might be promising, the quality and cost of the procedure need to balance these reductions in complication rates. In general the mean operation time was 245 min, which is considerably longer than the classical open procedure. At the same time, the median number of harvested lymph nodes was seven, and local recurrence rates were 6.6% [87]. Therefore, this technique should be evaluated in a prospective trial.

A first report has been published on the use of robot-assisted video endoscopic inguinal lymphadenectomy, but this was concerning one single case [88]. This too should be evaluated by a prospective trial.

References

1. Balch CM, Gershenwald JE, Soong SJ, et al. Final version of 2009 AJCC melanoma staging and classification. J Clin Oncol. 2009;27:6199–206.
2. Veronesi U, Adamus J, Bandiera DC, et al. Inefficacy of immediate node dissection in stage 1 melanoma of the limbs. N Engl J Med. 1977;297:627–30.
3. Sim FH, Taylor WF, Ivins JC, et al. A prospective randomized study of the efficacy of routine elective lymphadenectomy in management of malignant melanoma. Preliminary results. Cancer. 1978;41:948–56.

4. Balch CM, Soong SJ, Bartolucci AA, et al. Efficacy of an elective regional lymph node dissection of 1 to 4 mm thick melanomas for patients 60 years of age and younger. Ann Surg. 1996;224:255–63. discussion 263–6
5. Cascinelli N, Morabito A, Santinami M, et al. Immediate or delayed dissection of regional nodes in patients with melanoma of the trunk: a randomised trial. WHO melanoma programme. Lancet. 1998;351:793–6.
6. Morton DL, Wen DR, Wong JH, et al. Technical details of intraoperative lymphatic mapping for early stage melanoma. Arch Surg. 1992;127:392–9.
7. Gershenwald JE, Thompson W, Mansfield PF, et al. Multi-institutional melanoma lymphatic mapping experience: the prognostic value of sentinel lymph node status in 612 stage I or II melanoma patients. J Clin Oncol. 1999;17:976–83.
8. van der Ploeg AP, van Akkooi AC, Schmitz PI, et al. EORTC melanoma group sentinel node protocol identifies high rate of submicrometastases according to Rotterdam criteria. Eur J Cancer. 2010;46:2414–21.
9. van Akkooi AC, Atkins MB, Agarwala SS, et al. Surgical management and adjuvant therapy for high-risk and metastatic melanoma. Am Soc Clin Oncol Educ Book. 2016;35:e505–14.
10. Morton DL, Thompson JF, Cochran AJ, et al. Sentinel-node biopsy or nodal observation in melanoma. N Engl J Med. 2006;355:1307–17.
11. Morton DL, Thompson JF, Cochran AJ, et al. Final trial report of sentinel-node biopsy versus nodal observation in melanoma. N Engl J Med. 2014;370:599–609.
12. Thomas JM. Prognostic false-positivity of the sentinel node in melanoma. Nat Clin Pract Oncol. 2008;5:18–23.
13. Torjesen I. Sentinel node biopsy for melanoma: unnecessary treatment? BMJ. 2013;346: e8645.
14. van Akkooi AC. Sentinel node followed by completion lymph node dissection versus nodal observation: staging or therapeutic? Controversy continues despite final results of MSLT-1. Melanoma Res. 2014;24:291–4.
15. Altstein L, Li G. Latent subgroup analysis of a randomized clinical trial through a semiparametric accelerated failure time mixture model. Biometrics. 2013;69:52–61.
16. Altstein LL, Li G, Elashoff RM. A method to estimate treatment efficacy among latent subgroups of a randomized clinical trial. Stat Med. 2011;30:709–17.
17. Scolyer RA, Murali R, McCarthy SW, et al. Pathologic examination of sentinel lymph nodes from melanoma patients. Semin Diagn Pathol. 2008;25:100–11.
18. van Akkooi AC, Eggermont AM. SLNB in melanoma—DFS a true and cost-effective benefit? Nat Rev Clin Oncol. 2014;11:PMID:25311352.
19. Hodi FS, O'Day SJ, McDermott DF, et al. Improved survival with ipilimumab in patients with metastatic melanoma. N Engl J Med. 2010;363:711–23.
20. Larkin J, Chiarion-Sileni V, Gonzalez R, et al. Combined nivolumab and ipilimumab or monotherapy in untreated melanoma. N Engl J Med. 2015;373:23–34.
21. Postow MA, Chesney J, Pavlick AC, et al. Nivolumab and ipilimumab versus ipilimumab in untreated melanoma. N Engl J Med. 2015;372:2006–17.
22. Schadendorf D, Hodi FS, Robert C, et al. Pooled analysis of long-term survival data from phase II and phase III trials of ipilimumab in unresectable or metastatic melanoma. J Clin Oncol. 2015;33:1889–94.
23. Wolchok JD, Kluger H, Callahan MK, et al. Nivolumab plus ipilimumab in advanced melanoma. N Engl J Med. 2013;369:122–33.
24. Chapman PB, Hauschild A, Robert C, et al. Improved survival with vemurafenib in melanoma with BRAF V600E mutation. N Engl J Med. 2011;364:2507–16.
25. Flaherty KT, Infante JR, Daud A, et al. Combined BRAF and MEK inhibition in melanoma with BRAF V600 mutations. N Engl J Med. 2012;367:1694–703.
26. Flaherty KT, Robert C, Hersey P, et al. Improved survival with MEK inhibition in BRAF-mutated melanoma. N Engl J Med. 2012;367:107–14.
27. Hauschild A, Grob JJ, Demidov LV, et al. Dabrafenib in BRAF-mutated metastatic melanoma: a multicentre, open-label, phase 3 randomised controlled trial. Lancet. 2012;380:358–65.

28. Long GV, Stroyakovskiy D, Gogas H, et al. Combined BRAF and MEK inhibition versus BRAF inhibition alone in melanoma. N Engl J Med. 2014;371:1877–88.
29. Robert C, Karaszewska B, Schachter J, et al. Improved overall survival in melanoma with combined dabrafenib and trametinib. N Engl J Med. 2015;372:30–9.
30. Robert C, Long GV, Brady B, et al. Nivolumab in previously untreated melanoma without BRAF mutation. N Engl J Med. 2015;372:320–30.
31. Ribas A, Puzanov I, Dummer R, et al. Pembrolizumab versus investigator-choice chemotherapy for ipilimumab-refractory melanoma (KEYNOTE-002): a randomised, controlled, phase 2 trial. Lancet Oncol. 2015;16:908–18.
32. Robert C, Ribas A, Wolchok JD, et al. Anti-programmed-death-receptor-1 treatment with pembrolizumab in ipilimumab-refractory advanced melanoma: a randomised dose-comparison cohort of a phase 1 trial. Lancet. 2014;384:1109–17.
33. Robert C, Schachter J, Long GV, et al. Pembrolizumab versus ipilimumab in advanced melanoma. N Engl J Med. 2015;372:2521–32.
34. Eggermont AM, Chiarion-Sileni V, Grob JJ, et al. Adjuvant ipilimumab versus placebo after complete resection of high-risk stage III melanoma (EORTC 18071): a randomised, double-blind, phase 3 trial. Lancet Oncol. 2015;16:522–30.
35. Eggermont AM, Chiarion-Sileni V, Grob JJ, et al. Prolonged survival in stage III melanoma with ipilimumab adjuvant therapy. N Engl J Med. 2016;375:1845–55.
36. Morton DL. Overview and update of the phase III multicenter selective lymphadenectomy trials (MSLT-I and MSLT-II) in melanoma. Clin Exp Metastasis. 2012;29:699–706.
37. Leiter U, Stadler R, Mauch C, et al. Complete lymph node dissection versus no dissection in patients with sentinel lymph node biopsy positive melanoma (DeCOG-SLT): a multicentre, randomised, phase 3 trial. Lancet Oncol. 2016;17:757–67.
38. Starz H, Balda BR, Kramer KU, et al. A micromorphometry-based concept for routine classification of sentinel lymph node metastases and its clinical relevance for patients with melanoma. Cancer. 2001;91:2110–21.
39. Ranieri JM, Wagner JD, Azuaje R, et al. Prognostic importance of lymph node tumor burden in melanoma patients staged by sentinel node biopsy. Ann Surg Oncol. 2002;9:975–81.
40. Carlson GW, Murray DR, Lyles RH, et al. The amount of metastatic melanoma in a sentinel lymph node: does it have prognostic significance? Ann Surg Oncol. 2003;10:575–81.
41. Reeves ME, Delgado R, Busam KJ, et al. Prediction of nonsentinel lymph node status in melanoma. Ann Surg Oncol. 2003;10:27–31.
42. Starz H, Siedlecki K, Balda BR. Sentinel lymphonodectomy and s-classification: a successful strategy for better prediction and improvement of outcome of melanoma. Ann Surg Oncol. 2004;11:162S–8S.
43. Cochran AJ, Wen DR, Huang RR, et al. Prediction of metastatic melanoma in nonsentinel nodes and clinical outcome based on the primary melanoma and the sentinel node. Mod Pathol. 2004;17:747–55.
44. Dewar DJ, Newell B, Green MA, et al. The microanatomic location of metastatic melanoma in sentinel lymph nodes predicts nonsentinel lymph node involvement. J Clin Oncol. 2004;22:3345–9.
45. Vuylsteke RJ, Borgstein PJ, van Leeuwen PA, et al. Sentinel lymph node tumor load: an independent predictor of additional lymph node involvement and survival in melanoma. Ann Surg Oncol. 2005;12:440–8.
46. Pearlman NW, McCarter MD, Frank M, et al. Size of sentinel node metastases predicts other nodal disease and survival in malignant melanoma. Am J Surg. 2006;192:878–81.
47. van Akkooi AC, de Wilt JH, Verhoef C, et al. Clinical relevance of melanoma micrometastases (<0.1 mm) in sentinel nodes: are these nodes to be considered negative? Ann Oncol. 2006;17:1578–85.
48. Govindarajan A, Ghazarian DM, McCready DR, et al. Histological features of melanoma sentinel lymph node metastases associated with status of the completion lymphadenectomy and rate of subsequent relapse. Ann Surg Oncol. 2007;14:906–12.

49. Satzger I, Volker B, Al Ghazal M, et al. Prognostic significance of histopathological parameters in sentinel nodes of melanoma patients. Histopathology. 2007;50:764–72.
50. Scheri RP, Essner R, Turner RR, et al. Isolated tumor cells in the sentinel node affect long-term prognosis of patients with melanoma. Ann Surg Oncol. 2007;14:2861–6.
51. Debarbieux S, Duru G, Dalle S, et al. Sentinel lymph node biopsy in melanoma: a micromorphometric study relating to prognosis and completion lymph node dissection. Br J Dermatol. 2007;157:58–67.
52. Guggenheim M, Dummer R, Jung FJ, et al. The influence of sentinel lymph node tumour burden on additional lymph node involvement and disease-free survival in cutaneous melanoma—a retrospective analysis of 392 cases. Br J Cancer. 2008;98:1922–8.
53. Roka F, Mastan P, Binder M, et al. Prediction of non-sentinel node status and outcome in sentinel node-positive melanoma patients. Eur J Surg Oncol. 2008;34:82–8.
54. Satzger I, Volker B, Meier A, et al. Criteria in sentinel lymph nodes of melanoma patients that predict involvement of nonsentinel lymph nodes. Ann Surg Oncol. 2008;15:1723–32.
55. Frankel TL, Griffith KA, Lowe L, et al. Do micromorphometric features of metastatic deposits within sentinel nodes predict nonsentinel lymph node involvement in melanoma? Ann Surg Oncol. 2008;15:2403–11.
56. van Akkooi AC, Nowecki ZI, Voit C, et al. Sentinel node tumor burden according to the Rotterdam criteria is the most important prognostic factor for survival in melanoma patients: a multicenter study in 388 patients with positive sentinel nodes. Ann Surg. 2008;248:949–55.
57. Gershenwald JE, Andtbacka RH, Prieto VG, et al. Microscopic tumor burden in sentinel lymph nodes predicts synchronous nonsentinel lymph node involvement in patients with melanoma. J Clin Oncol. 2008;26:4296–303.
58. van der Ploeg IM, Kroon BB, Antonini N, et al. Comparison of three micromorphometric pathology classifications of melanoma metastases in the sentinel node. Ann Surg. 2009;250:301–4.
59. van der Ploeg IM, Kroon BB, Antonini N, et al. Is completion lymph node dissection needed in case of minimal melanoma metastasis in the sentinel node? Ann Surg. 2009;249:1003–7.
60. Riber-Hansen R, Nyengaard JR, Hamilton-Dutoit SJ, et al. Metastatic melanoma volume in sentinel nodes: objective stereology-based measurement predicts disease recurrence and survival. Histopathology. 2009;54:796–803.
61. Riber-Hansen R, Nyengaard JR, Hamilton-Dutoit SJ, et al. Automated digital volume measurement of melanoma metastases in sentinel nodes predicts disease recurrence and survival. Histopathology. 2011;59:433–40.
62. Francischetto T, Spector N, Neto Rezende JF, et al. Influence of sentinel lymph node tumor burden on survival in melanoma. Ann Surg Oncol. 2010;17:1152–8.
63. Younan R, Bougrine A, Watters K, et al. Validation study of the s classification for melanoma patients with positive sentinel nodes: the Montreal experience. Ann Surg Oncol. 2010;17:1414–21.
64. Cadili A, McKinnon G, Wright F, et al. Validation of a scoring system to predict non-sentinel lymph node metastasis in melanoma. J Surg Oncol. 2010;101:191–4.
65. Cadili A, Scolyer RA, Brown PT, et al. Total sentinel lymph node tumor size predicts nonsentinel node metastasis and survival in patients with melanoma. Ann Surg Oncol. 2010;17:3015–20.
66. Meier A, Satzger I, Volker B, et al. Comparison of classification systems in melanoma sentinel lymph nodes--an analysis of 697 patients from a single center. Cancer. 2010;116:3178–88.
67. van der Ploeg AP, van Akkooi AC, Haydu LE, et al. The prognostic significance of sentinel node tumour burden in melanoma patients: an international, multicenter study of 1539 sentinel node-positive melanoma patients. Eur J Cancer. 2014;50:111–20.
68. van der Ploeg AP, van Akkooi AC, Rutkowski P, et al. Prognosis in patients with sentinel node-positive melanoma is accurately defined by the combined Rotterdam tumor load and Dewar topography criteria. J Clin Oncol. 2011;29:2206–14.
69. Fink AM, Weihsengruber F, Duschek N, et al. Value of micromorphometric criteria of sentinel lymph node metastases in predicting further nonsentinel lymph node metastases in patients with melanoma. Melanoma Res. 2011;21:139–43.

70. Murali R, Desilva C, Thompson JF, et al. Factors predicting recurrence and survival in sentinel lymph node-positive melanoma patients. Ann Surg. 2011;253:1155–64.
71. Egger ME, Bower MR, Czyszczon IA, et al. Comparison of sentinel lymph node micrometastatic tumor burden measurements in melanoma. J Am Coll Surg. 2014;218:519–28.
72. Kim C, Economou S, Amatruda TT, et al. Prognostic significance of microscopic tumor burden in sentinel lymph node in patients with cutaneous melanoma. Anticancer Res. 2015;35:301–9.
73. Namikawa K, Yamazaki N, Nakai Y, et al. Prediction of additional lymph node positivity and clinical outcome of micrometastases in sentinel lymph nodes in cutaneous melanoma: a multi-institutional study of 450 patients in Japan. J Dermatol. 2012;39:130–7.
74. Murali R, Cochran AJ, Cook MG, et al. Interobserver reproducibility of histologic parameters of melanoma deposits in sentinel lymph nodes: implications for management of patients with melanoma. Cancer. 2009;115:5026–37.
75. Hughes TM, Thomas JM. Combined inguinal and pelvic lymph node dissection for stage III melanoma. Br J Surg. 1999;86:1493–8.
76. Badgwell B, Xing Y, Gershenwald JE, et al. Pelvic lymph node dissection is beneficial in subsets of patients with node-positive melanoma. Ann Surg Oncol. 2007;14:2867–75.
77. van der Ploeg AP, van Akkooi AC, Schmitz PI, et al. Therapeutic surgical management of palpable melanoma groin metastases: superficial or combined superficial and deep groin lymph node dissection. Ann Surg Oncol. 2011;18:3300–8.
78. Allan CP, Hayes AJ, Thomas JM. Ilioinguinal lymph node dissection for palpable metastatic melanoma to the groin. ANZ J Surg. 2008;78:982–6.
79. Oude Ophuis CM, van Akkooi AC, Hoekstra HJ, et al. Risk factors for positive deep pelvic nodal involvement in patients with palpable groin melanoma metastases: can the extent of surgery be safely minimized? A retrospective, multicenter cohort study. Ann Surg Oncol. 2015;22(Suppl 3):S1172–80.
80. Soderman M, Thomsen JB, Sorensen JA. Complications following inguinal and ilioinguinal lymphadenectomies: a meta-analysis. J Plast Surg Hand Surg. 2016;50:315–20.
81. Stuiver MM, Westerduin E, ter Meulen S, et al. Surgical wound complications after groin dissection in melanoma patients – a historical cohort study and risk factor analysis. Eur J Surg Oncol. 2014;40:1284–90.
82. van Akkooi AC, Bouwhuis MG, van Geel AN, et al. Morbidity and prognosis after therapeutic lymph node dissections for malignant melanoma. Eur J Surg Oncol. 2007;33:102–8.
83. Henderson MA, Burmeister BH, Ainslie J, et al. Adjuvant lymph-node field radiotherapy versus observation only in patients with melanoma at high risk of further lymph-node field relapse after lymphadenectomy (ANZMTG 01.02/TROG 02.01): 6-year follow-up of a phase 3, randomised controlled trial. Lancet Oncol. 2015;16:1049–60.
84. Bartlett EK, Meise C, Bansal N, et al. Sartorius transposition during inguinal lymphadenectomy for melanoma. J Surg Res. 2013;184:209–15.
85. Weldrick C, Bashar K, O'Sullivan TA, et al. A comparison of fibrin sealant versus standard closure in the reduction of postoperative morbidity after groin dissection: a systematic review and meta-analysis. Eur J Surg Oncol. 2014;40:1391–8.
86. Faut M, Heidema RM, Hoekstra HJ, et al. Morbidity after inguinal lymph node dissections: it is time for a change. Ann Surg Oncol. 2017;24:330–9.
87. Sommariva A, Pasquali S, Rossi CR. Video endoscopic inguinal lymphadenectomy for lymph node metastasis from solid tumors. Eur J Surg Oncol. 2015;41:274–81.
88. Sanchez A, Sotelo R, Rodriguez O, et al. Robot-assisted video endoscopic inguinal lymphadenectomy for melanoma. J Robot Surg. 2016;10:369–72.

Chapter 7
Considerations for Nonmelanoma Skin Cancer: Clinical Presentation

Matthew C. Perez, Carrie Luu, Amod A. Sarnaik, and Jonathan S. Zager

Lymphadenopathy of the groin, or inguinal lymphadenopathy, may be a feature of a number of systemic diseases, both benign and malignant. Benign causes of inguinal lymphadenopathy include infection, as well as vascular and autoimmune disease. Malignant groin lymphadenopathy may derive from primary skin tumors (melanoma, squamous cell carcinoma, and Merkel cell carcinoma being the most common) of the lower extremity or lower trunk, perineum, reproductive tracts, and lower gastrointestinal tract and leukemias and lymphomas [1]. This chapter focuses on inguinal lymph node metastasis from nonmelanoma cutaneous malignancies.

When evaluating lymphadenopathy in the groin, a complete history and physical examination are paramount and often sufficient to identify the etiology. Risk factors for cutaneous malignancies often include a history of extensive sun exposure and/or immunosuppression, and patients will commonly report a history of prior skin malignancy or prior skin biopsies. Malignant groin adenopathy typically presents as solitary or multiple painless, palpable masses. If there is a history of prior cutaneous malignancy, it is important to examine and palpate the area between the primary lesion and groin for any evidence of in-transit disease.

A diagnosis can be established by fine needle aspiration (FNA) or core needle biopsy of a palpable lymph node. However, if biopsies are negative or inconclusive, an excisional biopsy is warranted. Once a diagnosis of metastatic disease is established, full-body cross-sectional imaging, if appropriate for the diagnosis at hand, should be

M.C. Perez, M.D.
Department of Surgery, Emory University School of Medicine, Atlanta, GA, USA
e-mail: Matthew.perez@moffitt.org

C. Luu, M.D. • A.A. Sarnaik, M.D., F.A.C.S. • J.S. Zager, M.D., F.A.C.S. (✉)
Department of Cutaneous Oncology, Moffitt Cancer Center, Tampa, Florida, USA
e-mail: carrie.luu@health.slu.edu; Amod.sarnaik@moffitt.org; Jonathan.zager@moffitt.org

K.A. Delman, V.A. Master (eds.), *Malignancies of the Groin*,
https://doi.org/10.1007/978-3-319-60858-7_7

obtained to evaluate for further evidence of systemic disease. This can include magnetic resonance imaging (MRI) of the brain; computed tomography (CT) of the chest, abdomen, and pelvis; or full-body positron emission tomography (PET).

Differential Diagnosis

Cutaneous Squamous Cell Carcinoma

Cutaneous squamous cell carcinoma (SCC) accounts for approximately 20% of all nonmelanoma skin cancer cases [2]. Derived from the superficial layer of the epidermis, it develops most frequently in elderly males and in sun-exposed areas, especially the head, neck, and upper extremities. Ultraviolet (UV) radiation is the most common cause of cutaneous SCC. Immunosuppression, chronic inflammation, and certain chemical agents have also been associated with increased risk for the development of these tumors [3]. Classically, lesions present as a painless, slow-growing nodule(s) that is often ulcerated or with thickened, plaque-like features. Typically, there is a precursor lesion, such as actinic keratosis or SCC in-situ, also known as Bowen's disease [2, 3].

For the most part, SCC is a treatable disease, with an excellent prognosis and cure rates reported as high as 97% [4]. However, certain subtypes of SCC and clinical scenarios are found to be more aggressive and have a higher risk for regional and distant spread. These higher-risk cases include tumors >2 cm in diameter or 2 mm in thickness, poorly differentiated tumors, tumors with perineural invasion, recurrent SCCs, and those in immunosuppressed patients [5–7]. These higher-risk lesions require more aggressive surgical resection and often the addition of radiation therapy.

Merkel Cell Carcinoma

Merkel cell carcinoma (MCC) is a rare neuroendocrine tumor of the skin, characterized by rapid growth and a high propensity for local, regional, and distant spread. The total estimated incidence in 2003 was 1300 cases, with one study reporting its incidence to have tripled from 0.15 cases per 100,000 in 1986 to 0.44 cases per 100,000 in 2001 [8–10]. Located in the dermo-epidermal junction, Merkel cells are thought to be neurotactile cells with neuroendocrine features [10]. Histologically, MCC is very similar to small-cell carcinoma of the lung; therefore, immunohistochemical analysis with numerous markers including CK20 and TTF-1 staining is needed to differentiate MCC from other small-cell tumors [9].

Clinically, MCC characteristically appears as a painless, erythematous to violaceous nodule most frequently in the head and neck of elderly individuals. As seen in cutaneous SCC, sun exposure and immunosuppression are established risk factors, with one study reporting a 23.8-fold increase in MCC tumor development after solid organ transplantation [11]. MCC has also been found to be associated with infection by the Merkel cell polyomavirus, possibly acting synergistically with previously mentioned risk factors [12].

MCC typically behaves aggressively, with a higher mortality rate than most cutaneous malignancies. Survival rates for this disease have been reported between 20 and 80% [13]. Local failure is common and can occur in up to one-third of patients. Regional nodal involvement is also frequent, estimated to be as high as 30% at the time of diagnosis [14–16].

Extramammary Paget's Disease

Extramammary Paget's disease (EMPD) is a rare form of intraepithelial adenocarcinoma that originates from apocrine glands of the skin. It is most commonly found in the perineum, perianal region, and vulva. Peak incidence occurs between 50 and 80 years of age [17]. Diagnosis is best made by biopsy and immunohistochemical staining, demonstrating large round cells with centrally situated nuclei and CK7 and CK20 positivity. [18] The natural history of this disease remains widely unknown. There is a subset of EMPD with dermal invasion; though it is not clear whether EMPD with invasion represents a different disease entity, depth of tumor invasion is the most significant prognostic factor, with >1 mm depth or invasion into the reticular dermis associated with a poorer prognosis [19]. EMPD can also be associated with an underlying adnexal or visceral malignancy. In one review, 12% of patients with EMPD had a concurrent underlying internal malignancy [20].

Treatment of the Primary Malignancy

Squamous Cell Carcinoma

Conventional surgical resection or Mohs micrographic surgery (MMS) is the treatment of choice for most cutaneous SCC. Cure rates as high as 97% with negative margins have been reported in the literature [4, 21]. Recommended margin size depends on the histologic characteristics of the lesion. For conventional excision, the recommended margin is 4–6 mm for low-risk lesions and wider margins of 1 cm typically for high-risk tumors [4]. Larger surgical margins may not be feasible in certain anatomic locations and in situations where there is a risk of severe functional and/or cosmetic morbidity. In this setting, MMS is recommended to confirm complete removal of the tumor.

Radiation therapy (RT) has been used as both a primary and adjuvant therapy in the treatment of SCC. In low-risk lesions or tumors of the lip and eyelid, results have been comparable to that of surgery for primary treatment [21]. However, RT usually requires an intensive treatment schedule for multiple weeks and does not allow for histologic confirmation of tumor margins. Therefore, RT as the sole treatment modality is currently only recommended for those patients who are not surgical candidates or in locations where surgical resection would result in poor cosmetic or functional outcomes.

Merkel Cell Carcinoma

For all patients with MCC, wide excision with 1–2 cm margins and routine sentinel lymph node biopsy are recommended. Factors traditionally associated with SLNB positivity and overall survival include increasing tumor diameter and the presence of lymphovascular or angiolymphatic invasion [15, 16]. The authors' experience of 191 patients with MCC undergoing SLNB showed a >10% chance of a positive SLN even in the thinnest and smallest tumors. Furthermore, increasing tumor diameter and increasing tumor depth were shown to be independent prognostic factors of worse overall survival [22].

Extramammary Paget's Disease

For EMPD, surgical treatment with wide excision is the treatment of choice. It is currently unclear what size the surgical margin should be to achieve negative margins; however, 1–2 cm margins are routinely obtained [23]. Determining a clear surgical margin at the time of operation can be challenging, as Paget cells are often difficult to recognize on frozen sections. In a series of 48 patients with EMPD, the false-negative rate for frozen sections was 10.4% [24]. However, when combined with intraoperative immunostaining of CK-7, MMS has shown to be of comparable value with wide excision [25].

Recent studies have shown conflicting data regarding the association of SLNB and survival. In a series of 151 patients with EMPD, Fujisawa and colleagues recently reported similar survival outcomes between patients with positive and negative SLNB [26]. In contrast, Ogata and colleagues conducted a similar study of 59 patients with EMPD and reported a 5-year survival rate of 100% among negative SLNB patients versus 24% for those patients with positive SLNB ($p < 0.005$) [27]. Further studies are needed to develop a consistent treatment algorithm and staging guidelines.

Superficial Inguinal Lymph Node Dissection

In addition to treating the primary lesion, the nodal basin has to be addressed in the case of locoregional spread. Superficial inguinal lymph node dissection (ILND) (inguinofemoral lymphadenectomy) involves removal of the node-bearing tissues in the femoral triangle and the lymphatic tissues superficial to the external oblique aponeurosis above the inguinal ligament. Our method of ILND is outlined as follows [28].

Patients are placed in supine position and the affected extremity is gently frog-legged. Antibiotics are given within 30 min of the skin incision with a first-generation cephalosporin or the equivalent in the penicillin or cephalosporin-allergic patient. Patients with a personal or family history of thrombosis should receive deep venous thrombosis prophylaxis with low molecular weight heparin, and all patients should have bilateral lower extremity sequential compression devices placed. General anesthesia is given, and long-acting paralytics during induction are avoided to allow for stimulation of motor nerves during the procedure.

The technique of inguinofemoral lymphadenectomy varies slightly among surgeons; there are also several steps that tend to be more controversial. Commonly, a lazy "S" incision overlying any palpable adenopathy and which includes a prior biopsy site, seroma cavity, or sentinel lymph node biopsy scar is made. If an ilioinguinal lymphadenectomy is performed, this incision can be extended cranially. In some patients, if a sentinel lymph node biopsy scar is present above the inguinal ligament, then a transverse incision to incorporate the scar and a vertical counter incision below the inguinal ligament to remove the nodes in the femoral triangle is made. Skin flaps are then raised to clear the boundaries of dissection, which include the sartorius muscle laterally, the adductor longus medially, and the apex of the femoral triangle where the sartorius and adductor longus cross distally. Care is taken to avoid excessively thin flaps to prevent skin necrosis. Superiorly, the lymphatic tissues which are superficial to the external oblique fascia and which are bounded by the pubic tubercle, the anterior superior iliac spine, and the inguinal ligament are also removed. The external oblique fascia is preserved, while the fascia of the sartorius and adductor longus are typically incised. Lymphatic vessels should be ligated or clipped to decrease seroma or lymphocele formation.

During the course of dissection, the saphenous vein is encountered near the "apex" of the triangle and ligated distally (and later at the saphenofemoral junction). Saphenous vein preservation may be performed in an effort to decrease the risk of lymphedema. Despite the theoretical benefit, there are no randomized studies that demonstrate decreased lymphedema rates with this practice. In retrospective studies of groin dissection for carcinoma of the vulva, saphenous vein preservation is associated with decreased short- and long-term complications including cellulitis, wound dehiscence, and chronic lymphedema [29, 30]. Sabel et al. compared outcomes for patients undergoing inguinal lymph node dissection for sentinel lymph node-positive disease versus for palpable lymphadenopathy [31]. When the saphenous vein was ligated, the wound complication rate was 20%, and the lymphedema rate was 30% compared to 7% and 13%, respectively, when the saphenous vein was preserved, though this did not reach statistical significance. In a meta-analysis, four trials reported the rates of lymphedema with saphenous vein sparing technique; pooled results demonstrated a lower rate of lymphedema, acute cellulitis, and wound breakdown [32]. Still, some authors propose saphenous vein preservation only in select cases, such as with lymphadenectomy for micrometastatic disease in which the saphenous vein is uninvolved [33].

As dissection continues proximally, skeletonization of the femoral vessels is performed. The femoral artery is identified at the apex, anterior to the femoral vein. As it courses proximally, the femoral artery moves lateral to the femoral vein. All the lymphatic tissues must be dissected off the anterior aspect of the femoral vessels. Care must be taken to prevent injury to the femoral nerve, which is not visualized directly. If the saphenous vein is to be sacrificed, once all of the soft tissues are dissected free and the specimen is only attached by the saphenofemoral junction, the saphenous vein is divided and then suture ligated and tied again, taking care not to narrow the femoral vein.

One step of considerable variability is sartorius muscle transposition to protect the femoral vessels in the event of a wound dehiscence. Some advocate routine sartorius muscle transposition, while others rarely perform this procedure as part of the lymph node dissection. In most cases we perform sartorius muscle transposition, with some exceptions, such as with an incision that is located higher on the groin and not overlying the femoral vessels and in younger and/or very active patients. To start, the tendinous insertion of the sartorius muscle is detached from its origin on the anterior superior iliac spine with electrocautery. The muscle is mobilized medially and laterally to create a tension-free pedicled flap, taking care not to devascularize the muscle belly. A few lateral feeding vessels usually have to be tied off. The muscle is transposed over the femoral vessels and secured to the external abdominal oblique aponeurosis/inguinal ligament with several interrupted horizontal mattress sutures in a staggered fashion to avoid weakening of the fascia.

A retrospective review of patients with vulvar malignancy compared patients with sartorius transposition and those without and found that the transposition group had lower rates of wound breakdown and cellulitis; sartorius transposition was also the only factor associated with decreased wound morbidity [34]. However, in a randomized controlled trial of inguinofemoral lymphadenectomy for vulvar squamous cell carcinoma, there were no differences in wound infection, wound dehiscence, lymphedema, or rehospitalization whether sartorius muscle transposition was performed or not [35].

Prior to wound closure, a closed suction drain is placed in the dissection field and brought out through a separate stab incision adjacent to the wound. The incision is closed with 3-0 Vicryl for the subcutaneous layer and 4-0 Monocryl for a subcuticular layer. The skin is dressed with Dermabond adhesive or similar dressing. The extremity is wrapped in ACE bandages and SCDs are placed on bilateral extremities.

Ilioinguinal (Pelvic) Lymphadenectomy

The same lazy S type of incision may be used if a pelvic lymphadenectomy will be performed in the same setting, though it will need to be lengthened above the inguinal crease for adequate exposure. Lymph nodes removed in a pelvic lymphadenectomy include the obturator and external iliac nodes up to the bifurcation of the internal and external iliac artery; sometimes common iliac nodes can be retrieved, but para-aortic nodes are beyond the field of dissection.

Once the skin incision is extended, external and internal oblique aponeuroses are incised, parallel to the direction of the fibers of each muscle. The peritoneum is retracted superiorly and medially to expose the external iliac artery and vein. The ureter is usually mobilized medially out of the dissection field with this maneuver.

To retrieve the iliac nodal tissues, the external iliac vessel is skeletonized anteriorly from the inguinal ligament to the bifurcation proximally. To remove the obturator nodes, retract the external iliac vein laterally to expose the obturator space. Care must be taken to prevent injury to the obturator nerve, as well as the obturator vein medially, as injury to the vein or its tributaries may be difficult to control. Palpation of the obturator foramen and the pubic rami at the conclusion of the procedure ensure that abnormal nodes are not left behind. We generally do not dissect any tissue deep to the level of the obturator nerve.

After lymph node dissection is complete, a closed suction drain may be placed in the wound bed. The fascia of the internal and external oblique muscles are closed with separate running sutures. The wound is closed in the same fashion as described above for the inguinofemoral portion.

Cloquet's Node

In the melanoma literature, some authors advocate the use of Cloquet's node to determine if pelvic lymph node dissection is necessary. Cloquet's node is the highest node in the inguinal basin at the level of the inguinal ligament; it was first described by the surgeon-anatomist Jules Germain Cloquet [36]. Some surgeons advocate biopsy of Cloquet's node during groin dissection; if positive, deep pelvic lymphadenectomy would then be performed. However, the utility of Cloquet's node may be limited in the era of sentinel lymph node biopsy. In their review of patients undergoing groin dissection for sentinel lymph node-positive disease, Chu et al. argue that in the era of SLNB, groin dissections are often performed for an early microscopic disease that is not likely to be Cloquet's node positive, unlike in prior studies where ILND was performed for palpable adenopathy. In their study, the incidence of a positive Cloquet's node in patients undergoing groin dissection for SLNB was only 3.8% [37]. In addition, in patients whose Cloquet's node was positive, pelvic node dissection was already being performed for other indications. In various studies, the positive predictive value of Cloquet's node ranges from 27 to 79%, though the negative predictive value was higher at 80–97% [38–40].

In cutaneous squamous cell carcinoma, sentinel lymph node biopsy is not widely practiced. According to the National Comprehensive Cancer Network (NCCN), sentinel lymph node biopsy can be performed in certain high-risk lesions [41]. A review of 692 patients with anogenital and non-anogenital SCC patients reported positive sentinel lymph nodes in 24% and 21% of patients, respectively, with low false-negative rates [42].

Similarly, there are no guidelines for sentinel lymph node biopsy in extramammary Paget's disease. Conversely, sentinel lymph node biopsy is routinely performed for Merkel cell carcinoma. For nonmelanoma skin cancers such as SCC, EMPD, and MCC, regardless of whether sentinel lymph node biopsy is performed or whether ILND is undertaken for microscopic or palpable disease, Cloquet's node is not evaluated, as in some cases of inguinal dissection for melanoma.

Other Approaches to Inguinofemoral and Ilioinguinal Lymphadenectomy

For superficial ILND, minimally invasive approaches are also being evaluated. Delman et al. reported on 45 videoscopic inguinal lymphadenectomies for various malignancies [43]. A three-port approach was used and any prior biopsy scar was left intact [44]. In their series, the median nodal yield was 11, the postoperative complication rate was 18%, and lymphedema occurred in 11% of patients over long-term follow-up. Similar results were reported by Abbott et al. in which 13 minimally invasive lymphadenectomies were performed for melanoma [45]. The lymph node yield was similar, 11, and short-term morbidity was low.

Likewise, at our institution, robotic-assisted pelvic lymph node dissection is increasingly being performed. Using the da Vinci robotic system, three robotic ports, a camera port, and an assistant port were used to retrieve iliac nodes from the inguinal ligament to the bifurcation as well as the obturator nodes [46]. In our experience with 13 robotic pelvic lymphadenectomies, there were no differences in operative times or nodal yield compared to open lymphadenectomy, but hospital length of stay was significantly decreased with the robotic technique [46].

Thus far, results from minimally invasive techniques have been encouraging. Longer follow-up and further studies need to be carried out to ascertain equivalent operative and oncologic outcomes to open ILND.

Postoperative Care

To decrease the risk of morbidity from inguinal lymph node dissection, appropriate postoperative care is essential. Patients are placed on bed rest overnight but encouraged to ambulate the following day. Unless there is concern for hemorrhage, DVT prophylaxis is administered the night of surgery. The affected extremity may be wrapped in ACE bandages, which can be replaced with fitted compression stockings at a later time. If a combined inguinofemoral and ilioinguinal lymphadenectomy is performed, patients are started on a clear liquid diet on the first operative diet and advanced if there are no signs of ileus. Length of hospital stay varies, but patients can be discharged once they are tolerating a diet and pain is controlled with oral analgesics. During the course of follow-up, drains are removed as soon as the output is less than 30–40 mL/day for several days. As soon as lymphedema is recognized, then lymphatic massage may be ordered, and compliance with leg elevation and compression stockings is ascertained.

Short-term morbidity after inguinal lymph node dissection has been well documented, including cellulitis/infection, wound breakdown, skin flap necrosis, and seroma and hematoma formation. Reported complication rates range from 12 to 60% [32, 47]. Several prospective trials were performed to evaluate whether the application of fibrin sealant decreased postoperative wound complications. One trial did not find any difference in overall complications or time to drain removal [48]. Another similarly sized trial also found no difference in drain duration with fibrin glue [49].

Other Approaches to Inguinofemoral Malignant Adenopathy

Radiation therapy is used as primary or adjuvant treatment in many types of melanoma and nonmelanoma skin cancers with regional lymph node disease. Nodal disease is associated with a higher risk of recurrence and poorer survival. For cutaneous SCC, rates of recurrence and metastasis are highest for the skin of the head and neck and also of the penis, scrotum, and anus [50–52]. Metastasis occurs to the regional nodes in 85% of the cases with the remaining 15% spreading to distant sites [3, 50, 53]. For patients with regional nodal metastasis, the 10-year survival rate is less than 20% [3]. For Merkel cell carcinoma, 55% of patients develop regional lymph node metastasis over the course of the disease [54]. The role of radiation in the management of groin metastasis differs somewhat for cutaneous SCC versus Merkel cell carcinoma and will be discussed below.

Squamous Cell Carcinoma

Nodal metastasis significantly increases the risk of recurrence in cutaneous squamous cell carcinoma. Surgery is the first line of treatment for inguinal nodal metastasis. The literature regarding adjuvant radiation is conflicting. A retrospective review of patients with cutaneous SCC of the head and neck with metastatic regional nodes found that multimodality treatment with surgery and adjuvant radiation was associated with a lower rate of recurrence (20% vs. 43%) and higher 5-year disease-free survival rate (73% vs. 54%) compared to surgery alone [55]. Other studies are conflicting, however, and do not demonstrate improved locoregional control with adjuvant radiation [56–58]. Due to lack of consistent, high-level evidence on the utility of adjuvant radiation to the surgically treated inguinal nodal basin, adjuvant radiation is recommended only in cases where there are multiple involved nodes or extracapsular extension.

Merkel Cell Carcinoma

Unlike squamous cell carcinoma, Merkel cell carcinoma is a very uncommon and aggressive cutaneous malignancy. Due to the high rate of occult metastasis, the management of MCC routinely includes sentinel lymph node biopsy to more accurately stage the disease; regional lymph node metastasis is an important predictor of survival [59]. In a meta-analysis, a positive sentinel lymph node was detected in 32% of patients who had clinically negative nodal disease [60]. For patients with a positive sentinel lymph node biopsy, appropriate treatment options consist of a completion lymph node dissection *or* radiation to the nodal basin. In the same study, the 3-year relapse-free survival for patients who had a positive sentinel lymph node biopsy and received adjuvant therapy was 60% compared to 0% ($p < 0.01$) for those who did not receive any therapy [60]. Fang et al. reported a 100% 2-year regional recurrence-free survival in patients with microscopically positive nodal disease whether definitive lymph node irradiation or completion lymph node dissection ± adjuvant radiation was performed [61]. Based on available data, either definitive radiation or completion lymph node dissection is adequate in providing locoregional control in the setting of microscopic nodal disease.

For patients with clinically positive nodal disease, a multimodal approach is preferred in some clinical settings. Reports on advanced Merkel cell carcinoma are largely limited to small series that have been conflicting regarding single versus multimodal therapy for the nodal basin. In a retrospective series of 136 patients, 46 presented with clinical node-positive disease; one-fourth received lymph node dissection alone, one-fourth received radiation therapy alone, and half received both forms of therapy [62]. Radiation was associated with improved disease-free survival, though not overall survival; however, patients with nonmetastatic disease who received adjuvant radiation to the primary site were also included in the analysis, so the effect of radiation to the clinically positive nodal basin could not be adequately determined in this study. Due to the lack of available data, adjuvant radiation after therapeutic lymph node dissection for palpable lymphadenopathy is recommended in the case of multiple positive nodes or extracapsular extension [63, 64].

Conclusion

The inguinal nodal basin is an important anatomical area for many nonmelanoma cutaneous malignancies and is a common site of metastasis. In addition to meticulous surgical technique, appropriate postoperative care is necessary to decrease morbidity. Certain technical aspects of the procedure remain controversial. The impact of radiation on morbidity must also be taken into consideration as it is often used as primary therapy or adjuvant therapy in the management of inguinal nodal metastasis. More research is needed in this field to better improve patient morbidity, locoregional recurrence, and possibly survival.

References

1. Zaren HA, Copeland EM III. Inguinal node metastases. Cancer. 1978;41:919–23.
2. Prieto-Granada C, Rodriguez-Waitkus P. Cutaneous squamous cell carcinoma and related entities: epidemiology, clinical and histological features, and basic science overview. Curr Probl Cancer. 2015;39:206–15.
3. Alam M, Ratner D. Cutaneous squamous-cell carcinoma. N Engl J Med. 2001;344:975–83.
4. Parikh SA, Patel VA, Ratner D. Advances in the management of cutaneous squamous cell carcinoma. F1000Prime Rep. 2014;(6):70.
5. Mullen JT, Feng L, Xing Y, et al. Invasive squamous cell carcinoma of the skin: defining a high-risk group. Ann Surg Oncol. 2006;13:902–9.
6. Rowe DE, Carroll RJ, Day CL Jr. Prognostic factors for local recurrence, metastasis, and survival rates in squamous cell carcinoma of the skin, ear, and lip. Implications for treatment modality selection. J Am Acad Dermatol. 1992;26:976–90.
7. Farasat S, Yu SS, Neel VA, et al. A new American Joint Committee on Cancer staging system for cutaneous squamous cell carcinoma: creation and rationale for inclusion of tumor (T) characteristics. J Am Acad Dermatol. 2011;64:1051–9.
8. Agelli M, Clegg LX. Epidemiology of primary Merkel cell carcinoma in the United States. J Am Acad Dermatol. 2003;49:832–41.
9. Zhan FQ, Packianathan VS, Zeitouni NC. Merkel cell carcinoma: a review of current advances. J Natl Compr Cancer Netw. 2009;7:333–9.
10. Eng TY, Boersma MG, Fuller CD, et al. A comprehensive review of the treatment of Merkel cell carcinoma. Am J Clin Oncol. 2007;30:624–36.
11. Clarke CA, Robbins HA, Tatalovich Z, et al. Risk of Merkel cell carcinoma after solid organ transplantation. J Natl Cancer Inst. 2015;107
12. Locke FL, Rollison DE, Sondak VK. Merkel cell carcinoma and immunosuppression: what we still need to know. J Natl Cancer Inst. 2015;107
13. Strom T, Carr M, Zager JS, et al. Radiation therapy is associated with improved outcomes in Merkel cell carcinoma. Ann Surg Oncol. 2016;23:3572–8.
14. Allen PJ, Bowne WB, Jaques DP, Brennan MF, Busam K, Coit DG. Merkel cell carcinoma: prognosis and treatment of patients from a single institution. J Clin Oncol. 2005;23:2300–9.
15. Fields RC, Busam KJ, Chou JF, et al. Five hundred patients with Merkel cell carcinoma evaluated at a single institution. Ann Surg. 2011;254:465–73. discussion 73–5
16. Santamaria-Barria JA, Boland GM, Yeap BY, Nardi V, Dias-Santagata D, Cusack JC Jr. Merkel cell carcinoma: 30-year experience from a single institution. Ann Surg Oncol. 2013;20:1365–73.
17. Carbotta G, Sallustio P, Prestera A, Laforgia R, Lobascio P, Palasciano N. Perineal Paget's disease: a rare disorder and review of literature. Ann Med Surg (Lond). 2016;9:50–2.
18. Haydon N, Ting F, Southwell-Keely J. An important differential: extramammary Paget's disease. Eplasty. 2015;15:ic16.
19. Cohen JM, Granter SR, Werchniak AE. Risk stratification in extramammary Paget disease. Clin Exp Dermatol. 2015;40:473–8.
20. Chanda JJ. Extramammary Paget's disease: prognosis and relationship to internal malignancy. J Am Acad Dermatol. 1985;13:1009–14.
21. Fu T, Aasi SZ, Hollmig ST. Management of high-risk squamous cell carcinoma of the skin. Curr Treat Options in Oncol. 2016;17:34.
22. Smith FO, Yue B, Marzban SS, et al. Both tumor depth and diameter are predictive of sentinel lymph node status and survival in Merkel cell carcinoma. Cancer. 2015;121:3252–60.
23. van der Linden M, Meeuwis K, Bulten J, Bosse T, van Poelgeest M, de Hullu J. Paget disease of the vulva. Crit Rev Oncol Hematol. 2016;101:60–74.
24. Chan JY, Li GK, Chung JH, Chow VL. Extramammary Paget's disease: 20 years of experience in Chinese population. Int J Surg Oncol. 2012;2012:416418.

25. O'Connor WJ, Lim KK, Zalla MJ, et al. Comparison of Mohs micrographic surgery and wide excision for extramammary Paget's disease. Dermatol Surg. 2003;29:723–7.
26. Fujisawa Y, Yoshino K, Kiyohara Y, et al. The role of sentinel lymph node biopsy in the management of invasive extramammary Paget's disease: multi-center, retrospective study of 151 patients. J Dermatol Sci. 2015;79:38–42.
27. Ogata D, Kiyohara Y, Yoshikawa S, Tsuchida T. Usefulness of sentinel lymph node biopsy for prognostic prediction in extramammary Paget's disease. Eur J Dermatol. 2016;26:254–9.
28. Sarnaik AA, Puleo CA, Zager JS, Sondak VK. Limiting the morbidity of inguinal lymphadenectomy for metastatic melanoma. Cancer Contr. 2009;16:240–7.
29. Zhang SH, Sood AK, Sorosky JI, Anderson B, Buller RE. Preservation of the saphenous vein during inguinal lymphadenectomy decreases morbidity in patients with carcinoma of the vulva. Cancer. 2000;89:1520–5.
30. Dardarian TS, Gray HJ, Morgan MA, Rubin SC, Randall TC. Saphenous vein sparing during inguinal lymphadenectomy to reduce morbidity in patients with vulvar carcinoma. Gynecol Oncol. 2006;101:140–2.
31. Sabel MS, Griffith KA, Arora A, et al. Inguinal node dissection for melanoma in the era of sentinel lymph node biopsy. Surgery. 2007;141:728–35.
32. Abbas S, Seitz M. Systematic review and meta-analysis of the used surgical techniques to reduce leg lymphedema following radical inguinal nodes dissection. Surg Oncol. 2011;20:88–96.
33. Sarnaik AA, Puleo CA, Zager JS, Sondak VK. Limiting the morbidity of inguinal lymphadenectomy for metastatic melanoma. Cancer Control. 2009;16:240–7.
34. Paley PJ, Johnson PR, Adcock LL, et al. The effect of sartorius transposition on wound morbidity following inguinal–femoral lymphadenectomy. Gynecol Oncol. 1997;64:237–41.
35. Judson PL, Jonson AL, Paley PJ, et al. A prospective, randomized study analyzing sartorius transposition following inguinal-femoral lymphadenectomy. Gynecol Oncol. 2004;95:226–30.
36. Loukas M, El-Sedfy A, Tubbs RS, Wartman C. Jules Germain Cloquet (1790–1883)—drawing master and anatomist. Am Surg. 2007;73:1169–72.
37. Chu CK, Zager JS, Marzban SS, et al. Routine biopsy of Cloquet's node is of limited value in sentinel node positive melanoma patients. J Surg Oncol. 2010;102:315–20.
38. Illig L, Aigner KR, Biess B, et al. Diagnostic excision of the Rosenmuller's node. Screening for occult metastases before elective regional lymph node dissection in patients with lower limb melanoma? Cancer. 1988;61:1200–6.
39. Essner R, Scheri R, Kavanagh M, Torisu-Itakura H, Wanek LA, Morton DL. Surgical management of the groin lymph nodes in melanoma in the era of sentinel lymph node dissection. Arch Surg. 2006;141:877–82. discussion 82–4
40. Coit DG, Brennan MF. Extent of lymph node dissection in melanoma of the trunk or lower extremity. Arch Surg. 1989;124:162–6.
41. NCCN clinical practice guidelines in oncology: squamous cell skin cancer, version 1, 2016. 2016. NCCN.org. Accessed 19 Sept 2016.
42. Ross AS, Schmults CD. Sentinel lymph node biopsy in cutaneous squamous cell carcinoma: a systematic review of the English literature. Dermatol Surg. 2006;32:1309–21.
43. Delman KA, Kooby DA, Rizzo M, Ogan K, Master V. Initial experience with videoscopic inguinal lymphadenectomy. Ann Surg Oncol. 2011;18:977–82.
44. Delman KA, Kooby DA, Ogan K, Hsiao W, Master V. Feasibility of a novel approach to inguinal lymphadenectomy: minimally invasive groin dissection for melanoma. Ann Surg Oncol. 2010;17:731–7.
45. Abbott AM, Grotz TE, Rueth NM, Hernandez Irizarry RC, Tuttle TM, Jakub JW. Minimally invasive inguinal lymph node dissection (MILND) for melanoma: experience from two academic centers. Ann Surg Oncol. 2013;20:340–5.
46. Dossett LA, Castner NB, Pow-Sang JM, et al. Robotic-assisted transperitoneal pelvic lymphadenectomy for metastatic melanoma: early outcomes compared with open pelvic lymphadenectomy. J Am Coll Surg. 2016;222:702–9.

47. Stuiver MM, Djajadiningrat RS, Graafland NM, Vincent AD, Lucas C, Horenblas S. Early wound complications after inguinal lymphadenectomy in penile cancer: a historical cohort study and risk-factor analysis. Eur Urol. 2013;64:486–92.
48. Mortenson MM, Xing Y, Weaver S, et al. Fibrin sealant does not decrease seroma output or time to drain removal following inguino-femoral lymph node dissection in melanoma patients: a randomized controlled trial (NCT00506311). World J Surg Oncol. 2008;6:1.
49. Neuss H, Raue W, Koplin G, Schwenk W, Reetz C, Mall J. A prospective randomized trial: the influence of intraoperative application of fibrin glue after radical inguinal/iliacal lymph node dissection on postoperative morbidity. Eur J Surg Oncol. 2009;35:884–9.
50. Kwa RE, Campana K, Moy RL. Biology of cutaneous squamous cell carcinoma. J Am Acad Dermatol. 1992;26:1–26.
51. Brodland DG, Zitelli JA. Mechanisms of metastasis. J Am Acad Dermatol. 1992;27:1–8.
52. Committee on Guidelines of Care. Task Force on Cutaneous Squamous Cell Carcinoma. Guidelines of care for cutaneous squamous cell carcinoma. J Am Acad Dermatol. 1993;28:628–31.
53. Johnson TM, Rowe DE, Nelson BR, Swanson NA. Squamous cell carcinoma of the skin (excluding lip and oral mucosa). J Am Acad Dermatol. 1992;26:467–84.
54. Hitchcock CL, Bland KI, Laney RG III, Franzini D, Harris B, Copeland EM III. Neuroendocrine (Merkel cell) carcinoma of the skin. Its natural history, diagnosis, and treatment. Ann Surg. 1988;207:201–7.
55. Veness MJ, Morgan GJ, Palme CE, Gebski V. Surgery and adjuvant radiotherapy in patients with cutaneous head and neck squamous cell carcinoma metastatic to lymph nodes: combined treatment should be considered best practice. Laryngoscope. 2005;115:870–5.
56. Kraus DH, Carew JF, Harrison LB. Regional lymph node metastasis from cutaneous squamous cell carcinoma. Arch Otolaryngol Head Neck Surg. 1998;124:582–7.
57. Khurana VG, Mentis DH, O'Brien CJ, Hurst TL, Stevens GN, Packham NA. Parotid and neck metastases from cutaneous squamous cell carcinoma of the head and neck. Am J Surg. 1995;170:446–50.
58. Weinberg AS, Ogle CA, Shim EK. Metastatic cutaneous squamous cell carcinoma: an update. Dermatol Surg. 2007;33:885–99.
59. McAfee WJ, Morris CG, Mendenhall CM, Werning JW, Mendenhall NP, Mendenhall WM. Merkel cell carcinoma: treatment and outcomes. Cancer. 2005;104:1761–4.
60. Gupta SG, Wang LC, Penas PF, Gellenthin M, Lee SJ, Nghiem P. Sentinel lymph node biopsy for evaluation and treatment of patients with Merkel cell carcinoma: the Dana-Farber experience and meta-analysis of the literature. Arch Dermatol. 2006;142:685–90.
61. Fang LC, Lemos B, Douglas J, Iyer J, Nghiem P. Radiation monotherapy as regional treatment for lymph node-positive Merkel cell carcinoma. Cancer. 2010;116:1783–90.
62. Howle JR, Hughes TM, Gebski V, Veness MJ. Merkel cell carcinoma: an Australian perspective and the importance of addressing the regional lymph nodes in clinically node-negative patients. J Am Acad Dermatol. 2012;67:33–40.
63. NCCN clinical practice guidelines in oncology: Merkel cell carcinoma, version 1, 2016. 2016. NCCN.org. Accessed 19 Sept 2016.
64. Bichakjian CK, Lowe L, Lao CD, et al. Merkel cell carcinoma: critical review with guidelines for multidisciplinary management. Cancer. 2007;110:1–12.

Chapter 8
Considerations for Vulvar Cancer

Vasileios D. Sioulas and Yukio Sonoda

Vulvar Cancer

Epidemiology

Vulvar cancer is the fourth most common gynecologic cancer, accounting for approximately 5% of the genital malignancies in women [1, 2]. In 2016, an estimated 5950 women in the United States were diagnosed with vulvar cancer, while 1110 died of this disease [2]. Although vulvar cancer is usually diagnosed in postmenopausal women (median age at diagnosis, 68 years), an increasing incidence in younger women has been noted over the last decades [3, 4]. This could be attributed to the higher rates of human papillomavirus (HPV)-associated vulvar intraepithelial neoplasia (VIN) recorded in this population during the same time period [4, 5]. The widespread implementation of HPV vaccination programs may reverse this trend in the future. Studies indicate that the quadrivalent HPV vaccine can reduce the risk of any high-grade vulvar lesions by approximately 50% [6].

Histology/Etiology

Risk factors for vulvar cancer include cigarette smoking, immunosuppression, vulvar or cervical intraepithelial neoplasia, northern European ancestry, vulvar dystrophy (e.g., lichen sclerosus), and HPV infection [7]. The latter is responsible for 40–60% of vulvar cancers, while HPV 16, 18, and 33 have been shown to be the predominant subtypes [1, 8, 9].

V.D. Sioulas, M.D., Ph.D. • Y. Sonoda, M.D. (✉)
Gynecology Service, Department of Surgery, Memorial Sloan Kettering Cancer Center, New York, NY, USA
e-mail: vsioulas@gmail.com; sonoday@mskcc.org

K.A. Delman, V.A. Master (eds.), *Malignancies of the Groin*,
https://doi.org/10.1007/978-3-319-60858-7_8

Squamous cell carcinoma (SCC) accounts for greater than 90% of vulvar cancer cases [7, 10]. Melanoma, the second most common histologic type, is diagnosed in approximately 9% of these patients [4]. Rarely, verrucous carcinomas, basal cell carcinomas, soft tissue sarcomas, extramammary Paget's disease, adenocarcinomas of the Bartholin's gland, and neuroendocrine and neuroectodermal tumors can occur [4, 7].

Two different etiologies have been described for the pathogenesis of SCC of the vulva. Basaloid or warty carcinomas are encountered in younger women and linked to HPV infection [7, 11, 12]. In contrast, keratinizing squamous cell tumors are found in older women and are associated with inflammatory vulvar skin diseases such as lichen sclerosus [1, 12]. Mutations in the TP53 tumor suppressor gene have been proposed as the pathway possibly connecting vulvar dystrophies to SCC development [1, 13].

Clinical Manifestations/Diagnosis

The majority of women with vulvar cancer present with a visible or palpable lesion. Associated symptoms may include itching, a burning sensation, bleeding, discharge, dysuria, and dyspareunia, although patients can also be asymptomatic at the time of diagnosis [1, 7]. A delayed diagnosis is not uncommon. This is due to the lack of specific symptoms, the reluctance of many elderly women to report their symptoms or undergo routine gynecologic examinations, and the low biopsy rates of suspicious lesions [1, 14].

Vulvar cancer is most commonly localized in the labia (80%), the clitoris (10%), and the lower commissure (10%). Most of the tumors are unilateral; however, bilateral, as well as multifocal, lesions can be found. On visual inspection, any whitish, brownish, reddish, ulcerated, elevated, or thickened area is considered suspicious. If no lesions are recognized macroscopically in a symptomatic patient, colposcopic examination after applying 5% acetic acid solution can be employed [1]. All concerning lesions should be biopsied under local analgesia. Punch biopsies to allow for depth of invasion assessment are encouraged [1, 4]. The area(s) of the lesion that appears most abnormal should be preferentially sampled. In case of a discrepancy between the degree of clinical suspicion and pathologic results, a repeat biopsy should be considered [4, 7]. Primary care physicians should avoid excising the primary tumor to facilitate accurate evaluation of its size/location, as well as allowing for sentinel lymph node (SLN) mapping by a gynecologic oncologist [4, 7].

Staging

Vulvar cancer spreads by (1) direct extension into adjacent organs, (2) lymphatic embolization to regional lymph nodes in the groin, and (3) hematogenous dissemination to distant sites (e.g., bone, liver, lung), usually at the time of recurrence [4].

The initial evaluation of a patient with vulvar cancer includes a detailed medical history and physical/pelvic examination. The size of primary tumor and its proximity to the midline or surrounding structures (e.g., urethra, vagina, anus, bones) should be carefully assessed. Given that synchronous lower genital tract neoplasia is diagnosed in 10–15% of patients with vulvar cancer, a thorough examination of the entire vulva, perianal skin surface, vagina, and cervix, along with cervical cytology and colposcopy, should be performed [4, 7]. Bilateral groins and supraclavicular lymph nodes should be palpated to rule out enlarged lymph nodes [7]. In case of palpable groin lymph nodes, a biopsy should be obtained. Diagnostic imaging is not required in women with small primary tumors and body habitus not precluding clinical inguinal lymph node evaluation. In patients with large or fixed tumors and those who experience significant discomfort, exam under anesthesia may be beneficial. In advanced cases, cystourethroscopy and proctosigmoidoscopy can be considered to determine the extent of disease [1, 4, 7].

Radiographic studies may be helpful in patients with bulky tumors, clinically suspicious lymph nodes, and symptoms suggestive of distant metastasis or those who are poor surgical candidates due to comorbidities. Depending on the indication, ultrasonography (US), computed tomography (CT), magnetic resonance imaging (MRI), positron emission tomography (PET), or PET/CT are the available options [4, 15]. A small prospective study showed that there was no significant difference between CT/MRI and PET or PET/CT in detecting metastatic inguinal lymph nodes. However, CT/MRI was more efficacious than PET or PET/CT in the identification of pelvic lymph node or distant metastasis [16]. More recently, a study from Memorial Sloan Kettering Cancer Center underscored the important role that PET/CT may play in the management of vulvar cancer [17].

The staging system for vulvar cancer was last revised in 2009. Its current version, which is adopted by the International Federation of Gynecology and Obstetrics (FIGO), the American Joint Committee on Cancer (AJCC), and the International Union Against Cancer (UICC), is illustrated in Table 8.1 [18, 19]. Importantly, the depth of invasion (DOI) is defined as the measurement of the tumor from the epithelial-stromal junction of the adjacent most superficial dermal papilla to the deepest point of stromal penetration. Data pointing out that the risk of lymph node involvement in patients with DOI ≤1 mm is <1%, as compared to a risk of 15–20% in case of DOI between 3.1 and 5 mm, highlights its value in tailoring the treatment plan [4, 7, 20]. The new staging system reflects the prognostic value of the number of positive lymph nodes, the size of their largest metastasis, as well as the presence of extracapsular extension. Nonetheless, the number of lymph nodes that should be removed in order for the evaluation to be considered adequate is not clarified. Of note, this staging classification does not apply to vulvar malignant melanomas [4].

Table 8.1 Staging vulvar cancer [18, 19]

TNM categories	FIGO stages	Definition
Primary tumor (T)		
TX		Primary tumor cannot be assessed
T0		No evidence of primary tumor
Tis		Carcinoma in situ (preinvasive carcinoma)
T1a	IA	Lesions 2 cm or less in size, confined to the vulva or perineum and with stromal invasion 1.0 mm or less
T1b	IB	Lesions more than 2 cm in size or any size with stromal invasion more than 1.0 mm, confined to the vulva or perineum
T2	II	Tumor of any size with extension to adjacent perineal structures (lower/distal 1/3 urethra, lower/distal 1/3 vagina, anal involvement)
T3	IVA	Tumor of any size with extension to any of the following: upper/proximal 2/3 urethra, upper/proximal 2/3 vagina, bladder mucosa, rectal mucosa, or fixed to the pelvic bone
Regional lymph nodes (N)		
NX		Regional lymph nodes cannot be assessed
N0		No regional lymph node metastasis
N1		One or two regional lymph nodes with the following features
N1a	IIIA	One or two lymph node metastases, each 5 mm or less
N1b	IIIA	One lymph node metastasis 5 mm or greater
N2	IIIB	Regional lymph node metastasis with the following features
N2a	IIIB	Three or more lymph node metastases, each less than 5 mm
N2b	IIIB	Two or more lymph node metastases 5 mm or greater
N2c	IIIC	Lymph node metastasis with extracapsular spread
N3	IVA	Fixed or ulcerated regional lymph node metastasis
Distant metastasis (M)		
M0		No distant metastasis
M1	IVB	Distant metastasis (including pelvic lymph node metastasis)

Management of the Primary Tumor

Historically, the gold standard for vulvar cancer staging and treatment was en bloc radical vulvectomy and bilateral inguinofemoral lymphadenectomy through a butterfly incision. Despite the favorable survival outcomes, this procedure was associated with remarkable morbidity, including a high rate of wound infections, lymphedema, and major problems regarding patient's sexual function, body image, and self-assurance. Therefore, over the last decades, the excision of primary tumor and inguinofemoral lymph nodes has been accomplished through three separate incisions [8, 21]. This approach results in fewer complications, without compromising survival [22].

The technique used to resect the vulvar tumor depends on its size and extension [23]. Radical vulvectomy implies removal of the entire vulva down to the level of the deep fascia of the thigh, the periosteum of the pubis, and the inferior fascia of the urogenital diaphragm [8]. In current practice, the radical local excision and modified radical vulvectomy (also known as radical hemivulvectomy) are typically used [23]. Although data from prospective trials are lacking, the oncologic outcomes of the radical local excision, modified radical vulvectomy, and radical vulvectomy seem to be comparable [24, 25]. It should be stressed that the depth of resection is the same in all the aforementioned techniques (e.g., to the urogenital diaphragm) [26]. Importantly, T1 and T2 lesions ≤4 cm not infiltrating the urethra, vagina, or anus can be treated with wide local excision [8, 23, 26].

Independently of the selected surgical approach, every effort should be made to obtain normal tissue margins of 1–2 cm at primary surgery as a means to decrease the risk of local recurrence [23, 27]. If final pathology is consistent with closer (<8 mm) or positive surgical margins, re-excision or adjuvant radiation therapy (RT) should be offered. The decision is highly individualized and should take into account multiple parameters, including a patient's desire and comorbidities, lymph node status, and proximity to the urethra, vagina, and anus [23]. If the lesion involves the urethra, the distal 1 cm of the urethra can be excised without affecting continence [8].

Patients with locally advanced disease (T2 tumors >4 cm or with involvement of the urethra, vagina, or anus and T3 tumors) may benefit from neoadjuvant RT with concurrent platinum-based radiosensitizing chemotherapy [23, 28]. 5-Fluouracil in combination with cisplatin or mitomycin-C has been also used in this setting [29–32]. In case of residual tumor deemed to be resectable, surgical excision is favored. If the patient is a poor surgical candidate or the tumor is unresectable, additional individualized RT and/or chemotherapy or supportive care is recommended [10, 23].

Wound infection and breakdown are the most common complications that may follow surgery to the vulva. Their incidence ranges from 9 to 58%, while extensive surgery, increased age, obesity, diabetes, smoking, and prior RT are recognized as predisposing factors [33]. When large areas of the vulva are resected, tissue mobility is poor, or in case of neoadjuvant RT, primary closure of the vulvar defects may not be feasible. In these patients, vulvar reconstruction with fasciocutaneous or myocutaneous flaps, most commonly by plastic surgeons, may be needed [4, 8]. Urinary incontinence and vaginal prolapse are also listed among vulvar surgery complications [33]. Greater than 50% of the women who undergo vulvectomy report sexual dysfunction, including dyspareunia, decreased desire, or inability to orgasm, as well as resultant psychological issues. Conservative surgical techniques, when indicated, may result in better sexual and cosmetic outcomes [33].

Prognosis

Several factors have been reported to affect the prognosis of patients with vulvar cancer, including their age, the stage of disease, tumor size, DOI, capillary lymphatic space invasion, and lymph node status [34–37]. Among them, the presence of positive inguinofemoral lymph nodes is the single most significant determinant of disease-specific mortality [4, 8, 10, 34]. Further aspects of lymph node involvement (number, size of metastasis, and capsule infiltration) that exhibit a prognostic role are incorporated into the current staging system, presented in Table 8.1 [18, 19]. As mentioned, the adequacy of surgical margins is an important parameter that influences the risk of local recurrence. Heaps et al. found that a microscopic pathologic margin <8 mm corresponding to <1 cm in fresh tissue was associated with a 50% chance of recurrence [27]. Their results were supported by more recent studies [26, 36]. Lastly, a retrospective study published in 2016 revealed that among patients with SCC who received RT with or without surgical resection, the presence of HPV or its surrogate of p16 immunostaining was associated with better progression-free survival (PFS) and lower local recurrence rates [38]. In contrast, a series of 201 cases revealed increased risk of local relapse, either close or remote to the tumor margins, when vulvar SCC arose in a field of lichen sclerosis [39].

The prognosis of patients with vulvar cancer is quite good when appropriate and timely treatment is offered [8]. According to FIGO statistics, the 5-year overall survival (OS) rates per FIGO stage are I, 78.5%; II, 58.8%; III, 43.2%; and IV, 13.0% [40].

Follow-Up

According to the recently released National Comprehensive Cancer Network (NCCN) guidelines on SCC of the vulva, surveillance includes visits every 3–6 months for the first 2 years, followed by visits every 6–12 months until the completion of 5 years after treatment. Thereafter, the patients can be examined on an annual basis, although individual risk factors for disease recurrence may modify the schedule [23]. Of note, it was shown that relapses occur ≥5 years after the initial therapy in 35% of patients; these findings underscore the value of long-term surveillance [41, 42]. Surveillance visits should include a patient's history and detailed physical examination, with emphasis on the vulva, skin bridge, and groins [23, 42]. Distant areas of potential cancer recurrence (e.g., supraclavicular lymph nodes, lungs, brain, bones), as well as possible sites for neoplasia (cervix, vagina, perianal area), also should be evaluated [7, 42]. Regular cervical/vaginal cytology screening should be considered. In case of symptoms or clinical findings indicative of recurrence, appropriate laboratory workup and imaging studies should be ordered [23, 42]. The assessment for possible long-term complications from vulvar cancer treatment, including sexual health issues, should be part of the follow-up visit [10, 33].

Recurrence

Most treatment failures are diagnosed within 2 years after the initial surgical therapy [41]. Vulva constitutes the most common site of recurrence. In a series of 502 patients, 53.4% of them were diagnosed with local relapse. Inguinal, pelvic, distant, and multiple recurrences were found in 18.7%, 5.7%, 7.9%, and 14.2% of the patients, respectively [37]. According to a Gynecologic Oncology Group (GOG) study on patients who had undergone conservative therapy for vulvar cancer, the median time to recurrence in the vulva and groin was 35.9 and 7.0 months, respectively. Patients with local recurrence had better prognoses; the median survival after vulvar and groin relapse was 52.4 and 9.4 months, respectively [43]. Similarly, Maggino et al. showed that patients with recurrence in the vulva had favorable survival outcomes, compared to those with regional or distant recurrence. In their report, the 5-year survival rates were 60% for local, 27% for inguinal and pelvic, 15% for distant, and 14% for recurrence at multiple sites [37]. However, local relapse at the site of the primary tumor or skin bridge confers higher risk of cancer-related death compared with other perineal areas [44].

In recurrent cases, the treatment intent (curative vs. palliative) and plan depend on the patient's performance status, site of recurrence, and previous management. In patients not previously irradiated, recurrences confined to vulva with clinically negative lymph nodes are treated with radical re-excision. Unilateral or bilateral inguinofemoral lymphadenectomy (IFLD) is performed if neglected at the time of initial therapy [23]. Pelvic exenteration can be considered in case of local, central recurrence [23]. In this setting, PET/CT should be preoperatively ordered to rule out distant metastases [42]. Patients not previously exposed to pelvic RT can be scheduled for surgical resection if isolated pelvic lymph node recurrence is diagnosed. In case of multiple pelvic lymph node involvement, distant recurrence, or history of pelvic RT, management options include systemic chemotherapy, supportive care, or enrollment in a clinical trial [23]. The treatment recommendations for patients with groin recurrence are outlined in the relevant section of this chapter.

Management of Inguinofemoral Lymph Nodes

Inguinofemoral Lymphadenectomy (IFLD) in Patients with SCC of the Vulva

Full or complete IFLD refers to the removal of all lymph nodes of the Scarpa's triangle; the superficial chain lies between the superficial and femoral fascia, while the deep lymph nodes are situated within the fossa ovalis medial to the femoral vein. This procedure can be performed with the preservation of the femoral fascia [45].

Landmark studies published over the last decades determined the indications of performing IFLD in vulvar cancer patients that guide the current practice. As aforementioned, the risk of lymph node spread is <1% when DOI ≤1 mm [4, 7, 46]. In 1979, DiSaia et al. proposed that superficial IFLD could substitute for full IFLD in select patients. Their single-institution trial on women with stage I disease, DOI ≤5 mm, and negative nodes on frozen section supported this approach [21, 47]. However, the GOG-74 protocol demonstrated a disappointingly high rate of groin and/or vulva recurrence in women treated with radical local excision and ipsilateral superficial IFLD compared to historic controls having undergone radical vulvectomy and bilateral IFLD (15.6% vs. 6.7%, respectively) [48]. The GOG-88 protocol was designed to test whether groin RT was superior to and less morbid than IFLD in patients with clinically nonsuspicious nodes. The study was closed prematurely due to an unacceptably increased rate of groin recurrence in the RT group [49]. In terms of laterality, numerous studies pointed out that the risk of contralateral groin node metastasis in case of patients with early-stage, lateral disease and negative nodes after unilateral IFLD is <3% [48, 50, 51]. More recently, Gonzalez Bosquet et al. showed that in tumors located >1 cm from the midline with size ≤2 cm and DOI ≤5 mm, the risk of contralateral metastasis is zero [52].

Taken together, the NCCN recommends that IFLD can be safely omitted in patients with stage IA SCC of the vulva. In contrast, women with IB-II disease should undergo IFLD. For a tumor that is <2 cm in maximum dimension, located 2 cm or more from the vulvar midline and in the setting of clinically negative groin nodes, ipsilateral IFLD is appropriate. If positive lymph nodes are identified on pathology, contralateral IFLD or RT of the contralateral groin is recommended. Tumors closer than 2 cm from or crossing the vulvar midline should be treated with bilateral IFLD. Women with positive lymph nodes after bilateral IFLD should be offered RT with or without chemotherapy, especially when two or more lymph nodes are involved or the metastatic deposits are larger than 2 mm [23]. Homesley et al. demonstrated that for patients with positive groin nodes, adjuvant radiation is preferred over pelvic lymphadenectomy [53].

The management of bulky inguinofemoral lymph nodes in the setting of an unresectable or T3 primary vulvar tumor is more controversial. Platinum-based chemoradiation to the primary tumor, bilateral groins, and pelvis with or without prior debulking of the positive inguinofemoral lymph nodes is among the available options [23].

IFLD leads to significant morbidity; it is estimated that more than half of patients undergoing IFLD will experience at least one complication associated with the procedure. Lymphedema is a chronic condition that may be evident in almost 50% of these patients. It most commonly presents within 12 months of the groin dissection and may result in decreased mobility, severe limitations in daily activities, and psychological distress [33]. Obesity, a large number of lymph nodes removed, extensive surgery, postoperative infection or deep venous thrombosis (DVT), and RT to the groins are listed as risk factors for lymphedema development [33, 54, 55]. Lymphocele, a term used to describe collections of lymphatic fluid into the dead space resulting from lymph node dissection, may be diagnosed in up to 40% of patients after IFLD. Lymph leakage along with impaired lymphatic reabsorption seems to be the underlying mechanism for their formation. Importantly, lymphocele confers a higher risk for infection, edema, pain, and DVT [33]. Lastly, despite the remarkable decrease in their incidence after the implementation of the separate groin incisions, wound secondary events are common in patients undergoing IFLD. Wound infection, cellulitis, or breakdown may complicate 21–39%, 21–57%, or 17–39% of the cases, respectively [33].

Sentinel Lymph Node (SLN) Biopsy

Rationale, Diagnostic Accuracy, and Oncologic Safety

SLN mapping is an image-guided procedure used in the treatment of multiple types of cancer, primarily melanoma and breast cancers [56, 57]. This approach is based on the concept that lymph drains in an orderly pattern away from the tumor through the lymphatic system. Consequently, if the SLN is negative for metastasis, then the remaining nodes should also be negative and can be left behind [56]. The desire to limit the incidence and severity of the complications associated with complete lymphadenectomy largely accounts for SLN biopsy's increasing popularity over the last decades [57].

Lymph node involvement will be diagnosed in only 20–30% of early-stage vulvar cancer patients. Therefore, the vast majority of these women will be exposed to the risks of IFLD without, most likely, gaining a benefit from it [51]. Given that the need to exclude lymph node metastasis in the preoperative setting cannot be reliably met by imaging studies, including US, CT, MRI, and PET, the role of SLN biopsy has been extensively studied [15]. The encouraging results from several small studies led to the design of two large prospective trials, which validated the accuracy and established the safety of the procedure: the Groningen International Study on Sentinel nodes in Vulvar cancer (GROINSS-V) and GOG-173 [51, 58, 59].

The GROINSS-V study was a multicenter observational study of 403 women with early-stage vulvar cancer, in which full IFLD was omitted in patients with a negative SLN. Squamous cell histology, primary tumor smaller than 4 cm, DOI >1 mm, and nonsuspicious groin nodes at palpation served as the eligibility criteria. By utilizing a radioactive tracer and blue dye, the false-negative rate (FNR) was 5.9% (4.6% in patients with unifocal disease), and the false-negative predictive value was 2.9%. Among 259 patients with unifocal disease and negative SLN, a groin recurrence rate of 2.3% over a median follow-up time of 35 months and a 3-year disease-specific survival (DSS) rate of 97% were recorded [57, 58]. This rate of groin relapse was at least comparable to that reported for patients with early-stage vulvar cancer and negative lymph nodes following any type of formal IFLD [60]. In contrast, groin recurrence was diagnosed in 11.8% of patients with multifocal disease; this high rate resulted in the protocol amendment and exclusion of patients with multifocal disease. The rates of complications in patients with SLN biopsy and full IFLD, as indicated by positive SLN, were as follows: wound breakdown, 11.7% vs. 34.0%; cellulitis, 4.5% vs. 21.3%; recurrent erysipelas, 0.4% vs. 16.2%; and lymphedema, 1.9% vs. 25.2%, respectively [58]. In 2016, after a median follow-up time of 105 months, study outcomes were published. Among SLN-negative patients, the rate of isolated groin recurrence was 2.5%, and the 10-year DSS was 91% [61].

The GOG protocol 173 was a multi-institutional observational study of 452 women with early-stage vulvar cancer. Eligible women should have had SCC, DOI $\geq$1 mm, tumor size $\geq$2 and $\leq$6 cm, as well as clinically benign lymph nodes.

All patients underwent SLN biopsy, followed by full IFLD. The FNR for SLNs identified by blue dye alone, radiocolloid alone, and combination of blue dye and radiocolloid was 2.0%, 7.8%, and 1.6%, respectively. As far as the size of primary tumor was concerned, the FNR dropped from 8% in the study population to 5.6%, when the analysis was restricted to patients with tumor size <4 cm. The false-negative predictive value for SLN biopsy was 3.7%. In patients with tumors measuring <4 cm, the false-negative predictive value for SLN biopsy was 2.0%. In case of tumors measuring 4–6 cm, the false-negative predictive value for SLN biopsy climbed up to 7.4% [59].

The meta-analyses of the literature on SLN mapping in early-stage vulvar cancer patients confirmed the above findings. Hassanzade et al. found a detection rate per groin of 84.6% [62]. More recently, an FNR of 5% was reported after using the combination of radiocolloid and blue dye for detection and pathological examination with ultrastaging and immunohistochemistry [63]. Lastly, Covens and coworkers from Canada showed that the overall detection rate per groin with the combination of blue dye and radiocolloid was 86.9%. The SLN detection rate per groin with a radioactive tracer was much higher, compared to blue dye alone (85% vs. 63%). Similarly, the FNR in cases of combined technique was lower than that calculated for the use of radiocolloid or blue dye alone (6.6% vs. 10.4% vs. 9.3%, respectively). The groin recurrence rate in patients treated with SLN biopsy was estimated to be 3.4%, as opposed to 1.4% in those undergoing full IFLD [64]. It is worth mentioning that a repeat SLN procedure in patients with recurrent vulvar cancer is technically more challenging and leads to a lower SLN detection rate [65].

Selection Criteria, Technique, and Pathologic Processing

In light of the above data, NCCN guidelines incorporated SLN biopsy in the management of select patients with early-stage vulvar cancer. The candidates for this technique should have SCC, unifocal tumor measuring less than 4 cm, negative clinical groin examination and imaging, as well as no prior surgeries to the vulva that may have impacted lymphatic flow to the inguinal region [23]. The best imaging technique (e.g., US, CT, MRI, PET) for the preoperative assessment of groins has not been determined; the decision largely depends on local expertise and availability [57].

In terms of SLN technique, the combination of both radiocolloid and dye is favored as a means to increase the sensitivity of SLN detection. The radioactive tracer—most commonly technetium-99m sulfur colloid—is usually injected 2–4 h prior to the surgical procedure [23]. A preoperative lymphoscintigraphy (LSG) may be performed to aid in anatomically locating the SLN, although this strategy is more controversial [23, 57, 64]. Importantly, according to GOG-173 results, patients with lateral tumors (>2 cm from the midline) or lesions within 2 cm from the midline without involving it and only unilateral drainage on preoperative LSG may safely undergo unilateral SLN biopsy [66]. Approximately, 3–4 cc of dye—most commonly isosulfan blue 1%—is injected intradermally at the margins of the tumor

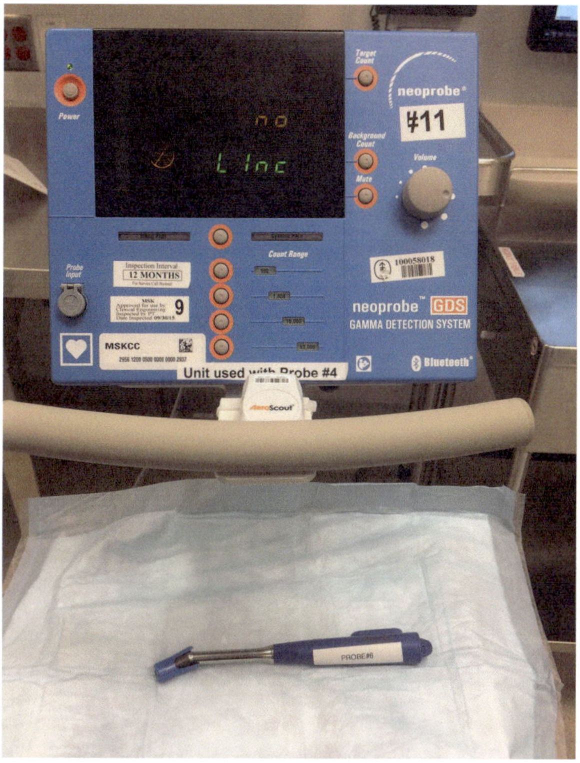

Fig. 8.1 Gamma probe used to detect sentinel lymph nodes by radiocolloid injection

(2, 5, 7, and 10 o'clock) within 15–30 min of initiating the SLN dissection. It should be stressed that the localization of blue dye in the nodal group of interest is transient (e.g., for 30–60 min). In order for the surgeons to explore the groins in a timely fashion and preserve the lymphatic channels connecting the vulvar tumor to the inguinal lymph node basin, the NCCN recommends that the SLN procedure be performed prior to the excision of the primary tumor. The location and size of the groin incision is more accurately determined after detecting the radiocolloid in the inguinofemoral region with the use of a gamma probe [23] (Fig. 8.1). A node that has >5 times the background radioactivity should be used to identify an SLN. Once the incision is made, the surgeon should also look for blue lymphatic channels and follow their course [64] (Fig. 8.2). It has been shown that 84% of SLNs are found in the superficial inguinal chain, while the remaining ones are deep femoral [67]. If an SLN is not identified, a side-specific, complete IFLD is required [23]. Of interest, a limited number of studies on the role of near-infrared (NIR) fluorescence imaging and indocyanine green (ICG) tracer in SLN mapping for vulvar cancer have been conducted, showing promising results [68–71] (Fig. 8.3).

The value of frozen-section analysis in these cases is unclear. In the GROINSS-V study, frozen sectioning was performed in 78% of the patients and showed a sensitivity of 48%, specificity of 100%, negative predictive value of 78%, and positive

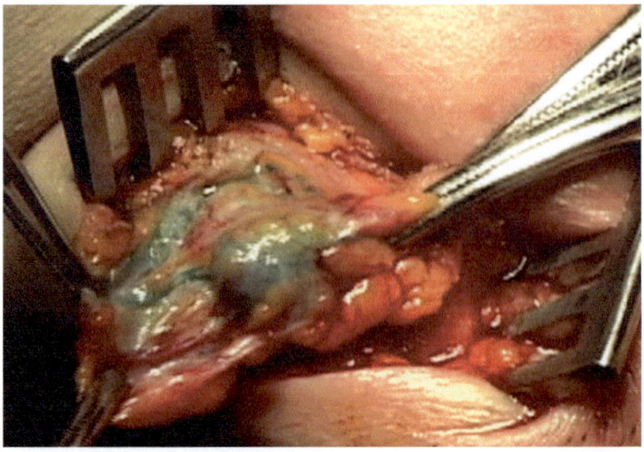

Fig. 8.2 Blue dye used to detect sentinel lymph nodes

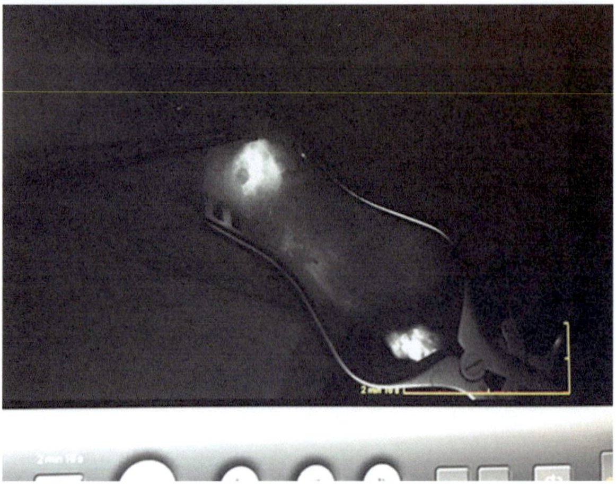

Fig. 8.3 A sentinel lymph node detected by fluorescent imaging using indocyanine green

predictive value of 100% [72]. Other researchers found a diagnostic accuracy of 98% for frozen-section analysis; 2 of 98 nodes were classified as false negative [67]. In their meta-analysis and guidelines, Covens and coworkers found insufficient evidence to make a recommendation for or against the use of frozen-section analysis. Until future research gives us further insight, the advantage of possibly avoiding a second procedure and the drawback of limiting the amount of available tissue for permanent analysis should be carefully weighed [64].

Ultrastaging of SLNs in patients with vulvar cancer refers to the evaluation of more sections, as compared to routine pathology. By using hematoxylin and eosin (H&E) staining and immunohistochemical (IHC) cytokeratin staining on paraffin-embedded SLN tissue cut at 400- to 500-μm intervals (as opposed to the 2- to 3-mm intervals used for traditional lymph node evaluation), pathologists may identify micrometastases in SLNs that would have been otherwise considered as benign [57, 64]. In the GROINSS-V study, 41% of positive SLNs were detected by ultrastaging: 14% with H&E and 27% with IHC staining. The risk of non-SLN metastases was higher when the SLN was found to be positive with traditional pathology than with ultrastaging. Likewise, in patients with positive SLNs detected by ultrastaging and routine microscopy, the 5-year DSS was 92.1% and 64.9%, respectively [72]. In GOG-173, 23% of all positive SLNs were detected by immunohistochemistry, while routine H&E staining had failed to reveal metastatic disease [59]. Despite the above results, consensus regarding the standards of histopathology and the need for ultrastaging is still lacking [57].

Management of Positive SLN

Patients with SLN metastasis, independent of size, have been treated with bilateral complete IFLD, followed by adjuvant RT with or without concurrent chemotherapy [23, 51]. However, accumulating evidence has questioned this approach. Woelber et al. revealed that none of 28 patients with unilateral positive SLNs had contralateral non-SLN metastasis when the SLN on that side was negative. Subsequently, the omission of contralateral IFLD to reduce surgical morbidity was proposed [73]. Moreover, as ultrastaging allows for the detection of more and smaller metastases, the risk of overtreatment, along with the resultant side effects and costs, should be taken into consideration [51]. The GROINSS-V study highlighted that the size of SLN metastases has significant prognostic implications. The rate of non-SLN metastases per size of SLN deposits was isolated tumor cells, 4.2%; ≤2 mm, 10.5%; >2 mm and ≤5 mm, 13.3%; and >5 mm, 47.6%. The 5-year DSS rates were 97%, 88%, 70%, and 69%, respectively. SLN metastases ≤2 mm were associated with significantly better 5-year DSS, as compared to SLN deposits >2 mm (94.4% vs. 69.5%), indicating that the cutoff of 2 mm could be used to define micrometastasis [72]. Nonetheless, the prognostic significance of micrometastasis or isolated tumor cells is not reflected in the current staging system, while no size of SLN metastases has been proposed below whose complete IFLD could be omitted without compromising oncologic outcomes [51].

More data on the optimal management of positive SLNs in early-stage vulvar cancer are expected upon the completion of the GROINSS-V-II/GOG-270 study. The researchers aim to investigate the safety of replacing full IFLD with adjuvant RT when the size of SLN metastasis is ≤2 mm. In addition, they will explore the efficacy, safety, and short- and long-term morbidity of IFLD and RT with or without chemotherapy in patients with SLN metastasis >2 mm. Lastly, this multicenter study will further evaluate the safety of observation alone for women with no metastases detected on SLN biopsy [57]. Of note, the NCCN has already incorporated RT (±chemotherapy) without full IFLD as an acceptable option in the treatment armamentarium for positive SLNs [23].

Quality of Life

The high complication rates of IFLD can have a major negative impact on quality of life, especially when adjuvant RT with or without chemotherapy is administered [57]. Although the GROINSS-V study showed that an SLN procedure is associated with lower morbidity compared to full IFLD, a follow-up survey sent to its participants did not record any differences in the overall quality of life between the two groups, as assessed by the European Organization for Research and Treatment of Cancer (EORTC) QLQ-C30 questionnaire. Only the financial difficulties scale was significantly worse in the IFLD group. When the Functional Assessment of Cancer Therapy-Vulvar (FACT-V) questionnaire was used, patients who had undergone IFLD had significantly worse results with respect to the contentment functional scale and edema, complaints, and stocking symptom scales [74]. Novackova et al. observed increased fatigue and impaired lymphedema in patients who had had full IFLD. These women also exhibited significantly worse outcomes in body image and cognitive functioning compared to patients treated with SLN biopsy [75]. Similarly, Forner et al. found that SLN mapping was linked to better sexual function than IFLD [76].

Given the relatively limited data, additional studies are needed to establish the favorable outcomes of SLN biopsy over IFLD in terms of quality of life [57].

Cost-Effectiveness

The cost-effectiveness of SLN biopsy in patients with early-stage vulvar cancer was explored in a few studies. Sutton et al. showed that SLN biopsy using technetium-99m and blue dye with ultrastaging may be considered the most cost-effective strategy based on the outcome of survival free of morbidity for 2 years [77]. In the United States, Erickson et al. found that the annual cost of SLN biopsy was $65.2 million compared with $76.8 million for IFLD. The lower cost-effectiveness ratio of SLN biopsy compared with IFLD was even more evident when complication costs were added ($23,711 vs. $31,198) [78]. Similarly, McCann et al. reported that SLN biopsy was less costly ($13,449 vs. $14,261) and more effective (quality-adjusted life years; 4.16 vs. 4.00) compared with complete IFLD. These differences were primarily attributed to the increased incidence of lymphedema after IFLD [79].

In summary, the shorter operating time and hospitalization, along with the fewer complications, offset the additional costs related to the procedure (e.g., radiocolloid, blue dye, intraoperative mapping, imaging) and make SLN biopsy the most cost-effective option for women with early-stage vulvar cancer [57, 78].

Learning Curve

Surgeon's familiarity with the procedure is a key factor for identifying SLNs in patients with early-stage vulvar cancer. Inadequate surgical experience may have accounted, at least partly, for the high FNR of SLN biopsy concluded by some studies [51, 80, 81]. Moreover, in the GOG-173 protocol, the rate of failure to identify an SLN was 16% during the first 2 years of the study, as compared to 7% for subsequent years [57, 59]. To optimize patients' outcomes and safety, it is recommended that surgeons perform at least ten cases of SLN biopsy followed by complete IFLD without any false negatives before implementing SLN biopsy alone in their practice

[57, 64]. This is in line with data from the GROINSS-V protocol [58]. However, given the rarity of vulvar cancer, achieving an adequate caseload for competence in SLN biopsy is harder for vulvar cancer than other malignancies [57]. Therefore, current guidelines encourage referral of patients to high-volume SLN surgeons in specialized centers [23, 64].

Special Populations

Vulvar Melanoma

Vulvar melanomas are rare tumors; only 644 cases were identified within the Surveillance, Epidemiology, and End Results (SEER) database of the US National Cancer Institute (NCI) between 1973 and 2003 [82]. Elective lymphadenectomy is not therapeutic in melanoma [83]. On this basis, elective IFLD should not be routinely considered in patients with vulvar melanoma and clinically nonsuspicious inguinofemoral lymph nodes [84]. Similarly to cutaneous melanoma, SLN mapping is the standard approach for these patients [83–85]. In cases of negative SLNs, completion lymphadenectomy is not required [83]. In contrast, completion lymphadenectomy has been employed in patients with positive SLNs despite the lack of a clear advantage in terms of OS [84, 85]. The Multicenter Selective Lymphadenectomy Trial II (MSLT-II) is expected to further elucidate whether a completion or delayed lymphadenectomy with closer surveillance is the appropriate treatment strategy when metastasis is identified on SLN biopsy [84]. When inguinofemoral lymph nodes are clinically suspicious, the patients should undergo lymphadenectomy with or without excision of primary tumor in the hopes of improved locoregional control [84].

Pregnancy

Vulvar cancer during pregnancy is extremely rare. A recent systematic review identified 36 published case reports. Squamous histology and FIGO stage I disease were found in 47.2% and 60% of the cases, respectively. Vulvectomy and IFLD were performed in 97.1% and 63.9% of women, respectively. Full-term delivery was recorded in 74.0% of the cases. Delay in diagnosis and advanced-stage disease were associated with decreased disease-free survival (DFS) and OS [86, 87].

In 2014, a group of international experts released guidelines on the management of gynecologic cancers during pregnancy. According to them, standard surgical treatment should be offered to patients with vulvar cancer, depending on the tumor diameter, stage of disease, and gestational age. In those with clinically negative nodes, radical local excision or radical vulvectomy with unilateral or bilateral IFLD or SLN biopsy should be performed. The surgical margins should be wide enough (2 cm macroscopically and 8 mm microscopically) to avoid the postoperative delivery of RT. Patients with positive SLNs require additional treatment. However, if isolated tumor cells are identified on SLN biopsy and the non-SLNs are negative, adjuvant RT might be omitted. When nodal involvement is evident after IFLD, pregnancy is terminated or delivery is planned, depending on the gestational age, and postpartum RT is administered. Delay of the latter by 6–8 weeks can be considered safe. When preoperative examination suggests inguinal lymph node involvement, the prognosis is less favorable. In these cases, immediate inguinal RT is vital, so termination of the pregnancy in the first and second trimester is favored. In the third trimester, cesarean delivery is indicated, followed by standard treatment. Neoadjuvant chemotherapy to reduce tumor size for locally advanced disease remains experimental [88].

Isolated Groin Recurrence

Recurrence in the groins has been traditionally considered as a fatal event [64]. In the era of full IFLD by separate incisions, its incidence was estimated to be approximately 6% [26]. In cases of negative nodes following IFLD, groin recurrence occurred in approximately 2% of patients [58]. The median time to recurrence in the groin was found to be 7.0 months [43]. Importantly, the prognosis in these cases was poor; a median survival of 9.4 months and a 5-year survival rate of 27% were reported [37, 43].

As far as SLN biopsy is concerned, the long-term follow-up of GROINSS-V study participants revealed an isolated groin recurrence rate of 2.5% when an SLN was not involved. The median time for its diagnosis was 13.5 months, while all patients died of the disease. In women with metastasis on SLN biopsy, the rate of isolated groin recurrence was 8.0%. In these cases, the median time to recurrence was 8 months. Among 8 patients, 7 died of vulvar cancer and 1 of intercurrent disease [61]. However, a retrospective study on 30 patients who had primarily undergone either IFLD or SLN biopsy showed a 50% OS rate 7 years after the groin recurrence diagnosis; the best results were noted in patients treated with surgery and RT (±chemotherapy) for disease relapse [89].

According to NCCN guidelines, the management of patients with inguinofemoral nodal recurrence and prior exposure to RT includes systemic chemotherapy, supportive care, or enrollment in a clinical trial. In RT-naive women, resection of positive inguinofemoral lymph nodes (± full IFLD) followed by RT with or without concurrent chemotherapy is favored, when feasible. In contrast, fixed lymph nodes or large recurrences are treated with chemoradiation alone [23].

Conclusion

Positive inguinofemoral lymph nodes remain the strongest prognostic factor for vulvar cancer. Indications for evaluation have been identified by clinicopathologic studies; however, traditional evaluation with IFLD is associated with significant morbidity. The use of SLN mapping has become more popular, and its use is supported by two large observational studies. The management of positive SLNs will continue to evolve as new information from ongoing trials becomes available.

References

1. Dittmer C, Fischer D, Diedrich K, Thill M. Diagnosis and treatment options of vulvar cancer: a review. Arch Gynecol Obstet. 2012;285(1):183–93.
2. Siegel RL, Miller KD, Jemal A. Cancer statistics, 2016. CA Cancer J Clin. 2016;66(1):7–30.
3. Hampl M, Deckers-Figiel S, Hampl JA, Rein D, Bender HG. New aspects of vulvar cancer: changes in localization and age of onset. Gynecol Oncol. 2008;109(3):340–5.
4. Moroney Jw KC, Wilkinson EJ, Levenback CF. Vulva. In: Barakat RR, Berchuck MM, Randall ME, editors. Principles and practice of gynecologic oncology. 6th ed. Philadelphia, PA: Lippincott Williams & Wilkins; 2013. p. 523–56.
5. Judson PL, Habermann EB, Baxter NN, Durham SB, Virnig BA. Trends in the incidence of invasive and in situ vulvar carcinoma. Obstet Gynecol. 2006;107(5):1018–22.
6. Munoz N, Kjaer SK, Sigurdsson K, et al. Impact of human papillomavirus (HPV)-6/11/16/18 vaccine on all HPV-associated genital diseases in young women. J Natl Cancer Inst. 2010;102(5):325–39.
7. Zweizig S, Korets S, Cain JM. Key concepts in management of vulvar cancer. Best Pract Res Clin Obstet Gynaecol. 2014;28(7):959–66.
8. Alkatout I, Schubert M, Garbrecht N, et al. Vulvar cancer: epidemiology, clinical presentation, and management options. Int J Women Health. 2015;7:305–13.
9. Monk BJ, Burger RA, Lin F, Parham G, Vasilev SA, Wilczynski SP. Prognostic significance of human papillomavirus DNA in vulvar carcinoma. Obstet Gynecol. 1995;85(5 Pt 1):709–15.
10. Stehman FB, Look KY. Carcinoma of the vulva. Obstet Gynecol. 2006;107(3):719–33.
11. De Koning MN, Quint WG, Pirog EC. Prevalence of mucosal and cutaneous human papillomaviruses in different histologic subtypes of vulvar carcinoma. Mod Pathol. 2008;21(3):334–44.
12. Brinton La SV, Trabert B. Epidemiology of gynecologic cancers. In: Barakat RR, Berchuck MM, Randall ME, editors. Principles and practice of gynecologic oncology. 6th ed. Philadelphia, PA: Lippincott Williams & Wilkins; 2013. p. 1–29.
13. Rolfe KJ, Maclean AB, Crow JC, Benjamin E, Reid WM, Perrett CW. TP53 mutations in vulval lichen sclerosus adjacent to squamous cell carcinoma of the vulva. Br J Cancer. 2003;89(12):2249–53.
14. Jones RW, Joura EA. Analyzing prior clinical events at presentation in 102 women with vulvar carcinoma. Evidence of diagnostic delays. J Reprod Med. 1999;44(9):766–8.
15. Kim KW, Shinagare AB, Krajewski KM, et al. Update on imaging of vulvar squamous cell carcinoma. AJR Am J Roentgenol. 2013;201(1):W147–57.
16. Lin G, Chen CY, Liu FY, et al. Computed tomography, magnetic resonance imaging and FDG positron emission tomography in the management of vulvar malignancies. Eur Radiol. 2015;25(5):1267–78.
17. Robertson NL, Hricak H, Sonoda Y, et al. The impact of FDG-PET/CT in the management of patients with vulvar and vaginal cancer. Gynecol Oncol. 2016;140(3):420–4.
18. Pecorelli S. Revised FIGO staging for carcinoma of the vulva, cervix, and endometrium. Int J Gynaecol Obstet. 2009;105(2):103–4.
19. Hacker NF. Revised FIGO staging for carcinoma of the vulva. Int J Gynaecol Obstet. 2009;105(2):105–6.
20. Yoder BJ, Rufforny I, Massoll NA, Wilkinson EJ. Stage IA vulvar squamous cell carcinoma: an analysis of tumor invasive characteristics and risk. Am J Surg Pathol. 2008;32(5):765–72.
21. Disaia PJ, Creasman WT, Rich WM. An alternate approach to early cancer of the vulva. Am J Obstet Gynecol. 1979;133(7):825–32.
22. Hacker NF, Leuchter RS, Berek JS, Castaldo TW, Lagasse LD. Radical vulvectomy and bilateral inguinal lymphadenectomy through separate groin incisions. Obstet Gynecol. 1981;58(5):574–9.
23. NCCN clinical practice guidelines in oncology. 2016.
24. Ansink A. VDVJ: surgical interventions for early squamous cell carcinoma of the vulva. Cochrane Database Syst Rev. 2000;2:CD002036.

25. Desimone CP, Van Ness JS, Cooper AL, et al. The treatment of lateral T1 and T2 squamous cell carcinomas of the vulva confined to the labium majus or minus. Gynecol Oncol. 2007;104(2):390–5.
26. De-Hullu JA, Hollema H, Lolkema S, et al. Vulvar carcinoma. The price of less radical surgery. Cancer. 2002;95(11):2331–8.
27. Heaps JM, Fu YS, Montz FJ, Hacker NF, Berek JS. Surgical-pathologic variables predictive of local recurrence in squamous cell carcinoma of the vulva. Gynecol Oncol. 1990;38(3):309–14.
28. Moore DH, Ali S, Koh WJ, et al. A phase II trial of radiation therapy and weekly cisplatin chemotherapy for the treatment of locally-advanced squamous cell carcinoma of the vulva: a gynecologic oncology group study. Gynecol Oncol. 2012;124(3):529–33.
29. Berek JS, Heaps JM, Fu YS, Juillard GJ, Hacker NF. Concurrent cisplatin and 5-fluorouracil chemotherapy and radiation therapy for advanced-stage squamous carcinoma of the vulva. Gynecol Oncol. 1991;42(3):197–201.
30. Wahlen SA, Slater JD, Wagner RJ, et al. Concurrent radiation therapy and chemotherapy in the treatment of primary squamous cell carcinoma of the vulva. Cancer. 1995;75(9):2289–94.
31. Deppe G, Mert I, Winer IS. Management of squamous cell vulvar cancer: a review. J Obstet Gynaecol Res. 2014;40(5):1217–25.
32. Tans L, Ansink AC, Van Rooij PH, Kleijnen C, Mens JW. The role of chemo-radiotherapy in the management of locally advanced carcinoma of the vulva: single institutional experience and review of literature. Am J Clin Oncol. 2011;34(1):22–6.
33. Wills A, Obermair A. A review of complications associated with the surgical treatment of vulvar cancer. Gynecol Oncol. 2013;131(2):467–79.
34. Burger MP, Hollema H, Emanuels AG, Krans M, Pras E, Bouma J. The importance of the groin node status for the survival of T1 and T2 vulval carcinoma patients. Gynecol Oncol. 1995;57(3):327–34.
35. Landrum LM, Lanneau GS, Skaggs VJ, et al. Gynecologic Oncology Group risk groups for vulvar carcinoma: improvement in survival in the modern era. Gynecol Oncol. 2007;106(3):521–5.
36. Chan JK, Sugiyama V, Pham H, et al. Margin distance and other clinico-pathologic prognostic factors in vulvar carcinoma: a multivariate analysis. Gynecol Oncol. 2007;104(3):636–41.
37. Maggino T, Landoni F, Sartori E, et al. Patterns of recurrence in patients with squamous cell carcinoma of the vulva. A multicenter CTF study. Cancer. 2000;89(1):116–22.
38. Lee LJ, Howitt B, Catalano P, et al. Prognostic importance of human papillomavirus (HPV) and p16 positivity in squamous cell carcinoma of the vulva treated with radiotherapy. Gynecol Oncol. 2016;142(2):293–8.
39. Yap JK, Fox R, Leonard S, et al. Adjacent Lichen sclerosis predicts local recurrence and second field tumour in women with vulvar squamous cell carcinoma. Gynecol Oncol. 2017;142(3):420–6.
40. Beller U, Quinn MA, Benedet JL, et al. Carcinoma of the vulva. FIGO 26th annual report on the results of treatment in gynecological cancer. Int J Gynaecol Obstet. 2006;95(Suppl 1):S7–27.
41. Gonzalez Bosquet J, Magrina JF, Gaffey TA, et al. Long-term survival and disease recurrence in patients with primary squamous cell carcinoma of the vulva. Gynecol Oncol. 2005;97(3):828–33.
42. Salani R, Backes FJ, Fung MF, et al. Posttreatment surveillance and diagnosis of recurrence in women with gynecologic malignancies: Society of Gynecologic Oncologists recommendations. Am J Obstet Gynecol. 2011;204(6):466–78.
43. Stehman FB, Bundy BN, Ball H, Clarke-Pearson DL. Sites of failure and times to failure in carcinoma of the vulva treated conservatively: a Gynecologic Oncology Group study. Am J Obstet Gynecol. 1996;174(4):1128–1132.; discussion 1132–1123.
44. Rouzier R, Haddad B, Plantier F, Dubois P, Pelisse M, Paniel BJ. Local relapse in patients treated for squamous cell vulvar carcinoma: incidence and prognostic value. Obstet Gynecol. 2002;100(6):1159–67.
45. Micheletti L, Preti M. Surgery of the vulva in vulvar cancer. Best Pract Res Clin Obstet Gynaecol. 2014;28(7):1074–87.
46. Sedlis A, Homesley H, Bundy BN, et al. Positive groin lymph nodes in superficial squamous cell vulvar cancer. A Gynecologic Oncology Group study. Am J Obstet Gynecol. 1987;156(5):1159–64.

47. Berman ML, Soper JT, Creasman WT, Olt GT, Disaia PJ. Conservative surgical management of superficially invasive stage I vulvar carcinoma. Gynecol Oncol. 1989;35(3):352–7.
48. Stehman FB, Bundy BN, Dvoretsky PM, Creasman WT. Early stage I carcinoma of the vulva treated with ipsilateral superficial inguinal lymphadenectomy and modified radical hemivulvectomy: a prospective study of the Gynecologic Oncology Group. Obstet Gynecol. 1992;79(4):490–7.
49. Stehman FB, Bundy BN, Thomas G, et al. Groin dissection versus groin radiation in carcinoma of the vulva: a Gynecologic Oncology Group study. Int J Radiat Oncol Biol Phys. 1992;24(2):389–96.
50. Burke TW, Levenback C, Coleman RL, Morris M, Silva EG, Gershenson DM. Surgical therapy of T1 and T2 vulvar carcinoma: further experience with radical wide excision and selective inguinal lymphadenectomy. Gynecol Oncol. 1995;57(2):215–20.
51. Oonk MH, Hollema H, Van Der Zee AG. Sentinel node biopsy in vulvar cancer: Implications for staging. Best Pract Res Clin Obstet Gynaecol. 2015;29(6):812–21.
52. Gonzalez Bosquet J, Magrina JF, Magtibay PM, et al. Patterns of inguinal groin metastases in squamous cell carcinoma of the vulva. Gynecol Oncol. 2007;105(3):742–6.
53. Homesley HD, Bundy BN, Sedlis A, Adcock L. Radiation therapy versus pelvic node resection for carcinoma of the vulva with positive groin nodes. Obstet Gynecol. 1986;68(6):733–40.
54. Cormier JN, Rourke L, Crosby M, Chang D, Armer J. The surgical treatment of lymphedema: a systematic review of the contemporary literature (2004-2010). Ann Surg Oncol. 2012;19(2):642–51.
55. Warren AG, Brorson H, Borud LJ, Slavin SA. Lymphedema: a comprehensive review. Ann Plast Surg. 2007;59(4):464–72.
56. Abu-Rustum NR. Sentinel lymph node mapping for endometrial cancer: a modern approach to surgical staging. J Natl Compr Cancer Netw. 2014;12(2):288–97.
57. Slomovitz BM, Coleman RL, Oonk MH, Van Der Zee A, Levenback C. Update on sentinel lymph node biopsy for early-stage vulvar cancer. Gynecol Oncol. 2015;138(2):472–7.
58. Van Der Zee AG, Oonk MH, De Hullu JA, et al. Sentinel node dissection is safe in the treatment of early-stage vulvar cancer. J Clin Oncol. 2008;26(6):884–9.
59. Levenback CF, Ali S, Coleman RL, et al. Lymphatic mapping and sentinel lymph node biopsy in women with squamous cell carcinoma of the vulva: a gynecologic oncology group study. J Clin Oncol. 2012;30(31):3786–91.
60. Cibula D, Oonk MH, Abu-Rustum NR. Sentinel lymph node biopsy in the management of gynecologic cancer. Curr Opin Obstet Gynecol. 2015;27(1):66–72.
61. Te Grootenhuis NC, Van Der Zee AG, Van Doorn HC, et al. Sentinel nodes in vulvar cancer: long-term follow-up of the GROningen INternational Study on Sentinel nodes in Vulvar cancer (GROINSS-V) I. Gynecol Oncol. 2016;140(1):8–14.
62. Hassanzade M, Attaran M, Treglia G, Yousefi Z, Sadeghi R. Lymphatic mapping and sentinel node biopsy in squamous cell carcinoma of the vulva: systematic review and meta-analysis of the literature. Gynecol Oncol. 2013;130(1):237–45.
63. Meads C, Sutton AJ, Rosenthal AN, et al. Sentinel lymph node biopsy in vulval cancer: systematic review and meta-analysis. Br J Cancer. 2014;110(12):2837–46.
64. Covens A, Vella ET, Kennedy EB, Reade CJ, Jimenez W, Le T. Sentinel lymph node biopsy in vulvar cancer: systematic review, meta-analysis and guideline recommendations. Gynecol Oncol. 2015;137(2):351–61.
65. Van Doorn HC, Van Beekhuizen HJ, Gaarenstroom KN, et al. Repeat sentinel lymph node procedure in patients with recurrent vulvar squamous cell carcinoma is feasible. Gynecol Oncol. 2016;140(3):415–9.
66. Coleman RL, Ali S, Levenback CF, et al. Is bilateral lymphadenectomy for midline squamous carcinoma of the vulva always necessary? An analysis from Gynecologic Oncology Group (GOG) 173. Gynecol Oncol. 2013;128(2):155–9.
67. Rob L, Robova H, Pluta M, et al. Further data on sentinel lymph node mapping in vulvar cancer by blue dye and radiocolloid Tc99. Int J Gynecol Cancer. 2007;17(1):147–53.
68. Schaafsma BE, Verbeek FP, Peters AA, et al. Near-infrared fluorescence sentinel lymph node biopsy in vulvar cancer: a randomised comparison of lymphatic tracers. BJOG. 2013;120(6):758–64.

69. Buda A, Dell'anna T, Vecchione F, Verri D, Di Martino G, Milani R. Near-infrared sentinel lymph node mapping with indocyanine green using the VITOM II ICG exoscope for open surgery for gynecologic malignancies. J Minim Invasive Gynecol. 2016;23(4):628–32.
70. Matheron HM, Van Den Berg NS, Brouwer OR, et al. Multimodal surgical guidance towards the sentinel node in vulvar cancer. Gynecol Oncol. 2013;131(3):720–5.
71. Verbeek FP, Tummers QR, Rietbergen DD, et al. Sentinel lymph node biopsy in vulvar cancer using combined radioactive and fluorescence guidance. Int J Gynecol Cancer. 2015;25(6):1086–93.
72. Oonk MH, Van Hemel BM, Hollema H, et al. Size of sentinel-node metastasis and chances of non-sentinel-node involvement and survival in early stage vulvar cancer: results from GROINSS-V, a multicentre observational study. Lancet Oncol. 2010;11(7):646–52.
73. Woelber L, Eulenburg C, Grimm D, et al. The risk of contralateral non-sentinel metastasis in patients with primary vulvar cancer and unilaterally positive sentinel node. Ann Surg Oncol. 2016;23(8):2508–14.
74. Oonk MH, Van Os MA, De Bock GH, De Hullu JA, Ansink AC, Van Der Zee AG. A comparison of quality of life between vulvar cancer patients after sentinel lymph node procedure only and inguinofemoral lymphadenectomy. Gynecol Oncol. 2009;113(3):301–5.
75. Novackova M, Halaska MJ, Robova H, et al. A prospective study in the evaluation of quality of life after vulvar cancer surgery. Int J Gynecol Cancer. 2015;25(1):166–73.
76. Forner DM, Dakhil R, Lampe B. Quality of life and sexual function after surgery in early stage vulvar cancer. Eur J Surg Oncol. 2015;41(1):40–5.
77. Sutton AJ, Barton P, Sundar S, et al. Cost-effectiveness of sentinel lymph node biopsy vs inguinofemoral lymphadenectomy in women with vulval cancer. Br J Cancer. 2013;109(10):2533–47.
78. Erickson BK, Divine LM, Leath CA 3rd, Straughn JM Jr. Cost-effectiveness analysis of sentinel lymph node biopsy in the treatment of early-stage vulvar cancer. Int J Gynecol Cancer. 2014;24(8):1480–5.
79. Mccann GA, Cohn DE, Jewell EL, Havrilesky LJ. Lymphatic mapping and sentinel lymph node dissection compared to complete lymphadenectomy in the management of early-stage vulvar cancer: a cost-utility analysis. Gynecol Oncol. 2015;136(2):300–4.
80. Radziszewski J, Kowalewska M, Jedrzejczak T, et al. The accuracy of the sentinel lymph node concept in early stage squamous cell vulvar carcinoma. Gynecol Oncol. 2010;116(3):473–7.
81. Hampl M, Hantschmann P, Michels W, Hillemanns P. Validation of the accuracy of the sentinel lymph node procedure in patients with vulvar cancer: results of a multicenter study in Germany. Gynecol Oncol. 2008;111(2):282–8.
82. Sugiyama VE, Chan JK, Shin JY, Berek JS, Osann K, Kapp DS. Vulvar melanoma: a multivariable analysis of 644 patients. Obstet Gynecol. 2007;110(2 Pt 1):296–301.
83. Garbe C, Peris K, Hauschild A, et al. Diagnosis and treatment of melanoma. European consensus-based interdisciplinary guideline—update 2012. Eur J Cancer. 2012;48(15):2375–90.
84. Leitao MM Jr, Cheng X, Hamilton AL, et al. Gynecologic Cancer InterGroup (GCIG) consensus review for vulvovaginal melanomas. Int J Gynecol Cancer. 2014;24(9 Suppl 3):S117–22.
85. Wong SL, Balch CM, Hurley P, et al. Sentinel lymph node biopsy for melanoma: American Society of Clinical Oncology and Society of Surgical Oncology joint clinical practice guideline. J Clin Oncol. 2012;30(23):2912–8.
86. Han SN, Verheecke M, Vandenbroucke T, Gziri MM, Van Calsteren K, Amant F. Management of gynecological cancers during pregnancy. Curr Oncol Rep. 2014;16(12):415.
87. Matsuo K, Whitman SA, Blake EA, et al. Feto-maternal outcome of pregnancy complicated by vulvar cancer: a systematic review of literature. Eur J Obstet Gynecol Reprod Biol. 2014;179:216–23.
88. Amant F, Halaska MJ, Fumagalli M, et al. Gynecologic cancers in pregnancy: guidelines of a second international consensus meeting. Int J Gynecol Cancer. 2014;24(3):394–403.
89. Frey JN, Hampl M, Mueller MD, Gunthert AR. Should groin recurrence still be considered as a palliative situation in vulvar cancer patients? A brief report. Int J Gynecol Cancer. 2016;26(3):575–9.

Chapter 9
Clinical Considerations and Surgical Management of Groin Disease in Penile and Scrotal Cancer

Christine Ibilibor, Pranav Sharma, and Philippe E. Spiess

Groin Disease in Penile Cancer

Introduction

Epidemiology

Penile cancer is rare, with the global incidence reported at 26,000 cases per year [1]. The majority of these cases occur in areas of South America, Africa, and Asia, accounting for up to 10% of all malignancies in the non-Western world, while penile cancer makes up only 0.4–0.6% of all cancers in men in the United States [2, 3]. It has been well reported in the literature that penile cancer typically plagues those men aged 50 years and older, with a reported incidence as high as 39% in patients greater than 66 years of age based on a Brazilian study [2, 4].

Risk Factors

The risk factors for penile carcinoma can largely be divided into behavioral/cultural practices versus infectious causes. Phimosis with its accompanying chronic inflammation and balanitis has been known to have a strong association with penile carcinoma with rates of malignancy reported to be significantly lower in populations and religious groups that practice neonatal circumcision [1, 2]. Poor hygiene and

C. Ibilibor, M.D. • P. Sharma, M.D.
Department of Urology, Texas Tech University Health Sciences Center, Lubbock, TX, USA
e-mail: christine.ibilibor@ttuhsc.edu; pranav.sharma@ttuhsc.edu

P.E. Spiess, M.D., M.S., F.R.C.S., F.A.C.S. (✉)
Department of Genitourinary Oncology, Moffitt Cancer Center,
12902 Magnolia Drive, Tampa, FL 33612, USA
e-mail: philippe.spiess@moffitt.org

K.A. Delman, V.A. Master (eds.), *Malignancies of the Groin*,
https://doi.org/10.1007/978-3-319-60858-7_9

cigarette smoking are other behavioral risk factors for penile cancer with smokers being at a 2.8-fold greater risk than nonsmokers [1, 2]. The human papillomavirus (HPV) is by far the most important and common infectious risk factor for penile carcinoma with its presence reported in 40–50% of cases and HPV-16 and HPV-18 being implicated as the most commonly found subtypes [1, 2, 5].

Nodal Involvement and Survival

While 80% of patients are diagnosed with localized penile cancer with a 5-year cancer-specific survival reported at 95%, those patients who present with a primary penile tumor and nodal involvement have a much more dismal prognosis [2, 6, 7]. It has been well documented, therefore, that the presence and extent of lymph node metastasis is by far the most important prognostic factor in determining long-term oncologic outcomes and patient survival [7, 8]. Graafland et al. detected nodal involvement in 16% of cases of penile squamous cell carcinoma (SCC) with a 38% overall survival at 9 years and 0% survival at 10 years in these patients compared to a 90% 10-year overall survival in patients without nodal metastases [7].

The risk of lymph node metastasis is influenced by both the stage of the primary penile tumor and the histologic subtype with an 89% risk of lymph node metastasis reported in sarcomatoid tumors and a 20–30% risk of lymph node metastasis reported in patients with pathologic T2–T3 disease in the primary lesion [2, 9]. Additionally, the number of inguinal lymph nodes involved, extranodal extension, and the presence of pelvic lymph node involvement further dictate disease-specific survival, making accurate determination of lymph node status of paramount importance [6, 10].

Three positive inguinal lymph nodes have been proposed in the literature as an appropriate cutoff point for risk-stratifying patients based on the extent of lymph node involvement [11]. Pandey et al. reported a 75.6% versus 8.4% overall survival at 5 years in patients with one to three versus four to five positive inguinal lymph nodes, respectively [6]. Similarly, Ravi et al. reported a 5-year overall survival of 75% in patients with one to three positive inguinal lymph nodes [12]. Leijte et al. found no statistical difference in survival between penile cancer patients with one versus two or more positive inguinal nodes, but for patients with between one to three positive inguinal lymph nodes versus four or more, the survival difference was statistically better with less than four positive inguinal nodes [13].

Extranodal extension and pelvic lymph node metastases are also very poor prognostic factors [6, 7, 14]. Multiple positive locoregional lymph nodes have been associated with both extranodal extension and pelvic node involvement [11]. Pandey et al. found 5-year survival rates of 8.9% and 0% in penile SCC patients with extranodal extension and positive pelvic lymph node disease [6].

Finally, inguinal lymph node laterality has been reported to have some prognostic value, with bilateral inguinal lymph node disease portending a worse survival rate than unilateral involvement [6, 11, 12]. A study by Zhu et al. found a 3-year recurrence-free survival rate of 59.2% versus 26.7% in unilateral versus bilateral inguinal lymph node disease [11].

Anatomic Considerations

Lymphatics and Inguinal Lymph Node Anatomy

The lymphatic drainage of the penis originates from a dense and intricate network of lymphocapillaries within the penile skin, mucous membrane and submucosa of the urethra, septum of the glans, tunica albuginea of the corpora cavernosa, and Buck's fascia [15]. Thus, penetration of Buck's fascia or the tunica albuginea by the primary penile tumor allows for dissemination of tumor cells into the lymphatic system [16]. The regional penile lymphatics can be divided into superficial and deep systems, which drain the penile skin, corporal bodies, and glans into the superficial and deep inguinal lymph nodes, respectively [15, 17].

Classically, the superficial inguinal lymph nodes have been divided into five quadrants or zones named the superomedial, superolateral, inferomedial, inferolateral, and central zones [18]. Cadaveric and anatomic studies have shown that there are approximately 4–25 superficial inguinal lymph nodes in these five regions with more recent studies reporting 5–17 lymph nodes in this area [17, 18]. In a study of the lymphangiographic patterns of penile cancer patients, Cabanas et al. showed that injection into the dorsal penile vasculature consistently drained into a node anterior or medial to the superficial epigastric vein or superomedial to the sapheno-epigastric junction with subsequent drainage into the deep inguinal lymph nodes and iliac nodes. Each patient in the study showed metastasis to this sentinel node [19]. A study using single-photon emission computed tomography (SPECT) scanning to assess lymphatic drainage patterns found sentinel lymph nodes to be present in the superior and central zones, which suggested that all modified inguinal lymph node dissections should include these two regions [20].

The deep inguinal lymph nodes are separated from the superficial nodes by the fascia lata, and they have been noted to be fewer in number with one cadaveric study reporting only zero to five lymph nodes in the deep system located medial to the saphenous vein and draining into the pelvic nodes (namely, the external iliac, internal iliac, and obturator nodes) [17, 21]. There have been no anatomic or lymphangiographic studies demonstrating direct lymphatic drainage to the pelvic lymph nodes from the penis, which is evidenced by the lack of metastatic spread to the pelvic lymph nodes from a primary penile tumor in the absence of metastatic spread to the inguinal lymph nodes [15, 20, 21]. It has been well documented, however, that lymphatic drainage of penile SCC can be unilateral or bilateral to the groins with lymphatic spread of penile carcinoma to unilateral or bilateral inguinal lymph nodes seen [20].

Diagnosing Lymph Node Involvement

Palpable inguinal lymph nodes in the setting of penile SCC is considered to be secondary to metastatic spread as a trial of antibiotics is no longer advised since it can delay oncologic diagnosis and treatment [22]. Surgical staging with radical inguinal lymph node dissection is warranted in the setting of fixed inguinal lymph nodes (cN3) on physical examination, often requiring neoadjuvant chemotherapy prior to lymphadenectomy [22]. Palpable inguinal lymph nodes have been noted to harbor metastatic spread in 60–80% of cases, and diagnostic modalities such as fine needle aspiration or 18F-flurodeoxyglucose positron emission tomography/computerized tomography (PET/CT) can be used to confirm the presence of penile carcinoma in suspicious regions, but this does not obviate the need for inguinal lymph node dissection for adequate oncologic diagnosis and staging [22–24].

Fine needle aspiration has been shown to have a sensitivity of 93% in detecting carcinoma in clinically positive lymph nodes [23]. A meta-analysis also reported PET/CT to have a pooled sensitivity of 96.4% for patients with palpable or clinically positive inguinal lymph nodes [25]. Similarly, a study by Schlenker et al. found PET/CT to have a sensitivity and specificity of 88.2% and 98.1%, respectively, in all clinically node-positive penile cancer patients with a false-negative rate of 11.8% [24]. While PET/CT can be used as an adjunct in diagnosing metastatic involvement in clinically positive lymph nodes, it has shown less utility in clinically negative inguinal lymph node cases [26, 27]. A study by Rosevear et al. reported PET/CT to have a sensitivity of 0% and a false-negative rate of 100% in detecting carcinoma patients with clinically negative inguinal lymph nodes [26]. While the study had a small population, it demonstrated that PET/CT had limited utility in the setting of nonpalpable lymph nodes that may harbor micrometastatic disease, thus ought not to be used solely to rule out the presence of inguinal nodal involvement. It must be noted, however, that in this same study, the patients who later developed palpable lymph nodes had positive PET/CT scans, which were confirmed histologically [26]. Thus, it is the management of clinically negative inguinal lymph nodes with the identification of micrometastasis that is the most problematic aspect of penile cancer staging.

Approximately 25% of clinically normal groins have micrometastatic disease, and cross-sectional imaging studies such as CT and magnetic resonance imaging (MRI) are unable to accurately detect these cases and thus are only largely used to assess for the presence of pelvic lymph node involvement [22, 26]. Patients with noninvasive, well-differentiated penile tumors (i.e., low-grade or pathologic stage Ta, Tis, or T1 without LVI (lymphovascular invasion) [T1a]) and clinically negative groins warrant surveillance as this group is at low risk for inguinal nodal spread and are unlikely to benefit from lymphadenectomy [22].

Dynamic sentinel lymph node biopsy (DSNB) is a minimally invasive diagnostic tool which can serve as an intermediary between noninvasive imaging modalities and surgical resection when identifying those patients with clinically negative groins who would benefit from inguinal lymphadenectomy [28]. Dynamic sentinel lymph node biopsy is based on the assumption that penile cancer cells will initially spread unilaterally or bilaterally to a single inguinal lymph node before disseminating to adjoining lymph nodes and that this sentinel lymph node can have a variable position among individuals [29]. DSNB involves injecting technetium-99m-labeled nanocolloids and patent blue dye around the primary penile tumor usually on the same day as lymph node biopsy. Lymphoscintigraphic images are obtained using a gamma-ray detection probe intraoperatively to identify the first lymph node that drains the primary tumor (i.e., the sentinel lymph node), which is subsequently resected [28, 30]. The sensitivity of dynamic sentinel lymph node biopsy has been reported at 93% by Leijte et al. and 88% by Lam et al. with false-negative rates of 7% and 5%, respectively. Additionally, the morbidity of dynamic sentinel lymph node biopsy is significantly less than that of a modified inguinal lymph node dissection or a standard lymphadenectomy, with complication rates ranging between 7.6 and 4.7% [28, 30]. The use of fine needle aspiration with ultrasonographic imaging has been shown to further improve the diagnostic yield of DSNB and decrease the rate of false negatives [29, 30]. Although its utility has been well documented, widespread use of DSNB remains limited and generally restricted to high-volume centers [30].

Stage and grade of the primary penile tumor have been known to dictate the management of penile SCC with nonpalpable inguinal lymph nodes. The 2014 European Association of Urology penile cancer guidelines recommend the surgical staging of clinically negative inguinal lymph nodes in penile cancer patients with high-grade, pT1 with LVI (pT1b), or pT2–T4 tumors using a bilateral modified inguinal lymph node dissection or DSNB since this group is considered to be intermediate to high risk for locoregional metastasis [22]. Graafland et al. reported occult nodal metastases in 23% of patients with clinically negative groins who had undergone complete inguinal lymphadenectomy and who had high-grade disease, pT2–T4 penile carcinoma, or when lymphovascular invasion was noted within the primary tumor [31].

A modified inguinal lymph node dissection has been described as a method to stage clinically negative groins in penile SCC and diagnose nodal involvement with less morbidity compared to standard radical inguinal lymph node dissection, but it is important to note that a modified dissection should be converted to a radical inguinal lymphadenectomy if positive inguinal lymph nodes are present on frozen section [22, 32].

Surgical Considerations

Indications for Lymphadenectomy and Templates

Clinically positive groins (cN1 or cN2) warrant bilateral radical inguinal lymphadenectomy, while fixed or bulky groin disease (i.e., >4 cm inguinal lymph nodes) is managed with neoadjuvant chemotherapy and subsequent radical lymphadenectomy in those patients that respond clinically [21, 22, 33]. The boundaries of a radical inguinal lymph node dissection are the inguinal ligament superiorly, the sartorius muscle laterally, the adductor longus muscle medially, and the junction at which these two muscles cross serving as the inferior boundary [21, 34]. Draseler et al. also described the superior boundary as a 12-cm line parallel to that of the inguinal ligament beginning from the pubic tubercle and extending laterally, the lateral boundary as a 20-cm line beginning at the anterior superior iliac spine extending inferiorly, and the medial border beginning at the pubic tubercle and extending inferiorly 15 cm [18].

A skin incision is made 2–3 cm below the inguinal crease that is parallel to the inguinal ligament, and skin flaps are subsequently raised. The dissection begins with exposing the inferior border of the inguinal ligament, spermatic cord, and external oblique aponeurosis. Lymphatic tissue is then removed above the fascia lata. The dissection is deepened through the fascia lata overlying the sartorius muscle laterally and the adductor longus muscle medially. Additionally, lymphatic tissue is removed around the femoral artery and vein until these vessels are skeletonized. The saphenous vein is ligated at the saphenofemoral junction. The sartorius muscle is then released from its attachment to the anterior superior iliac spine and transposed over the femoral vessels to serve as a myocutaneous flap for coverage. A suction drain is then placed in the wound with reapproximation of the skin and subcutaneous tissues [34, 35].

Catalona et al. first described a modified inguinal lymph node dissection involving resection of lymphatic tissue from the superomedial quadrant of the groin, preservation of the saphenous vein, limiting resection of deep inguinal nodes to those medial to the femoral vein, avoiding dissection lateral to the femoral vein or caudal to the fossa ovalis, and eliminating transposition of the sartorius muscle [32, 36]. This technique can also preserve vasculature to the groin that runs parallel to the inguinal ligament, which can decrease the risk of flap necrosis. Jacobellis et al. described minimizing the dissection beneath the superficial layer of the fascia lata since the superficial branches of the inferior epigastric, external pudendal, and circumflex iliac arteries run in this region and supply the skin of the groin [37]. Avoiding the removal of lymphatic tissue in this area can further preserve arterial blood supply and reduce risk of skin breakdown and wound dehiscence [37].

Surgical Techniques: Minimally Invasive Inguinal Lymph Node Dissection

Traditionally, both standard and modified inguinal lymph dissections were performed via an open approach. Minimally invasive techniques, however, have been described and utilized. Video endoscopic inguinal lymph node dissection (VEIL) involves the placement of trocars at the vertex, medial, and lateral boundaries of the femoral triangle [38]. The boundaries of inguinal lymph node dissection are similar to that of an open inguinal lymphadenectomy with preservation of the saphenous vein and creation of a working space beneath the skin and subcutaneous tissues via insufflation [38]. Due to the decreased morbidity of this minimally invasive technique, it has been generally best suited for patients with clinically negative groins with a high-grade or T1b or greater primary tumor that have an intermediate to high risk for locoregional metastatic spread. These patients undergoing a prophylactic groin dissection with VEIL may be spared the complications associated with the standard open approach [38, 39]. For example, Tobias-Machado et al. reported a significantly lower rate of skin-related complications with VEIL in their prospective study comparing it to traditional open radical groin dissection with 5% of patients in the VEIL group developing skin-related events compared to 50% in the open group [38]. Additionally, the overall complication rate with VEIL was 20% compared to 70% in the open group [38]. Pahwa et al. asserted similar findings with an overall reported morbidity of 20% in their VEIL cohort [39].

Robot-assisted VEIL has also been reported by Josephson et al. to be a viable minimally invasive approach with retrieval of lymph node numbers comparable to that of open surgery with improved visualization, precision, and greater degrees of freedom than that of standard laparoscopic instruments [40]. Large prospective studies with long-term follow-up, however, are necessary to demonstrate oncologic equivalence to standard open procedures.

Complications

Open radical inguinal lymphadenectomy is a fairly morbid procedure with complication rates reported to be as high as 50%. Some of the most commonly encountered complications are skin related including wound infection, skin flap necrosis, wound dehiscence, and seroma formation [38, 41]. One observational study reported that 58% of the patients in their cohort experienced one or more wound-related complications with 43% of patients being treated for wound infections [41].

Lymphatic complications are another common class of complications associated with radical groin dissection [22, 32, 41]. Lymphatic complications such as lymphocele or scrotal/lower extremity lymphedema have been reported in 10–20% of patients [32, 38]. This postoperative morbidity, however, has been reported to be markedly reduced with use of a modified inguinal lymphadenectomy template [32, 37, 38]. Yao et al. retrospectively observed an overall complication rate of 14% with a modified groin dissection compared to 25–50% reported for standard radical inguinal lymphadenectomy [22, 32]. Meticulous usage of clips, instead of electrocautery, to ligate lymphatic channels has been noted to help reduce lymphatic complications, and inguinal pressure dressings, antibiotic regimens, and stockings are additional maneuvers that have been described to reduce postoperative morbidity [22].

Outcomes

Early groin dissection in penile cancer patients with clinically negative inguinal lymph nodes has been associated with a survival benefit compared to delayed lymphadenectomy performed at the time of recurrence during surveillance. In a prospective study by Kroon et al. in a population of patients with pT2–T3 penile carcinoma, those that underwent immediate, early prophylactic groin dissection had a 3-year survival of 84% compared to 35% in those whom inguinal lymphadenectomy was performed for clinically positive groins detected during 6 months of surveillance [9]. Similarly, early groin dissection in patients with clinically negative groins has been reported to have a superior 5-year survival rate compared to patients that undergo immediate radiotherapy or surveillance based on a Kulkarni et al.

study in which the 5-year overall survival rates were 74%, 66%, and 63%, respectively, with lower recurrence rates noted in the early groin dissection cohort [42].

Although less lymphatic tissue is resected in a modified groin dissection, recurrence and false-negative rates have been reported to be low and comparable to that of the traditional radical template [43]. Yao et al. retrospectively observed a 0% inguinal recurrence rate after bilateral modified inguinal lymphadenectomy [32]. Similarly, d'Ancona reported a false-negative rate of only 5.5% in penile cancer patients undergoing modified lymphadenectomy during a 2-year period [43].

Patients with fixed inguinal lymph nodes, on the other hand, have a dismal prognosis as radical groin dissection is generally not curative. Such patients, therefore, are managed with neoadjuvant chemotherapy and postchemotherapy lymphadenectomy in clinical responders [22, 33]. Bermejo et al. reported a 40% 5-year survival rate with a median survival time of 26 months in their retrospective study of such penile SCC patients who were treated with neoadjuvant paclitaxel, ifosfamide, and cisplatin prior to surgery [33].

Groin Disease in Primary Scrotal Cancer

Introduction

Primary scrotal carcinoma is an exceedingly rare entity with an incidence reported at ten cases per ten million people in 2002 using the Surveillance, Epidemiology, End Results (SEER) cancer registry with the majority of cases being reported in patients over the age of 50 [44]. Several histologic subtypes of scrotal carcinoma have been identified. The most common subtype is squamous cell carcinoma, which accounts for 32–35% of all scrotal cancers, followed by extramammary Paget's disease (21%), basal cell carcinoma (18%), sarcoma (18%), and melanoma (8%) [44, 45]. Occupational exposure of the genitals to soot and coal in the eighteenth century was noted to be associated with the development of squamous cell carcinoma of the scrotum, while UV exposure and radiotherapy have been associated with basal cell carcinoma, but little is known regarding the risk factors for the development of extramammary Paget's disease of the scrotum [44, 46, 47].

Anatomic Considerations

Primary Tumor and Diagnosing Inguinal Node Involvement

The scrotum is divided into an anterior and posterior portion supplied by the external pudendal and perineal arteries, respectively, but both regions are drained by the superficial inguinal lymph nodes [45]. At the microscopic level, the scrotal skin is comprised of the epidermis, dermis, subcutaneous tissue, and underlying dartos fascia with the lymphatic capillaries residing within the reticular dermis [48, 49].

Anastomotic connections between the lymphatic networks of the right and left hemiscrotum at the level of the median raphe allow lymphatic drainage into bilateral groins, so inguinal metastatic spread can be unilateral or bilateral [45, 50].

The majority of cases of scrotal carcinoma are confined to the epidermis and are noninvasive [51]. Due to the rarity of this disease, there are limited reports in the literature with small cohort sizes. One clinical series, however, reported inguinal metastatic disease in 16.5% of all cases of extramammary Paget's disease (EMPD) of the scrotum and 35.1% of invasive cases [48]. Additionally, Andrews et al. reported that 36% of patients with squamous cell carcinoma of the scrotum had lymphadenopathy at presentation [51].

Depth of invasion of the primary tumor has been demonstrated to be associated with the risk of lymph node metastasis [48]. Tsutsumida et al. found no lymph node involvement in patients with EMPD confined to the epidermis (i.e., carcinoma in situ) or with microscopic invasion into the papillary dermis, while all EMPD patients with invasion into the subcutaneous tissue or dartos layer had pathologically positive inguinal lymph nodes [49]. Both depth of invasion and presence of inguinal lymph node metastasis have been noted to affect disease-specific survival with invasive primary scrotal tumors and nodal involvement portending a worse prognosis [48, 49, 52]. A retrospective study of patients with EMPD by Ito et al. reported a 5-year survival of 20% in patients with nodal involvement compared to 87% in patients with pathologically negative nodes [52]. Tsutsumida et al. also reported an overall survival of less than 50 months in EMPD patients with nodal involvement [49].

Considering the prognostic importance of inguinal lymph node involvement with regard to patient survival, identifying the presence of locoregional metastasis is of paramount importance. Sentinel lymph node biopsy has been noted to be useful in identifying nodal spread in patients with EMPD. Hatta et al. reported that patients with EMPD in their cohort who were found to have a negative sentinel lymph node biopsy were disease-free during the follow-up period with a false-negative rate of 0% [53]. Fine needle aspiration of clinically positive inguinal lymph nodes has also been described in EMPD and utilized as a method for determining the presence of nodal involvement [48].

Indications for Inguinal Lymph Node Dissection and Templates

Currently there are no formal guidelines for the management of inguinal disease in the setting of primary scrotal carcinoma. Historically, Ray and Whitmore performed and recommended ipsilateral radical inguinal groin dissection in the presence of biopsy-proven nodal involvement and delaying contralateral inguinal lymph node dissection unless clinical evidence of metastatic spread to the contralateral groin developed [54]. A study by Tsutsumida et al., however, suggested the use of inguinal lymph node dissection in patients with EMPD who had extension of carcinoma into the dermis or subcutaneous tissues even in the presence of clinically negative lymph nodes [49]. A study by Zhang et al. recommended groin dissection only in EMPD patients with clinically positive lymph nodes that are confirmed histologically [55]. Similar controversies exist with the management of the groins in the setting of squamous cell carcinoma of the scrotum [50].

The template for a radical groin dissection in patients with primary scrotal carcinoma is similar to that performed in patients with penile carcinoma. A combined inguinal and pelvic dissection, however, has been utilized in patients with EMPD and malignant melanoma in which the external iliac and obturator nodes are dissected along with the lymphatic tissue of the inguinofemoral region and femoral triangle [48, 56].

Surgical Considerations

Complications and Outcomes

Due to the rarity of EMPD of the scrotum, the majority of studies performed to investigate patient-related outcomes and survival after groin dissection are restricted to small case series. Also, there have been no randomized control trials comparing survival in patients with clinically negative nodes managed with groin dissection versus those managed with surveillance. The vast majority of studies, however, report lower rates of disease-specific and overall survival in patients with lymph node disease diagnosed after a prior inguinal lymph node dissection [48, 49, 52]. Hatta et al. described three patients with EMPD found to have inguinal spread after radical groin dissection one of whom passed away with distant metastases at 10 months, while another developed distant recurrence of disease 3 months postoperatively [53]. Similarly, Zhang et al. reported an overall survival of 26 months in an EMPD patient with nodal spread found after inguinal lymphadenectomy who passed away with systemic metastatic disease [55]. Koh et al. noted improved disease-free survival in their cohort of EMPD patients who had undergone a combined groin and pelvic lymph node dissection with an overall disease-free survival rate ranging from 20 to 60 months [48].

The postoperative morbidity associated with groin dissection for scrotal carcinoma is similar to that seen when it is performed for penile carcinoma, including risk of wound infection, skin necrosis, seroma, and chronic lymphedema [48, 53]. Adjuvant and palliative radiation as well as systemic chemotherapy have been utilized in scrotal cancer patients with iliofemoral lymph node involvement; however, their indications are not well defined [48, 52].

Conclusions

Significant advances have been made in our current state of knowledge on the most suitable diagnostic evaluation and management of penile and scrotal neoplasms. Key surgical techniques and approaches that are highlighted in this chapter optimize our anticipated perioperative outcomes, with present and evolving minimally invasive techniques offering great potential to establish a new benchmark for reduced rates of treatment-related morbidity.

References

1. Christodoulidou M, Sahdev V, Houssein S, Muneer A. Epidemiology of penile cancer. Curr Probl Cancer. 2015;39(3):126–36.
2. Pizzocaro G, Algaba F, Horenblas S, et al. EAU penile cancer guidelines 2009. Eur Urol. 2010;57(6):1002–12.
3. Siegel R, Ma J, Zou Z, Jemal A. Cancer statistics, 2014. CA Cancer J Clin. 2014;64(1):9–29.
4. Favorito LA, Nardi AC, Ronalsa M, Zequi SC, Sampaio FJ, Glina S. Epidemiologic study on penile cancer in Brazil. Int Braz J Urol. 2008;34(5):587–91. discussion 583–91
5. Flaherty A, Kim T, Giuliano A, et al. Implications for human papillomavirus in penile cancer. Urol Oncol. 2014;32(1):53 e51–8.
6. Pandey D, Mahajan V, Kannan RR. Prognostic factors in node-positive carcinoma of the penis. J Surg Oncol. 2006;93(2):133–8.
7. Graafland NM, Verhoeven RH, Coebergh JW, Horenblas S. Incidence trends and survival of penile squamous cell carcinoma in the Netherlands. Int J Cancer. 2011;128(2):426–32.
8. Horenblas S, van Tinteren H. Squamous cell carcinoma of the penis. IV. Prognostic factors of survival: analysis of tumor, nodes and metastasis classification system. J Urol. 1994;151(5):1239–43.
9. Kroon BK, Horenblas S, Lont AP, Tanis PJ, Gallee MP, Nieweg OE. Patients with penile carcinoma benefit from immediate resection of clinically occult lymph node metastases. J Urol. 2005;173(3):816–9.
10. Cubilla AL. The role of pathologic prognostic factors in squamous cell carcinoma of the penis. World J Urol. 2009;27(2):169–77.
11. Zhu Y, Ye DW. Lymph node metastases and prognosis in penile cancer. Chin J Cancer Res. 2012;24(2):90–6.
12. Ravi R. Correlation between the extent of nodal involvement and survival following groin dissection for carcinoma of the penis. Br J Urol. 1993;72(5 Pt 2):817–9.
13. Leijte JA, Gallee M, Antonini N, Horenblas S. Evaluation of current TNM classification of penile carcinoma. J Urol. 2008;180(3):933–8. discussion 938
14. Diorio GJ, Leone AR, Spiess PE. Management of penile cancer. Urology. 2016;96:15–21.
15. Wood HM, Angermeier KW. Anatomic considerations of the penis, lymphatic drainage, and biopsy of the sentinel node. Urol Clin North Am. 2010;37(3):327–34.
16. Pow-Sang MR, Ferreira U, Pow-Sang JM, Nardi AC, Destefano V. Epidemiology and natural history of penile cancer. Urology. 2010;76(2 Suppl 1):S2–6.
17. de Carvalho JP, Patricio BF, Medeiros J, Sampaio FJ, Favorito LA. Anatomic aspects of inguinal lymph nodes applied to lymphadenectomy in penile cancer. Adv Urol. 2011;2011:952532.
18. Daseler EH, Anson BJ, Reimann AF. Radical excision of the inguinal and iliac lymph glands; a study based upon 450 anatomical dissections and upon supportive clinical observations. Surg Gynecol Obstet. 1948;87(6):679–94.
19. Cabanas RM. An approach for the treatment of penile carcinoma. Cancer. 1977;39(2):456–66.
20. Leijte JA, Valdes Olmos RA, Nieweg OE, Horenblas S. Anatomical mapping of lymphatic drainage in penile carcinoma with SPECT-CT: implications for the extent of inguinal lymph node dissection. Eur Urol. 2008;54(4):885–90.
21. Sharma P, Zargar-Shoshtari K, Spiess PE. Current surgical management of penile cancer. Curr Probl Cancer. 2015;39(3):147–57.
22. Hakenberg OW, Comperat EM, Minhas S, et al. EAU guidelines on penile cancer: 2014 update. Eur Urol. 2015;67(1):142–50.
23. Saisorn I, Lawrentschuk N, Leewansangtong S, Bolton DM. Fine-needle aspiration cytology predicts inguinal lymph node metastasis without antibiotic pretreatment in penile carcinoma. BJU Int. 2006;97(6):1225–8.
24. Schlenker B, Scher B, Tiling R, et al. Detection of inguinal lymph node involvement in penile squamous cell carcinoma by 18F-fluorodeoxyglucose PET/CT: a prospective single-center study. Urol Oncol. 2012;30(1):55–9.

25. Sadeghi R, Gholami H, Zakavi SR, Kakhki VR, Horenblas S. Accuracy of 18F-FDG PET/CT for diagnosing inguinal lymph node involvement in penile squamous cell carcinoma: systematic review and meta-analysis of the literature. Clin Nucl Med. 2012;37(5):436–41.
26. Rosevear HM, Williams H, Collins M, Lightfoot AJ, Coleman T, Brown JA. Utility of (1)(8) F-FDG PET/CT in identifying penile squamous cell carcinoma metastatic lymph nodes. Urol Oncol. 2012;30(5):723–6.
27. Sharma P, Zargar H, Spiess PE. Surgical advances in inguinal lymph node dissection: optimizing treatment outcomes. Urol Clin North Am. 2016;43(4):457–68.
28. Leijte JA, Hughes B, Graafland NM, et al. Two-center evaluation of dynamic sentinel node biopsy for squamous cell carcinoma of the penis. J Clin Oncol. 2009;27(20):3325–9.
29. Neto AS, Tobias-Machado M, Ficarra V, et al. Dynamic sentinel node biopsy for inguinal lymph node staging in patients with penile cancer: a systematic review and cumulative analysis of the literature. Ann Surg Oncol. 2011;18(7):2026–34.
30. Lam W, Alnajjar HM, La-Touche S, et al. Dynamic sentinel lymph node biopsy in patients with invasive squamous cell carcinoma of the penis: a prospective study of the long-term outcome of 500 inguinal basins assessed at a single institution. Eur Urol. 2013;63(4):657–63.
31. Graafland NM, Lam W, Leijte JA, et al. Prognostic factors for occult inguinal lymph node involvement in penile carcinoma and assessment of the high-risk EAU subgroup: a two-institution analysis of 342 clinically node-negative patients. Eur Urol. 2010;58(5):742–7.
32. Yao K, Tu H, Li YH, et al. Modified technique of radical inguinal lymphadenectomy for penile carcinoma: morbidity and outcome. J Urol. 2010;184(2):546–52.
33. Bermejo C, Busby JE, Spiess PE, Heller L, Pagliaro LC, Pettaway CA. Neoadjuvant chemotherapy followed by aggressive surgical consolidation for metastatic penile squamous cell carcinoma. J Urol. 2007;177(4):1335–8.
34. Heyns CF, Fleshner N, Sangar V, Schlenker B, Yuvaraja TB, van Poppel H. Management of the lymph nodes in penile cancer. Urology. 2010;76(2 Suppl 1):S43–57.
35. Kean J, Hough M, Stevenson JH. Skin excision and groin lymphadenectomy: techniques and outcomes. Lymphology. 2006;39(3):141–6.
36. Catalona WJ. Modified inguinal lymphadenectomy for carcinoma of the penis with preservation of saphenous veins: technique and preliminary results. J Urol. 1988;140(2):306–10.
37. Jacobellis U. Modified radical inguinal lymphadenectomy for carcinoma of the penis: technique and results. J Urol. 2003;169(4):1349–52.
38. Tobias-Machado M, Tavares A, Silva MN, et al. Can video endoscopic inguinal lymphadenectomy achieve a lower morbidity than open lymph node dissection in penile cancer patients? J Endourol. 2008;22(8):1687–91.
39. Pahwa HS, Misra S, Kumar A, Kumar V, Agarwal A, Srivastava R. Video endoscopic inguinal lymphadenectomy (VEIL)—a prospective critical perioperative assessment of feasibility and morbidity with points of technique in penile carcinoma. World J Surg Oncol. 2013;11:42.
40. Josephson DY, Jacobsohn KM, Link BA, Wilson TG. Robotic-assisted endoscopic inguinal lymphadenectomy. Urology. 2009;73(1):167–70. discussion 170–61
41. Stuiver MM, Djajadiningrat RS, Graafland NM, Vincent AD, Lucas C, Horenblas S. Early wound complications after inguinal lymphadenectomy in penile cancer: a historical cohort study and risk-factor analysis. Eur Urol. 2013;64(3):486–92.
42. Kulkarni JN, Kamat MR. Prophylactic bilateral groin node dissection versus prophylactic radiotherapy and surveillance in patients with N0 and N1-2A carcinoma of the penis. Eur Urol. 1994;26(2):123–8.
43. d'Ancona CA, de Lucena RG, Querne FA, Martins MH, Denardi F, Netto NR Jr. Long-term followup of penile carcinoma treated with penectomy and bilateral modified inguinal lymphadenectomy. J Urol. 2004;172(2):498–501. discussion 501
44. Wright JL, Morgan TM, Lin DW. Primary scrotal cancer: disease characteristics and increasing incidence. Urology. 2008;72(5):1139–43.
45. Huang JH, Broggi M, Osunkoya AO, Master VA. Surgical management of primary scrotal cancer. Urol Clin North Am. 2016;43(4):531–44.

46. Melicow MM. Percivall Pott (1713–1788): 200th anniversary of first report of occupation-induced cancer scrotum in chimmey sweepers (1775). Urology. 1975;6(6):745–9.
47. Verhoeven RH, Louwman WJ, Koldewijn EL, Demeyere TB, Coebergh JW. Scrotal cancer: incidence, survival and second primary tumours in the Netherlands since 1989. Br J Cancer. 2010;103(9):1462–6.
48. Koh YX, Tay TK, Xu S, Lee CM, Teo MC. A clinical series and literature review of the management of inguinal nodal metastases in patients with primary extramammary Paget disease of the scrotum. Asian J Surg. 2015;38(1):40–6.
49. Tsutsumida A, Yamamoto Y, Minakawa H, Yoshida T, Kokubu I, Sugihara T. Indications for lymph node dissection in the treatment of extramammary Paget's disease. Dermatol Surg. 2003;29(1):21–4.
50. Azike JE. A review of the history, epidemiology and treatment of squamous cell carcinoma of the scrotum. Rare Tumors. 2009;1(1):e17.
51. Andrews PE, Farrow GM, Oesterling JE. Squamous cell carcinoma of the scrotum: long term followup of 14 patients. J Urol. 1991;146(5):1299–304.
52. Ito Y, Igawa S, Ohishi Y, Uehara J, Yamamoto AI, Iizuka H. Prognostic indicators in 35 patients with extramammary Paget's disease. Dermatol Surg. 2012;38(12):1938–44.
53. Hatta N, Morita R, Yamada M, et al. Sentinel lymph node biopsy in patients with extramammary Paget's disease. Dermatol Surg. 2004;30(10):1329–34.
54. Ray B, Whitmore WF Jr. Experience with carcinoma of the scrotum. J Urol. 1977;117(6):741–5.
55. Zhang ZM. Extramammary Paget's disease of the scrotum: an analysis of 26 cases. Zhonghua Yi Xue Za Zhi. 2011;91(28):1970–2.
56. Karakousis CP, Driscoll DL. Positive deep nodes in the groin and survival in malignant melanoma. Am J Surg. 1996;171(4):421–2.

Chapter 10
Soft Tissue Tumors of the Groin and Inguinal Region

Stefano Radaelli, Chiara Colombo, Marco Fiore, and Alessandro Gronchi

Introduction

Soft tissue sarcomas (STSs) are rare tumors accounting for 1% of all adult malignancies, with a global incidence of four/five new cases per 100,000 population per year [1–3]. The typical clinical presentation consists of a solid mass arising from the mesenchymal tissue of the limbs and trunk (75%), retroperitoneum (15%), and head and neck (10%) [4–6].

The whole family of STSs includes several histologies from low-grade to highly aggressive tumors with very poor prognosis. Histological distribution does correlate with the sites of origin and patient characteristics [7, 8]. Well-differentiated/dedifferentiated liposarcoma, leiomyosarcoma, undifferentiated pleomorphic sarcoma, and solitary fibrous tumor are the most frequent abdominal and retroperitoneal histologies [9–11]. Myxofibrosarcoma is the most common histotype affecting the extremities in elderly patients [12, 13], while myxoid liposarcoma and especially synovial sarcoma affect the limbs usually in younger patients [14–17]. Malignant peripheral nerve sheath tumors are ubiquitarian, usually deeply located and often related to von Recklinghausen's disease. They may arise from the principal motor nerves such as the femoral, ischiatic nerve in the lower limb or the ulnar, median, radial nerve in the upper limb [18, 19].

The perineum and ischiorectal fossa are usually the most typical sites of presentation of proximal epithelioid sarcoma, a tumor with a tendency for lymph node spread and early-stage metastatization [20, 21].

Mesenchymal lesions arising in the distal extremities are often found to be classic epithelioid sarcoma and acral myxoinflammatory fibroblastic sarcoma. These

S. Radaelli, M.D. • C. Colombo, M.D. • M. Fiore, M.D. • A. Gronchi, M.D. (✉)
Sarcoma Service, Department of Surgery, Fondazione IRCCS Istituto Nazionale dei Tumori, Via Venezian 1, 20133 Milan, Italy
e-mail: stefano.radaelli@istitutotumori.mi.it; chiara.colombo@istitutotumori.mi.it; marco.fiore@istitutotumoti.mi.it; Alessandro.Gronchi@istitutotumori.mi.it

K.A. Delman, V.A. Master (eds.), *Malignancies of the Groin*,
https://doi.org/10.1007/978-3-319-60858-7_10

are usually indolent tumors which could require a complex surgical approach including functional reconstructions for their peculiar site of presentation [21–23].

Low-grade fibromyxoid sarcoma may be ubiquitarious with a predilection for young adults' lower extremities and trunk wall. Multifocal presentation is not uncommon. Despite the bland histopathological features and the slow rate of growth, it may show a propensity for local recurrence and distant metastasis [24, 25].

Clear cell sarcoma and alveolar soft part sarcoma are generally located in the distal limbs, the former typically arising from tendons or aponeurosis and presenting as a superficial and tiny lesion predominantly in the third and fourth decades and behaving aggressively with a pattern of melanoma-like spread and the latter commonly presenting as a subfascial larger lesion, often being metastatic at its presentation but with a more indolent course and better medium-term prognosis [26–30].

The scalp is typically affected by superficial STSs; the majority of them are angiosarcoma with a peak incidence during the seventh or eighth decade of life and dermatofibrosarcoma protuberans especially in young adults [31–34].

A peculiar histology is represented by Ewing sarcoma. It mainly affects younger patients and may potentially originate from any anatomical site. Retroperitoneal and subcutaneous limb/trunk wall locations are common primary locations. Given its sensitivity to medical treatment, the conventional approach generally consists of numerous cycles of specific chemotherapy which differs from the first-line regimen used as standard of care in STSs [35, 36].

Although local control always represents the most relevant and challenging aspect of its treatment, STS may also meta stasize, predominantly to the lung and less frequently to liver, bone, and soft tissues [37–39].

Lymphatic spread is uncommon (<5% of cases) and may occur in certain histotypes which usually present epithelial aspects in their pattern of growth (clear cell sarcoma, epithelioid sarcoma) or particular biological aggressiveness (rhabdomyosarcoma, angiosarcoma) [40, 41].

Surgery is the mainstay of treatment of localized disease. External beam radiation therapy can be delivered in a neoadjuvant or adjuvant setting for most high-grade, larger, and deeply located STSs as it has been proved to be able to increase the chance of local control in this setting of disease with a comparable local control of an amputation. External beam radiation is the most frequently applied modality of radiotherapy (RT) in STS treatment [42–46].

The value of systemic chemotherapy in STS treatment has been largely debated over recent years given the contradictory outcomes of several consistent published studies [47, 48]. Indeed doxorubicin and ifosfamide have been shown to be the most active single agents in extremity and high-grade STS treatment, achieving response rates up to 30% in patients presenting with advanced setting of disease [49, 50]. Furthermore, a recent randomized multicentric study, recruiting patients with primary and localized high-grade extremity STSs, compared three preoperative cycles of epirubicin (120 mg/m^2) plus ifosfamide (9 g/m^2) versus three preoperative cycles of one of five histologically tailored regimens (gemcitabine + docetaxel in undifferentiated pleomorphic sarcoma, trabectedin in high-grade myxoid liposarcoma, high-dose prolonged-infusion ifosfamide in synovial sarcoma, etoposide + ifosfamide in malignant peripheral nerve sheath tumors, gemcitabine + dacarbazine leiomyosarcoma). The results of the study showed a significant advantage in terms of both disease-free and overall survival in the epirubicin plus ifosfamide group [52].

Sarcomas of the groin and inguinal region include different histotypes arising from the soft tissues of the inguinal canal, from the spermatic cord, and from the anatomical structures within the femoral triangle. Exceptionally giant retroperitoneal sarcoma may herniate through the abdo-inguinal ring mimicking an indirect inguino-scrotal hernia. All these locations account for between 5 and 10% of all soft tissue malignancies and are frequently misdiagnosed.

Groin STSs typically present as solid, painless, and irreducible groin or scrotal masses. To the general surgeon or urologist, the most likely differential diagnosis is therefore an incarcerated inguinal or crural hernia, locoregional lymphadenopathies, or testicular malignancy. Misdiagnosis frequently leads to inadequate surgery, where tumor excision can be incomplete with a subsequent risk of tumor recurrence. Surgery remains the mainstay of treatment for the largest part of STSs arising in the inguinal region, while chemotherapy and RT are usually proposed on an individualized basis. Because of the small number of case series and literature reports regarding this setting of disease, proper surgical expertise may be difficult to achieve, while prognostic factors and disease outcomes are not completely established to date [53–59].

Anatomical Aspects

In human anatomy, the inguinal region and the groin represent the junctional area between the abdomen and the thigh on both sides of the pubic bone [60–63]. The anterior abdominal wall superiorly, the inguinal canal anteriorly, the femoral triangle inferiorly, the hip bone posteriorly, and the adductor muscles originating from the pubic bones on the medial side represent its anatomical borders.

Due to the presence of embryologically different anatomical structures included within the groin area, several types of soft tissue neoplasms may arise from it.

These anatomical structures are the iliac-femoral artery and vein; the femoral and obturator nerves; the spermatic cord/round ligament in males and females, respectively; the locoregional lymph nodes; the muscle; and the subcutaneous fat of the anterior abdominal wall and of the upper third of the anterior aspect of the thigh. STSs may arise from all of these elements, and their histological and clinical features, diagnostic process, type of treatment, and prognosis will be described in the following paragraphs according to their site of origin or involvement.

Anterior Abdominal Wall

The contents of the abdomen are protected by a band of muscles with a little skeletal support. There are three layers of abdominal muscles that include five muscles: obliquus externus, transversus, obliquus internus, rectus, and pyramidalis.

From outside to inside the anterior abdominal wall is composed of:

The superficial fascia. The superficial fascia of the abdomen is the most external structure covering the greater part of the abdominal wall. It is composed of a single layer containing a variable quantity of subcutaneous fat, but near the groin, two layers are easily recognizable. Between these two layers, the superficial vessels, nerves, and inguinal lymph nodes are located.

The *superficial layer* (fascia of Camper) is thicker, passing over the inguinal ligament and continuing with the superficial fascia of the thigh. In the male, Camper's fascia is extended over the penis and helps to form the dartos. In the female, Camper's fascia is continued from the abdomen into the labia majora.

The *deep layer* (fascia of Scarpa) is thinner, adherent to the aponeurosis of the obliquus externus abdominis, and medially attached to the linea alba and to the pubic symphysis. It is prolonged downward where it helps to form the dartos and the labia majora in males and females, respectively.

The *aponeurosis of the obliquus externus abdominis* is a strong fibrous structure, the fibers of which are directed downward and medialward from the anterosuperior iliac spine to the pubic tubercle and joining the aponeurosis of the contralateral obliquus externus abdominis muscle along the midline, originating from the linea alba which extends from the xiphoid process to the pubic symphysis.

The lowest portion of the aponeurosis included between the anterosuperior iliac spine and the pubic tubercle is the *inguinal ligament*: a thick band reflected inward and prolonged below with the fascia lata.

Within the aponeurosis of the obliquus externus, immediately above the crest of the pubis, there is a triangular opening, the *subcutaneous inguinal ring*, formed by a separation of the fibers of the aponeurosis of the obliquus externus.

The *obliquus externus* is located on the lateral and anterior aspect of the abdomen and is the widest and the most superficial muscle in this region.

It arises, by eight digitations, from the inferior borders of the lower eight ribs, interlacing with the ones from the latissimus dorsi and serratus anterior to find their insertion on the iliac crest and on the anterosuperior iliac spine.

The *obliquus internus abdominis* lies beneath the obliquus externus and is thinner and smaller than the latter. It arises, by fleshy fibers, from the thoracolumbar fascia, from the lateral half surface of the inguinal ligament, and from the anterior aspect of the iliac crest. From this origin, the fibers partially diverge downward and medialward forming the conjoint tendon together with those of the transversus and partially passing in front of the linea semilunaris to be inserted into the linea alba.

The *transversus abdominis* is the most internal of the three flat muscles of the abdominal wall, placed immediately underneath the obliquus internus and separated from the peritoneum by the transversalis fascia. It arises, by fleshy digitations, from the lower six ribs, from the lateral aspect of the inguinal ligament, and from the anterior component of the iliac crest, interlacing with the diaphragm on its upper part. The lower fibers are directed downward and medialward interdigitating with those of the obliquus internus to be inserted into the inguinal ligament forming the conjoint tendon.

The transversalis fascia continues medially and inferiorly passing horizontally the midline and finding its insertion into the linea alba together with the fibers of the obliquus internus and the rectus. In the transversalis fascia, 1 cm above the inguinal ligament, midway between the anterosuperior iliac spine and the pubic symphysis, there is the *abdominal inguinal ring*, which lets the gonadal vessels together with the spermatic cord in males and the round ligament of the uterus in females pass through the abdominal wall.

The *rectus abdominis* is a long, quite thick muscle vertically extended along the entire length of the anterior abdominal wall, being detached from the contralateral one by the linea alba.

It originates by two tendons: the lateral one from the crest of the pubis and the medial one embracing the contralateral tendon on its opposite side to be connected into the pubic symphysis.

The insertion of the muscle is across the cartilages of the fifth, sixth, and seventh ribs.

The rectus is crossed by three fibrous intersections, called the tendinous inscriptions, and it is enveloped in a sheath by the aponeuroses of the oblique and transversus which enclose the muscle anteriorly and posteriorly and end to be inserted into the linea alba. From the umbilicus to the pubic symphysis, the posterior rectal sheath ends into a convex line, *the linea semilunaris*, which is separated from the peritoneum by the *transversalis fascia*.

The *pyramidalis* is a small triangular muscle originating from the pubic crest into the linea alba. It is placed in front of the rectum and enclosed by its sheath at the lower part of the abdomen.

Inguinal Canal

This is an oblique canal extending downward and medialward from the external to the internal inguinal ring. Its length is approximately 4 cm and it contains the gonadic vessels, the ilioinguinal nerve, and the spermatic cord or the uterine round ligament in males and females, respectively.

It runs parallel and above the inguinal ligament for a length of approximately 4 cm.

It is bounded, *in front*, by the aponeurosis of the obliquus externus throughout its whole length; *behind*, by the reflected inguinal ligament, the inguinal aponeurotic falx, the transversalis fascia, the extraperitoneal fat, and the peritoneum; *above*, by the arched fibers of obliquus internus and transversus abdominis; *below*, by the union of the transversalis fascia with the inguinal ligament; and at its medial end, by the lacunar ligament.

Femoral Triangle

The inguinal area belongs to the anterolateral abdominal wall and can be divided into two distinct regions: the inguino-abdominal region above the inguinal ligament and the inguinocrural one just below it. Beneath the inguinal ligament, the vascular and the muscular lacunae transmit muscles, vessels, and nerves from the retroperitoneum to the thigh. The *vascular lacuna* is the medial compartment and contains the common femoral artery and vein, the Cloquet or Rosenmuller lymph nodes, and the femoral branch of the genitofemoral nerve. It is separated by the iliopectineal arch from the *muscular lacuna*, the lateral compartment, for the passage of the iliopsoas muscle, the femoral nerve, and the lateral cutaneous nerve of the thigh.

On the proximal and anterior aspect of the thigh, the femoral nerve and vessels are contained within the *femoral triangle* (Scarpa's triangle). It is delimited superiorly by the inguinal ligament, medially by the medial border of the adductor longus muscle, and laterally by the medial border of the sartorius muscle. Its floor is formed by the pectineus and adductor longus muscles medially and iliopsoas muscle laterally. Its roof is formed by the fascia lata, except at the saphenous hiatus where it is formed only by the fascia cribrosa.

Muscles of the Proximal Medial Aspect of the Thigh

The medial border of the groin is delimited by the adductor muscles originating from the pubic bones.

The *gracilis* is the most superficial muscle on the medial aspect of the thigh.

It arises by a thin aponeurosis from the anterior margins of the lower half of the pubic symphysis and the upper half of the pubic arch. The muscle's fibers run longitudinally downward, ending in a thick tendon inserted into the upper part of the medial surface of the tibia immediately above that of the semitendinosus muscle, and its upper edge is overlapped by the tendon of the sartorius muscle. For this reason, the muscle is a lower limb adductor.

The *pectineus* is situated at the anterior part of the upper and medial aspect of the thigh. It arises from the pectineal line of the superior pubic ramus, and the fibers are directed downward, backward, and lateralward, to be inserted into the femur below the lesser trochanter.

The *adductor longus* is the most superficial of the three adductors, arising from the superior ramus of the pubis. It passes downward to be inserted by a common aponeurosis with the adductor magnus and vastus medialis into the linea aspera of the femur.

The *adductor brevis* lies behind the pectineus and the adductor longus arising by a narrow origin from the superior and inferior rami of the pubis. Its fibers pass backward and lateralward to find their insertion into the linea aspera of the femur.

The *adductor magnus* is the largest muscle of the medial aspect of the thigh. It originates partially from the inferior ramus of the pubis and partially from the ischium. Its fibers are inserted into the linea aspera; at their insertion, they form a series of tendinous arches attached to the bone. The upper four openings are small and give passage to the perforating branches of the profunda femoris artery and vein. The lowest is large and transmits the superficial femoral vessels to the popliteal fossa.

The *innervation* of the three adductors and the gracilis is provided by the third and fourth lumbar nerve roots through the obturator nerve; the adductor magnus receives an additional branch from the sacral plexus through the sciatic nerve. The pectineus is supplied by the second, third, and fourth lumbar nerves through the femoral nerve and by the third lumbar root through the accessory obturator when this latter exists.

The pectineus and three adductors adduct the thigh. The pectineus and adductors brevis and longus assist the psoas major and iliacus in flexing the thigh upon the pelvis.

The gracilis assists the sartorius in flexing the leg and rotating it inward; it is also an adductor of the thigh.

The Diagnostic Process

As largely documented, sarcomas of the soft tissue represent a very large family of neoplasms able to affect virtually any part of the human body. Involvement of the groin and the inguinal region has been reported in up to 10% of patients among the largest series ever published on this topic.

Given the heterogeneity of the anatomical structures present in this specific context, we may recognize several different histotypes of primary and metastatic STS.

Independently from its possible nature and site of origin, the diagnostic flowchart in case of a suspicious malignant lesion is always composed of clinical examination, radiological assessment, and pretreatment biopsy.

Physical Examination

The clinical presentation of patients with limb STSs is highly variable. STSs may not cause any signs or symptoms in their early stages. As the tumor grows, its appearance is of a palpable nodule, usually covered by healthy tissue [64–66].

In a minor population of patients with locally advanced disease, a direct infiltration of the skin might be present, sometimes resulting in ulcerated and bleeding lesions [67, 68].

STSs can initially be misdiagnosed as deep venous thrombosis or spontaneous intramuscular hematomas, leading to a late diagnosis. Spontaneous intramuscular hematomas in extremities are very uncommon and should be approached with a high degree of clinical suspicion [69, 70].

Unilateral, painless, and slow-growing inguinal masses can often be interpreted as inguinal hernia, while intrascrotal lumps with dubious characteristics are often diagnosed as testicular tumors or hydrocele/hematocele.

Due to the numerous lymphatic chains within the inguinocrural area, firm and solid nodules may also be confused for inflammatory or neoplastic lymphadenopathies [55, 56, 71, 72]. Although uncommon, certain STS histotypes with epithelial pathological features or very aggressive biology present a consistent risk (up to 5%) of lymphatic spread [40, 41].

Systemic symptoms (such as fever, weight loss, or malaise) and paraneoplastic syndromes rarely occur in STSs [73–75].

Imaging

Diagnostic imaging plays a significant role in detection and treatment planning in patients with musculoskeletal tumors. A multidisciplinary approach that includes active participation by expert radiologists is beneficial for optimal diagnostic evaluation and treatment of STSs [76]. Recent advances in diagnostic imaging include a better understanding of the practical roles of plain radiography, ultrasonography, contrast-enhanced computed tomography, and magnetic resonance imaging in the evaluation of primary and recurrent STSs. The use of positron emission tomography (PET) scanning is limited to specific clinical situations because of the consistent related risk of false negatives.

Follow-up imaging is scheduled differently according to tumor and patient's characteristics [77, 78].

Ultrasonography (US)

The initial evaluation should begin with an ultrasound scan of the mass or the region affected. Ultrasonography (US), in fact, is an easily available, noninvasive, and quite inexpensive procedure, and it is widely used in the early local staging of a soft tissue mass. Notable information such as tumor size, location, and consistency (cystic or solid lesion) can all be provided by US, particularly when STSs are located in extremities more than deeper in the abdomen or in the chest.

Ultrasound criteria considered highly suspicious of malignancy include larger size, irregular margins, tissue heterogeneity and architectural distortion, and deeper location. Either benign or malignant soft tissue masses may displace rather than invade noble anatomical structures. Direct infiltration or encasement of nerves, blood vessels, or bones is, in fact, quite unlikely, and happens more frequently in case of tumors directly arising from those anatomical structures; in the groin region, it could be the case of vascular leiomyosarcomas originating from the femoral or greater saphenous veins or malignant peripheral sheath tumors (MPNST) arising from the femoral and obturator nerves. A further element able to determine a consistent suspicion of malignancy is the vascular pattern of a soft tissue mass. Predominant peripheral blood flow with a necrotic central area is more typical of malignant tumors, while avascular lesions with weak power and color Doppler signal are more probably benign. Color Doppler US may also be a valid aid in the early postoperative period when distinguishing residual or recurrent tumor from seroma, hematoma, abscess, and granulation tissue may not be simple. Ultrasonography can also be used to guide percutaneous Tru-cut needle biopsy is useful especially when a suspected malignant lesion is not palpable because it is deeply located or it is very proximal to critical structures (major blood vessels, nerves, organs). Additionally, US-guided biopsy may identify the most viable part of a tumor, avoiding necrotic or hemorrhagic areas and providing representative tissue for an adequate histopathological examination [79–81].

Plain Radiography

Plain X-ray has a limited role in the diagnosis and staging of STSs, but it is quick to perform and inexpensive and allows evaluation of calcification within lesions and bone involvement.

In case of large and prominent masses, radiography may show the distortion of the superficial tissue planes. Some readily evident features on plain radiography can help in the diagnostic process, e.g., calcified phleboliths (hemangiomas), a cumulus cloud-like appearance (extraskeletal chondro-/osteosarcoma), and mature peripheral trabecular bone (myositis ossificans) [81, 82].

Magnetic Resonance Imaging (MRI)

MRI is considered the preferred imaging modality for diagnosis and staging of soft tissue tumors. It is able to give remarkable information regarding the precise anatomic location of the mass and its relationship to the adjacent neurovascular and skeletal structures. MRI provides excellent contrast detail of various soft tissue elements (fluids, fat, muscle, bone) anticipating information about the pathological nature of soft tissue masses; however, it has limited ability in defining patterns of soft tissue calcification and potential neoplastic bone involvement.

Lesions are generally considered benign when they are small in size with regular margins and homogeneous consistency. Larger and heterogeneous masses should be considered at least malignant until proved otherwise as only 5% of benign soft tissue tumors measure more than 5 cm.

Soft tissue masses, albeit malignant, generally appear encapsulated at MRI and tend to grow without invading the anatomic planes. Therefore, the presence of a peripheral rim which delimits the tumor is not distinctive only of benign lesions.

The standard MRI sequences used to evaluate a soft tissue tumor include axial and coronal T1- and T2-weighted sequences, fat saturation, and T1-weighted sequences with contrast.

These sequences provide the best means of determining the anatomic orientation of the mass. T1-weighted MRI allows excellent anatomic visualization, given its high spatial resolution. T2-weighted and fat suppression techniques show abnormal changes and highlight tissue edema.

MR signal intensity characteristics of the majority of the STSs are generic and have heterogeneous appearance on T1- and T2-weighted sequences. However, most soft tissue tumors are hypointense on T1-weighted images and hyperintense on T2-weighted images and show contrast enhancement on post-contrast magnetic resonance images. In selected cases, MRI may help the clinician to generate a differential diagnosis and an appropriate treatment plan for the patient with a soft tissue mass. However, it has a limited ability to provide specific tissue diagnosis except in the case of lipomatous tumors, peripheral nerve sheath tumors, and hemangiomas.

Administration of intravenous contrast agent (Gd-DTPA) has increased the potential of MRI. Though not specific, high-grade STSs frequently tend to present a peripheral enhancing zone with a non-enhancing necrotic center. In addition, gadolinium enhancement is helpful for assessing treatment response or can also help the physician in distinguishing a recurrent tumor or viable residual tumor tissue from granulation tissue in the surgical field. In contrast to the delayed enhancement seen with granulation tissue, viable residual tumor tissue displays vivid and early contrast enhancement [82–85].

Computed Tomography (CT)

CT scan plays different roles in the diagnostic process of STSs, although its use has decreased as the role of MRI has evolved. It is typically the imaging technique of choice for patients who cannot undergo MRI, including those with pacemakers or metal implants and those who are claustrophobic or too large to fit in the MRI scanner.

CT is the modality of choice for the local evaluation of retroperitoneal sarcomas as the bowel movements may interfere with the sharpness of abdomen MRI scan while they do not affect the quality of CT images. CT scan is also preferred in bone and vascular sarcoma assessment given its major sensitivity in detecting infiltration of the cortical layer and the vascular wall, respectively. It is also the favorite imaging technique to assess locoregional/lymph node spread or to identify distant (pulmonary) metastases both at the initial staging and during the follow-up scans.

CT is also important in aiding the radiologist in various intervention techniques such as aspiration and bioptic procedures [82, 86, 87].

Positron Emission Tomography (PET)

Positron emission tomography (PET), most commonly used in conjunction with computed tomography (CT) (PET-CT), may provide precise anatomic correlation and information about metabolic activity of a soft tissue tumor.

However, the use of PET in the diagnosis of sarcoma is still controversial and may be misleading in certain contexts. Typically, in fact, high-grade sarcomas show high rates of glucose metabolic activity, while low-grade or benign soft tissue lesions have a minor or absent uptake. Benign classic or ancient schwannoma is an exception since this subtype may reveal an increased F-18 fluorodeoxyglucose enhancement especially in the cellular histological variant.

PET-CT is generally not used for the initial evaluation of STSs, but it plays a remarkable role in assessing treatment response after systemic or radiation therapy or in identifying neoplastic recurrence in a previously radiated or surgical bed. It helps also to observe the tumor response after radiofrequency ablation or chemoembolization of metastatic lesions or to identify potential local and distant recurrences in case of dubious CT or MRI scan, especially in a previously radiated or surgical bed [82, 88, 89].

Biopsy

Initial biopsy of soft tissue mass is mandatory. The only exception may be lipoma-like fatty lesions or clinically evident sebaceous cysts. Although the procedure is not technically demanding, complications are not unusual and can ultimately compromise the final surgical outcome [90, 91]. In a study of complications following a biopsy performed by a non-oncologic surgeon, the Musculoskeletal Tumor Society reported a complication rate of 19% in 597 biopsies performed at a referral center; this rate was 12-fold greater than that of biopsies performed by a surgical oncologist.

The complications included 18 unnecessary amputations; any soft tissue biopsy, in fact, should be viewed as the first step of a successful limb salvage operation [92].

An incorrect biopsy may violate the compartmental anatomy, leading to the risk of tumor seeding, and may change a wide excision to an amputation. Ideally, the needle track should be placed in the plane of the future incision in order for it to be taken out during the operation [93].

Needle and open biopsies are the two modalities of biopsy for soft tissue tumors [94].

Core needle biopsy is usually the technique of choice since it allows the collection of a valuable amount of tissue with a very limited and less invasive procedure, usually performed under local anesthesia. Several passes of the Tru-cut needle are typically necessary to obtain consistent slices of pathological tissue, which allow the pathologist to perform a complete morphological analysis, also leaving tissue available for immunohistochemical or molecular studies.

This is a simple and quite inexpensive procedure, with a very low contamination risk, which can be done in office for easily palpable masses. For any deeper lesion or one located contiguously to critical structures, the same needle approach needs to be managed under radiological guidance. CT and US are the most commonly used tools to assist in guided biopsies, and the coaxial needle system may be used to decrease needle seeding [95–98].

Open biopsies can be divided into incisional and excisional biopsies.

With an *incisional biopsy*, only part of the tumor is removed in order to achieve a proper diagnosis. The incision line has to be placed longitudinally, possibly right above the tumor or just parallel to it such that the whole tract can be removed safely and en bloc during the time of definitive surgery [82–99].

Excisional biopsy implies a marginal resection of the tumor, and it is generally reserved for supposed benign lesions such as lipoma, schwannoma, or intramuscular angioma or in case of very small nodules (less than 3 cm) where the core-needle biopsy would probably fail in achieving enough tissue samples to reach a correct diagnosis.

Although excisional biopsy provides the highest level of diagnostic accuracy, it may be related to neoplastic contamination in case of intralesional resection. Because soft tissue masses are so common, clinicians of any medical specialty may encounter them. Being also variegate and challenging to treat, in order to optimize the chance of cure, it is mandatory to immediately refer any patient to the national reference center [82, 100, 101].

Clinical Presentations

Spermatic Cord Sarcomas

Respectively to the genitourinary (GU) tract, STSs are relatively rare tumors, accounting for 2.1% of STSs and 1–2% of all urological malignancies. The spermatic cord is the most commonly involved urological site (30% of cases), but it is often misdiagnosed. The most common histologic subtypes at this site are liposarcoma and leiomyosarcoma in older patients and rhabdomyosarcoma in children and young adults.

Spermatic cord sarcomas typically present as unilateral, hard, firm, slow-growing masses of the inguinal canal or the scrotum. Their clinical manifestations are usually nonspecific, with a sensation of compression and heaviness due to local mass effect, which is diagnosed after a long evolution, since they are generally asymptomatic. To the general surgeon or urologist, possible differential diagnosis could be an incarcerated inguinal hernia, inguinal lymphadenopathy, testicular malignancy, or hydroceles. Misdiagnosis frequently leads to inappropriate surgery, where tumor excision can be incomplete with a subsequent risk of tumor recurrence. The association of involved surgical margins with local recurrence is consistent with what has been extensively reported for STS at other sites [102, 103]. Other pathological features that increase the risk of recurrence include large tumor size, inguinal location, previous inadequate surgery, and manipulation of the tumor and the depth of invasion. Unexpectedly, tumor grade has little influence on the probability of local recurrence, and a low-grade sarcoma is as likely to recur locally as a high-grade sarcoma [104, 105]. In these circumstances, every attempt should be made to achieve always negative surgical margins in order to minimize the risk of local relapse: wide excision of the tumor en bloc with ipsilateral testicle and scrotum, spermatic cord, and vessel ligation at the level of the abdominal inguinal ring and resection of the adjacent soft tissue ring is the aim to obtain a higher chance of locoregional control. The iliac and femoral vessels must be exposed and controlled, and in locally advanced disease, they may need to be dissected or even excised and reconstructed. In selected cases, bone resection may be necessary. This usually includes an osteotomy of the ipsilateral superior pubic ramus. Obviously the specimen has to be removed en bloc from the surgical bed. The resulting anterior abdominal wall defect most frequently needs a combination of primary tension-free closure and placement of polyester or polypropylene synthetic mesh, sutured to the pubic tubercle, the inguinal ligament, and the surrounding abdominal wall aponeurosis by two hemi-running 2/0 Prolene sutures. In case of large skin/muscular excision, a pedicled (fasciocutaneous or myocutaneous) flap, or occasionally a free flap, has to be performed. In selected cases, when tumor location is completely intrascrotal, the abdominal wall might be limited, and its reconstruction would be performed just by a primary Prolene suture of the external oblique muscle aponeurosis avoiding any mesh placement [103, 106, 107] (Figs. 10.1 and 10.2).

The main pattern of spread of spermatic cord sarcomas is a contiguous extension from the cord through the inguinal canal and into the abdominal cavity. Although the majority of spermatic cord sarcoma types have low propensity for nodal spread, a significant incidence of inguino/iliac/retroperitoneal nodal relapse has been described in selected histotypes (rhabdomyosarcoma, pleomorphic liposarcoma, epithelioid sarcoma). However, given the low incidence of nodal metastasis from adult spermatic cord sarcomas and the potential morbidity of locoregional lymphadenectomy, the consensus is against prophylactic retroperitoneal lymph node dissection, while it is deemed mandatory in patients with preoperative evidence of retroperitoneal lymph node metastasis.

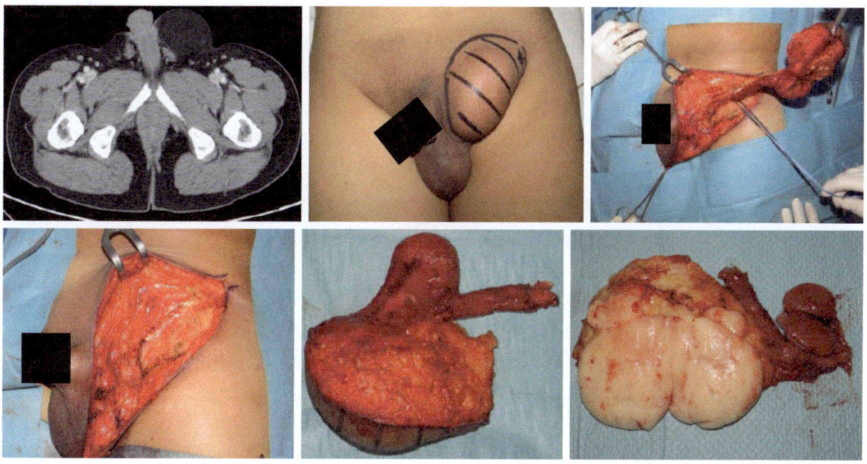

Fig. 10.1 A 41-year-old male, affected by primary, well-differentiated left paratesticular liposarcoma, excised en bloc with ipsilateral testicle and spermatic cord ligated and resected at the abdominal inguinal ring level. The complete intrascrotal location allowed a limited abdominal wall excision without requiring mesh reconstruction

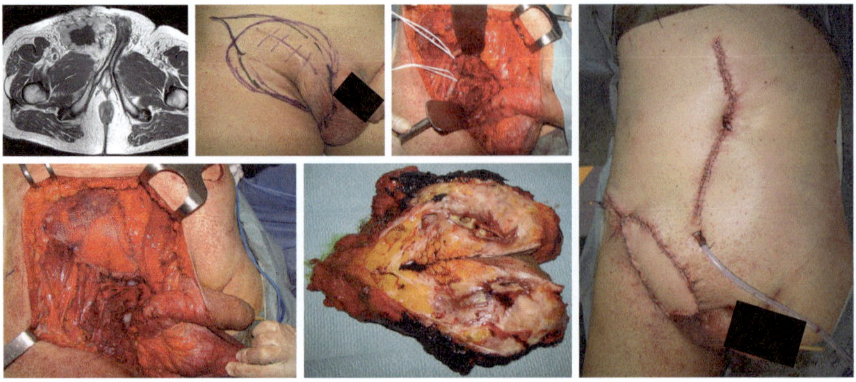

Fig. 10.2 A 60-year-old male, affected by recurrent high-grade dedifferentiated left spermatic cord liposarcoma. The locally advanced disease presentation required subadventital dissection of the iliac-femoral bundle and wider abdominal wall excision en bloc with ipsilateral testicle, spermatic cord, and portion of the superior pubic ramus. Reconstruction of the groin defect was performed by Prolene mesh repair and contralateral pedicled VRAM flap coverage

In contrast, retroperitoneal lymph node dissection is a controversial question in the rhabdomyosarcoma subtype; in these specific histologies, ipsilateral lymph node dissection has been advocated as the risk of metachronous lymphatic spread has been described up to 50% in same case series [108–110].

Although up to 90% of patients may be disease-free at 5 years and 15-year overall survival rate may exceed 52% of cases, still one-third of patients may present local or distant relapse 5 years after surgery. Thus, the sarcoma scientific community is left wondering about the opportunity to implement the surgical treatment with complementary therapies both in neoadjuvant and adjuvant settings [103, 111, 112].

Due to the limited experience in the treatment of spermatic cord sarcomas, the role of RT in their management remains controversial. The conventional approach in the treatment of high-grade STSs in the extremities and trunk wall includes performing wide en bloc excisions in combination with preoperative or postoperative radiation therapy, thus producing higher rates of local control rather than surgery alone. Unfortunately, experience with spermatic cord sarcoma is limited, albeit the medical literature has reported better locoregional control and disease-free survival after adjuvant RT. Therefore, a treatment combination with surgery and radiation therapy might be the most appropriate approach in order to decrease the local recurrence rates. To date, any relevant trial has been able to support this strategy, and looking forward any randomized study could ever be designed on this topic given the rarity of the tumor in object.

Differently regarding paratesticular rhabdomyosarcoma, RT can be more than a simple option given the sensitiveness toward ionizing radiations of this specific histotype.

In general, radiation therapy should supplement rather than replace a wide surgical excision and should be delivered postoperatively only after a complete resection has been performed. In selected cases when specific margins cannot be surgically improved by a re-excision, radiation therapy may find a role as definitive treatment. Furthermore, by improving the local control, radiation therapy directly impacts the final outcome as subsequent recurrences in the inguinal region may become unresectable leading to the death of the patient.

The potential benefit should be considered in light of the inherent side effects of inguinal irradiation.

Wounds in the groin are notorious for their high rate of tissue breakdown. The reasons for this are the difficulty in keeping the area clean and dry, the constant movement, and the potential for seroma formation, particularly if lymphadenectomy has been performed. Radiation therapy can exacerbate these wound problems; then once all the criticisms have been assessed and the risk of local recurrence determined, the decision to administer radiation therapy should be made on an individualized basis.

Currently, there is no definitive role of chemotherapy in the treatment of primary and localized spermatic cord sarcomas. Due to the rarity of this disease, neither randomized nor retrospective studies have been provided on this specific topic. Starting from the outcomes of STS in the extremities and trunk wall, we should begin to consider neoadjuvant chemotherapy in selected cases even in patients with spermatic cord sarcomas. This specific location, in fact, presents analogous characteristics to the extremities in terms of histological subtypes (mostly liposarcomas and leiomyosarcomas), but it is potentially worse in terms of surgical margin constraints and higher rate of local relapse, although most of them are in fact well-differentiated or intermediate-grade dedifferentiated liposarcomas, which carry a limited risk of distant spread and therefore are not good candidates for a neoadjuvant therapy.

Chemotherapy has a central role in childhood rhabdomyosarcomas, showing benefits in all stages of disease and delivering patients combined treatment based on vincristine, dactinomycin, and ifosfamide/cyclophosphamide.

To date, given the lack of published data about the efficacy of chemotherapy in adult spermatic cord sarcoma, a routine adjuvant systemic therapy is not justified, except in selected cases (high-grade and large-volume tumors) or in rhabdomyosarcoma patients [106, 109, 113–115].

Because of the high risk of locoregional relapse, patients affected by spermatic cord sarcomas should be strictly monitored during the first 5 years. Subsequently, periodic follow-up is still recommended since late recurrences are not infrequent (up to 45% at 15 years). Follow-up should include chest X-rays and US scan for low-grade tumors, while chest and full abdomen CT scan is suggested in case of high-grade disease with a major risk of intra-abdominal or lung spread [103, 105].

Retroperitoneal Sarcoma Extending to the Groin

A clinical finding of groin swelling or palpable lump within the inguinal region may generate multiple differential diagnoses which need to be cleverly examined in order to recognize the correct one among a wide range of possibilities.

Inguinal or crural hernias, reactive or neoplastic lymphadenopathies, hydrocele or varicocele are the most likely entities responsible for the arising of a groin or scrotal mass.

Patient medical history and a careful clinical examination may address the physician toward the most probable diagnosis, although misdiagnosing sometimes occurs leading to subsequent inappropriate treatments [102, 103].

Primary tumors, in fact, may affect the groin, but given their rarity (incidental tumors may be discovered in less than 0.1% patients treated for inguinal hernia repair) and the lack of clinical peculiar findings or laboratory abnormalities, identifying them usually represents a real challenge also for expert clinicians [71].

As already discussed and largely reported, the majority of STSs arising within the inguinal region are considered to originate from the spermatic cord [103, 105–108]. Interestingly, to date an increasing number of cases regarding primary and recurrent retroperitoneal sarcoma herniating through the deep inguinal ring and mimicking an inguinal hernia or a generic groin swelling have been recently described. According to this particular disease presentation, the clinical aspects and the pattern of symptoms tend to be more specific, showing during patient physical examination a firm and irreducible solid mass in the context of the inguinal canal or just below the inguinal ligament. The retroperitoneal cavity, in fact, communicates with the pelvic and the inguinal region following the gonadal vessels, and hence retroperitoneal STSs can occasionally extend through the inguinal canal into the scrotum. This possibility is further increased by the deep inguinal ring, which is a defect in the fascia transversalis through which cord structures enter the inguinal canal. This peculiar anatomical situation provides a favorable pathway of spread through which retroperitoneal sarcomas may invade the groin presenting as an indirect inguinal hernia or less frequently through the femoral and obturator foramen herniating into the thigh. The largest part of retroperitoneal STS usually shows an indolent pattern of growth, but due to the absence of anatomical barriers within the retroperitoneal cavity, they might grow indefinitely, becoming symptomatic only after the tumor has reached a remarkable size. Diffuse and a specific abdominal pain is the most common symptom, affecting approximately 50% of patients; less frequent symptoms may include general discomfort, fatigue, early satiety, nausea, vomiting, weight loss, lower extremity swelling, and GI hemorrhage or obstruction.

In the clinical scenario of discovering a groin lump, a retroperitoneal or pelvic tumor with inguinal or scrotal extension should be considered and preoperatively excluded. MRI and CT scan may be helpful either in the staging or diagnostic process and to define the most appropriate therapeutic strategy also including and planning the proper intervention.

In contrast, if the diagnosis of sarcoma was incidentally made in the operating room during the initial operation for an inguinal hernia or a lipoma removal, the suspicious of facing a retroperitoneal sarcoma should always be taken into consideration and surgery therefore suspended after collecting enough tissue sample to achieve the correct diagnosis. The patient would need to be evaluated and fully staged for retroperitoneal sarcoma and surgery rescheduled referring the patient to a sarcoma reference center [116–121].

Undoubtedly, surgery remains the cornerstone of treatment also in these locally advanced pelvic and retroperitoneal sarcomas. In order to improve the tumor resectability, an abdomino-inguinal incision is often preferred. It involves a lower midline incision, which is extended from 2 cm below the umbilicus transversely to the mid-inguinal point on the affected side and then vertically for few centimeters in the femoral triangle. The ipsilateral rectus abdominis and the anterior sheath are divided, as well as the obliquus externus, internus, and transversus muscles; the inguinal ligament is divided off the pubic tubercle, the inferior epigastric vessels are ligated and divided near their origin from the femoral vessels, which are exposed, and the lateral third of the inguinal ligament is detached from the iliac fascia. This approach provides better surgical exposure in one continuous field from the lower abdomen to the groin area on the side affected by the tumor, allowing safer proximal and distal control of the iliac-femoral vessels, easier identification of the femoral nerve lateral to the femoral artery, and easier disconnection of the iliac-psoas muscle from the lesser trochanter if required to achieve a wider resection.

This peculiar surgical approach, though very effective in the intraoperative management of locally advanced retroperitoneal and pelvic sarcomas, leaves the abdominal wall extremely weak. The integrity of the anterior abdominal wall could be difficult to restore by primary sutures, and for this purpose the use of a synthetic mesh is often required. The use of a nonadsorbable polypropylene or polyester prosthesis is preferable, which is more suitable to reinforce the iliac fossa anterior wall and to reconstruct the inguinal canal into the lacuna vasorum and lacuna musculorum. In case of sporadic larger soft tissue defect with a consequent harmful exposure of major vessels and nerves, a solid plastic reconstruction is often recommended. The more common alternatives may be the transposition of the ipsilateral sartorius muscle and the rotation of the contralateral rectus abdominis myocutaneous flap [122, 123] (Fig. 10.3).

Ultimately the incidental finding of STSs herniating through the deep inguinal ring to the inguinal canal may be an uncommon presentation of primary retroperitoneal STSs (most commonly liposarcoma). Given the rarity of this scenario, the diagnosis is not always immediate and may be incidentally detected during the repair of a suspected inguinal hernia or lipoma removal. Of course, it represents an important diagnostic challenge for a surgeon, due both to the surgical and oncological implications.

Any patient with the suspicion of a retroperitoneal sarcoma should be referred to a tertiary center to be properly staged and treated. In fact, although surgery represents the formal approach for retroperitoneal STS, the extension of the surgical resection or the potential administration of complementary treatments has to be assessed and eventually performed only in a high-volume sarcoma center.

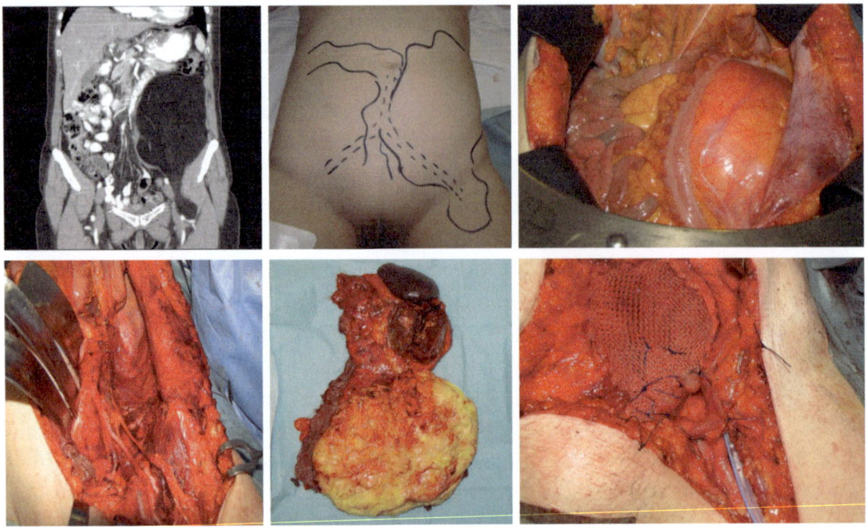

Fig. 10.3 A 69-year-old female, affected by primary left retroperitoneal well-differentiated liposarcoma extending to the groin through the inguinal canal. In order to achieve better exposure, an abdomino-inguinal incision was performed dividing the inguinal ligament. The tumor was removed en bloc with the spleen and the pancreatic tail, the sigmoid colon, the left kidney, the adrenal gland, and the ipsilateral iliopsoas muscle. The neurovascular iliac-femoral tree was entirely dissected. The abdominal wall defect was repaired by placing a polyester mesh, while the reconstruction of the inguinal canal into the lacuna vasorum and musculorum was achieved by shaping a Vicryl plug beneath the abdominal wall prosthesis

Soft Tissue Sarcomas of the Femoral Triangle

The management of STSs is generally a crucial problem when the tumor develops in constrained anatomical areas adjacent to noble or vital structures. The goal of R0 resection in these peculiar regions often results in complex defects including soft tissue layers with the overlying skin and major vessels and nerves.

Indeed, sarcomas usually respect anatomical boundaries, with local anatomy able to influence tumor growth by setting natural barriers to their extension, and in general, sarcomas follow the path of least anatomical resistance, initially growing within the anatomical compartment in which they arise. However, major vessels and nerves may sometimes be involved or even give rise to STSs, and their resection becomes mandatory in roughly 5% of all STSs. Undoubtedly, whether STSs arise from arterial or venous blood vessels or infiltrate or encase the vascular tree, the vessels must be resected in order to achieve adequate surgical margins. In contrast, STSs surrounded by a plane of normal tissue can be dissected from major blood vessels. By longitudinally splitting the adventitia opposite the tumor, a rim of normal tissue is preserved in the vessel-tumor interface. This thin layer of tissue, while not infiltrated by the tumor, is usually sufficient to provide microscopic negative margins without requiring any type of vascular resection and reconstruction (Fig. 10.4).

Recently, the continuous development of reconstructive techniques including flap coverage, limb revascularization, and nerve grafting has improved the surgical outcome for major defects and vascular involvement [124–128].

However, albeit resection of major vessels is feasible and has potentially facilitated local control, it cannot offset the high biological risk of these tumors. Therefore, although technically manageable, involvement of major vessels may be regarded as an added negative prognostic factor and possibly addressed by means of an effective systemic therapy, when the malignancy grade of the tumor is high [125, 126, 129].

Limb preservation is then possible, although given the high complexity of these surgical procedures the rate of postoperative complications may be substantial: disabling lymphedema, vascular graft thrombosis, wound dehiscence especially when flap coverage is required, and ultimately amputation are frequent complications of limb salvage with vascular reconstruction [124–126, 130].

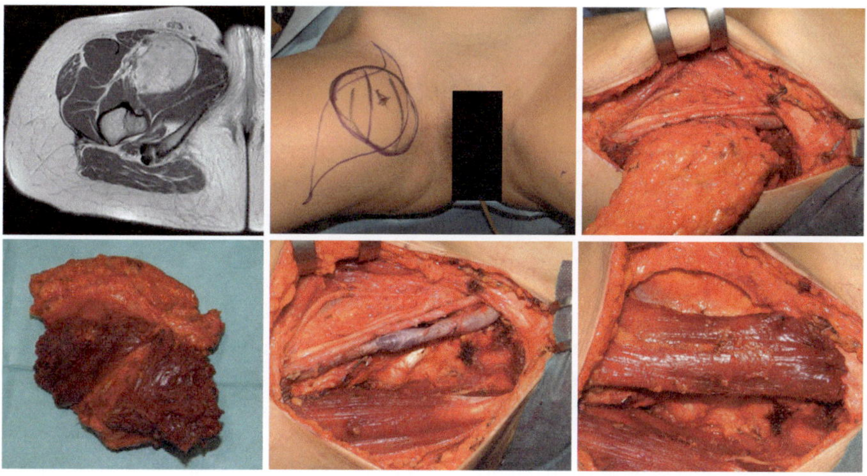

Fig. 10.4 A 34-year-old female, affected by primary classic solitary fibrous tumor of the proximal aspect of the right thigh. The tumor was excised en bloc with the surrounding soft tissue, and common femoral vessels were dissected underneath the adventital plane, an anatomical barrier which is seldom infiltrated by low-grade tumor. The rotation of the ipsilateral sartorius muscle allowed the exposed neurovascular bundle to be covered

When looking at the different histology subtypes growing within the inguino-femoral region, it was no surprise that leiomyosarcoma was identified as one of the most frequent tumor types given that leiomyosarcomas are the most common malignancy affecting the vascular system, directly arising from the smooth muscle cells of the major vessel wall. Indeed leiomyosarcomas of intravascular origin are extremely rare, presenting only 0.001% of all malignancies, which arise five times more commonly from the venous than from the arterial system, with a strong predilection for the larger veins. The inferior vena cava, in fact, is the most commonly involved vein, being the site of origin in over 50% of cases, while most of the reported sarcomas arising in the extremities affect the femoral bundle.

A direct infiltration of the vascular bundle is not a unique prerogative of vascular leiomyosarcoma. Any STS subtypes would be theoretically able to invade or encase the vascular tree. According to this peculiar anatomical location, synovial sarcoma and liposarcoma are the two most common histologies responsible for secondary vascular involvement.

Synovial sarcoma often affects girdles and the proximal aspect of upper and lower limbs. It may even arise within the proximal thigh close to the neurovascular bundle; synovial sarcoma, in fact, is the second most common histotype after malignant peripheral nerve sheath tumor originating from the peripheral nerves and in this context from the femoral nerve, then invading the nearby vascular axis.

Liposarcoma typically arises from the retroperitoneum/pelvis, and its usual massive extension may lead the tumor to herniate through the deep inguinal ring down to the inguinal or femoral canal, often encasing the iliac-femoral artery and vein.

Conversely, vascular involvement is exceedingly rare for myxofibrosarcoma aside from its typical infiltrative pattern of growth. This may be explained by the predilection for different sites which these tumors originate from; myxofibrosarcomas, in fact, are more common in superficial tissues or distal extremities and only occasionally approach the deep location close to the major vessels [125, 131–133].

Several methods for vascular reconstruction have been reported. Conduit selection, whether autologous venous graft or synthetic graft, is the most controversial aspect of complex groin reconstruction. While the need for arterial reconstruction is obvious, veins are reconstructed predominantly only if they were patent at the time of surgery and had no clinical or radiological evidence of collateralization. The need for vein replacement is controversial, mainly due to its related risk of developing graft thrombosis and subsequent pulmonary embolism. Indeed it is debatable whether it is worth considering reconstructing veins which are already occluded by the tumor, especially in the presence of collaterals. In contrast, when the vein is patent, it is beneficial to replace it in order to improve the short-to-medium-term functional outcome avoiding or delaying the onset of limb edema.

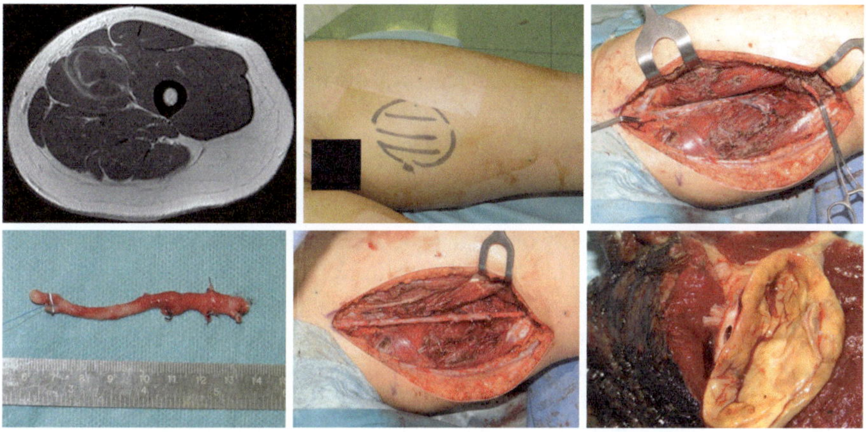

Fig. 10.5 A 26-year-old female, affected by primary synovial sarcoma of the left crural region. The patient received neoadjuvant chemo-/radiation therapy, and the tumor was excised en bloc with part of the anteromedial muscular compartment of the proximal thigh and the superficial femoral artery and vein, which were found to be encased at the pathological examination. The femoral artery was replaced with the contralateral greater saphenous vein, while the femoral vein was ligated only since the return flow from the leg would be provided by the superficial venous circulation

Autologous venous grafts are the ideal conduit because of their long-term superior patency rate without requiring any anticoagulation therapy. Among the wide range of possibilities, the most preferred autologous graft is the contralateral superficial femoral vein. Alternatively, when the diameter of the resected vascular stump is smaller, the contralateral greater saphenous vein is also a valid conduit (Fig. 10.5). Of course in any case of arterial replacement, the autologous venous graft needs to be reversed (to allow normal blood circulation avoiding the opposition of the vein valves), while it has to be kept straight in case of vein replacement.

A suitable option instead of autologous venous grafting is the use of a banked cadaver graft, which possesses all the advantages of the autologous graft, limiting any additional incision on the patient.

Unfortunately, the availability of homologous graft is generally limited, especially when surgery is not performed in a transplant center where tissue banking is usually available allowing the selection of the most appropriate cadaver graft for the receiving patient.

In contrast, when any kind of homologous vein graft is available, the most suitable choice is a PTFE graft with or without integrated rings. PTFE grafts are resistant to kinking and compression; in addition, their diameter can be properly selected to match the diameter of the arterial or venous stump.

The major drawbacks of PTFE grafts are the requirement for lifelong anticoagulation therapy, a higher risk of postoperative infection, and a less favorable long-term patency rate [125–128, 134, 135].

Primary STS originating from neural structures such as MPNST and synovial sarcoma or locally advanced tumors directly invading the femoral nerve may require its sacrifice when identifying a safe plane of dissection beneath the perineurium is not feasible. Perineurium nerves such as the adventitia of major vessels generally represent an acceptable margin which is seldom surmountable by neoplastic cells. Indeed, when the tumor breaks that anatomical barrier, the transection of the nerve with clear margins is mandatory. Although nerve reconstruction is not a lifesaving procedure and patients' quality of life may be acceptable even after the resection of the major peripheral nerve, the attempt to restore the neuromotor unit should be pursued especially in young patients with a short, possibly less than 10 cm, neural gap. Classically, the sural nerve has been the predominant source of nerve autograft, as sural donor site morbidity is minimal resulting in diminished sensation at the lateral foot and minimal, often invisible, scars [136–141].

In case of extensive groin resections, soft tissue and skin defect may also be challenging to reconstruct. Transferring a soft tissue flap into the surgical defect fills the dead space; incorporates healthy, well-vascularized tissue into the wound; and facilitates tension-free closure. Various flap coverage techniques are available, and flap selection is potentially a critical component for these complex defects, because multiple vital anatomical structures should be reconstructed simultaneously. Given its reliable vascularization, the large skin surface, and the considerable amount of tissue, pedicled vertical rectus abdominis myocutaneous (VRAM) flap is usually the preferred choice especially for remarkable defects of the inguinal region.

An ipsilateral VRAM would always be the ideal option, but it is seldom practicable since the ipsilateral deep inferior epigastric vessels are necessarily transected during the tumor removal. A contralateral VRAM flap is then a valid alternative as it can reach the contralateral distal thigh without excessive tension. The disadvantages of using contralateral VRAM flap depend on the consequent weakness of the anterior abdominal wall, although fixed by a synthetic mesh, which can lead to donor site bulging or herniation.

For patients with smaller groin soft tissue defects, pedicled anterolateral thigh (ALT) flap may be an appropriate reconstruction as long as the source vessel is not sacrificed proximally at the tumor ablation.

When local flaps are not available or adequate to cover the whole surgical defect, the transfer of a free flap remains the last option. The choice of which free flap depends on the size of the groin defect. For patients with small to moderate defects, an ALT free flap is normally indicated, while for wider defects, a latissimus dorsi free flap is usually more appropriate. The major obstacle to free flap transfer for complex groin defects is the availability of suitable recipient vessels when the external iliac or femoral vessels are transected [124, 142–144].

In conclusion, we confirm that locally advanced groin sarcoma can be removed with negative microscopic resection margins by a limb salvage approach. Involvement or infiltration of noble or vital structures does not represent a contraindication to perform or attempt surgery since major peripheral vessels, motor nerves, and soft tissue defects may be well reconstructed. However, this category of patients is at high risk of developing metastatic disease due to the underlying nature of their disease. This should be carefully considered when planning the treatment strategy; for example, the perioperative risks/postoperative functional impact should be adequately assessed against the limited benefit in overall survival. This is a rare indication for a rare tumor, and patients should definitely be treated only in STS reference centers where all the medical and surgical specialties are available in order to deliver the best standard of care.

Metastatic Disease (Lymph Node Involvement)

About 25–30% of extremity soft tissue sarcoma (ESTS) patients develop metastatic disease, mainly disseminating through the bloodstream, with the lungs being the primary site in 80% of patients with distant metastases.

In contrast, lymph node metastases have been identified in less than 5% of cases, although some series report overall rates of lymphatic spread in up to 10% of ESTS patients [145–147].

In specific histotypes, there is a strong propensity toward lymph node metastasis. Between 2000 and 2009, from the National Cancer Data Base of the United States, 27,536 patients with extremity STS were identified; 1924 (7.0%) underwent nodal evaluation and 25,612 (93.0%) did not have any nodes examined at surgery. Of 1924 patients with extremity STS who underwent nodal evaluation, 290 (15.1%) had nodal metastases. Assessment by histologic subtype revealed higher rates of lymph node metastases in patients with rhabdomyosarcoma (32.1%), angiosarcoma (24.1%), clear cell sarcoma (27.7%), and epithelioid sarcoma (31.8%). Lower rates of nodal metastases were seen in patients with fibrosarcoma (9.5%), leiomyosarcoma (7.5%), synovial sarcoma (6.0%), and liposarcoma (3.5%). Nodal metastasis rates were higher in patients with tumors 5–10 cm in size (18.6%) than in patients with tumors <5 cm in size (12.4%). Patients with high-grade tumors (18.4%) also had higher rates of nodal metastases than patients with low-grade tumors (5.3%) [148].

Neoplastic lymph node involvement has also been described in myxofibrosarcoma. In this last subtype, the tumor cells may spread into the lymph node parenchyma or peculiarly give rise to soft tissue metastases invading only the perilymph node stroma [12, 149].

Metastatic lymph node involvement is usually suspected when regional nodes are enlarged on clinical examination or have an abnormal appearance on ultrasound, computed tomography (CT), or magnetic resonance imaging (MRI). However, both clinical and radiographic assessments may overestimate the presence of metastatic disease in the case of normal reactive nodes or underestimate tumor involvement if the nodes are not pathologically enlarged. One cooperative group study reported that 17% of clinically and radiologically normal lymph nodes were found to contain neoplastic cells at the time of biopsy [150].

Various approaches have been proposed in order to try to identify the presence of metastatic lymph nodes in extremity STS patients.

Historically, random sampling of the anatomic regional lymph node basin has been performed, but this can lead to excessive dissection with a high morbidity rate such as infection, nerve and vascular damage, and chronic lymphedema. Sentinel lymph node biopsy (SLNB) was designed as a reasonable alternative to random sampling, hypothesizing that the first (sentinel) node receiving lymphatic drainage from a tumor site would be the most likely one to contain metastatic cells. The methylene blue and the gamma probe guide the surgeons to the identification of the sentinel lymph nodes, which are subsequently excised. It has quickly become the standard of care in selected types of cancer. Several studies, in fact, have supported this procedure in both melanoma and breast cancer.

Due to the potential advantages of SLNB, many physicians have explored, with mixed success, the use of lymphatic mapping for malignancies other than breast cancer or melanoma. These included lung cancer, colon cancer, head and neck squamous cell carcinoma, gynecologic cancers, thyroid cancer, Merkel cell carcinoma, upper gastrointestinal cancers, and non-small cell lung cancer.

Unfortunately, until date, SLNB has not been formally investigated in the management of sarcoma. This is not particularly surprising given that the majority of sarcomas spread by local extension or hematogenously. Regional lymph node metastases seldom occur, developing in less than 10% of patients with localized disease. In addition, recurrence within the locoregional nodal basin is rare, representing 4–10% of local recurrences [151–153].

Functional imaging using [18F]fludeoxyglucose positron emission tomography/computed tomography (FDG-PET-CT) represents a valid and noninvasive alternative to SLNB to assess the regional lymphatic basin for the presence of metastatic disease. It uses not just the size but also the metabolic characteristics of tissue to determine whether a metastatic tumor may be present. Although PET/CT scans have been found to be less reliable than SLNB in melanoma and breast cancer, it has proved to be superior to any other radiological method for staging of locoregional lymph nodes and the detection of skeletal metastases in sarcoma patients. However, what emerged from recent studies is that PET-CT cannot be considered specific enough for nodal metastases such that biopsy can be avoided. The positive predictive value of PET-CT was reported to be 29%, while the negative predictive value of PET-CT was 79%. These findings imply poor predictive power for PET to identify small-volume metastatic nodal disease in STSs.

Therefore, when clinical examination and imaging are not conclusive, fine-needle aspiration biopsy (FNAB) should be performed preoperatively to assess whether tumoral cells within the lymph node are present. It is, in fact, absolutely crucial to distinguish between metastatic and reactive enlarged lymph nodes before defining any patient treatment plan [150–153].

Given the unlikely event of nodal spread in ESTS patients and consequently the relatively limited experience in this area, data are lacking, and comprehensive recommendations on lymph node evaluation (SLNB-FNAB-PET/CT scan) or treatment when involved are not available in the National Comprehensive Cancer Network's guidelines for sarcoma. There is a formal consensus regarding the uselessness of performing prophylactic locoregional lymph node dissection as part of the initial treatment of these patients, at least in the absence of macroscopic lymphatic disease.

The confirmation of metastases in regional nodes is a clinical expression of the biological aggressiveness of the sarcoma, with 5-year survival rates between 10 and 23%, while 10-year survival rates have been reported to be approximately 3% [147, 154].

For STSs, isolated lymph node metastasis has been thought to carry a prognosis similar to distant metastatic disease, and, in fact, the 2002 American Joint Committee on Cancer staging guidelines classify lymph node metastasis as stage IV disease [147]. Nevertheless, according to this classification, patients with lymph node-positive sarcomas had worse overall survival when compared to patients with localized disease, but showed improved survival rates if compared to patients with distant metastases. Localized disease, regional lymph node metastatic disease, and distant metastatic disease had 5-year survival rates of 81%, 51%, and 22%, respectively [145]. Besides, as recently underlined in a large ESTS patients series, the outcome of patients presenting with simultaneous locoregional lymphatic disease and distant metastasis definitely had a worse long-term prognosis when compared to patients only presenting with nodal spread. The 1- and 2-year survival for patients with isolated regional lymph node metastasis (RLNM) was 77% and 47%, respectively. The 1- and 2-year survival for patients with distant metastases present at the time of presentation of RLNM was 36% and 21%, respectively. The 5-year survival for patients with isolated RLNM was 24%, while it was 0% for patients who presented with RLNM and distant metastasis.

There is a general agreement that lymph node metastasis detected at the time of diagnosis indicates a poorer outcome. The 1-year survival for metachronous and synchronous RLNM was 94% and 68%, respectively, and the 2-year survival was 56% and 42%, respectively. Metachronous RLNM then has a better outcome than synchronous RLNM at the time of diagnosis of primary STSs [147]. In contrast to this tendency, recent studies have supported the view of better survival for patients with lymph node metastases at the time of diagnosis. Post regional lymphadenectomy, disease-free survival was significantly longer in patients with regional lymph node metastases at the time of diagnosis than in patients with lymph node recurrence after prior curative surgery. Significantly, patients with initial and recurrent regional lymph node metastases showed longer disease-free survival than patients with distant metastases [146].

The surgical treatment of radical lymph node dissection is generally considered a palliative procedure, because it invariably indicates distant micrometastatic disease [145]. However, several studies demonstrated that aggressive treatment could bring about long-term survivors. It was, in fact, demonstrated that primary surgical treatment of lymph nodes gives a better survival if compared to patients treated by chemotherapy and RT [41, 146]. To date, the question about the extension of the lymphatic dissection is controversial, and the role of radical lymphadenectomy among node-positive patients remains to be defined. Although it is widely known that the presence of lymph node metastasis directly impacts overall survival, it has been recently reported that neither the degree of lymph node burden nor the extent of resected lymph nodes influenced survival, calling into question the role of radical lymphadenectomy in patients with isolated regional lymph node metastasis. In further case series of ESTS patients with limited nodal disease, any survival benefit was proved for those patients who underwent radical lymphadenectomy compared to patients who had a limited lymph node dissection consisting in the resection of the only macroscopically detectable tumor [145, 155, 156].

In contrast with these results, evidence in favor of extended lymph node dissection in sarcoma patients with isolated regional lymph node metastasis is primarily based on previous retrospective investigations and studies supporting the advantage of radical lymphadenectomy, such as axillary dissection including level I to III lymph nodes and ilioinguinal dissection, in order to improve the local control rate and prolong the overall survival in ESTS patients. In addition, when there is recurrence in the regional lymph node basin, performing radical lymphadenectomy would further increase survival [40, 41, 147, 157, 158].

In conclusion, lymph node metastases are uncommon in the majority of extremity STSs. In specific histotypes, with increased rates of lymph node metastasis such as rhabdomyosarcoma, clear cell sarcoma, and epithelioid sarcoma, heightened attention directed to the regional lymph node basin is warranted. In the absence of distant metastasis, in fact, lymph node status is the most relevant prognosticator of disease-specific survival.

The preoperative (FNAB-PET/CT scan) or intraoperative (sentinel node biopsy) evaluation of the locoregional lymph node status is performed only in case of clinical or radiological suspicious. The management of positive lymph nodes remains controversial, although an extended lymph node dissection would seem to improve at least the local control. However, definitive data are lacking, and future studies are needed to determine the optimal treatment strategy for extremity STS patients with isolated regional lymph node metastasis.

Conclusion

STSs are a rare and heterogeneous group of tumors potentially arising from any component of the human body. R0-surgery is the standard of care for primary neoplasms, although radiation therapy and to some extent chemotherapy are routinely administered in locally advanced disease settings or very sensitive histologies.

Additionally, the inguino-femoral location represents a further complex element to face in STS treatment since the anatomical constraints present herein may make it difficult to achieve adequate microscopic resection margins with a higher risk of locoregional recurrences which are often unresectable.

The involvement of the inguinal canal or the proximal anterior aspect of the thigh always requires massive soft tissue and muscular resection afterward repaired with tension-free synthetic mesh in order to reinforce the abdominal wall and to avoid postincisional hernias.

The encasement of the neurovascular bundle, although not a formal contraindication to surgery, often reveals consistent tumor aggressiveness; therefore, beneficial of a complex operation, albeit technically feasible, must be balanced over the high metastatic risk and the poor global outcome when these tumor presentations occur.

In case of vascular involvement, the replacement of the artery is always mandatory. The contralateral superficial femoral vein is the most preferred option, although cadaveric grafts or PTFE prosthesis can be used too. In contrast, vein reconstruction (which can lead to graft thrombosis and consequent pulmonary embolism) is controversial, especially if the greater saphenous vein or significant collateral vessels may be preserved.

Femoral nerve reconstruction by transposing the sural nerve is seldom required, unless the tumor originates from the nerve or grows beyond its perineurium. The indication of nerve reconstruction subsists only when the chance of recovery after nerve resection is substantial (young patients, short neural gap).

STSs predominantly metastasize through the bloodstream, the lung usually being the first site of distant spread. In contrast, lymph node metastases occur in roughly 5% of cases. Epithelioid sarcoma, clear cell sarcoma, rhabdomyosarcoma, and angiosarcoma are the histological subtypes with the higher incidence of lymphatic involvement. Sentinel node biopsy is not routinely performed in the clinical practice, while groin dissection is proposed only in the presence of macroscopic lymph node disease. In case of extensive groin resections, local or free flaps are often required in order to fill the surgical defect and cover viable anatomical structures such as major vessels or motor nerves.

Groin soft tissue tumors are historically associated with poor local control and higher rates of intra- and postoperative complications. The extreme variety of STS presentations within the inguinal region reflects the complexity of their treatment, which is often planned on an individualized basis. Therefore, the diagnostic and therapeutic approach may be very different from case to case, and in order to receive the most appropriate cure, patients should always be referred from the beginning to a tertiary cancer center.

References

1. ESMO/European Sarcoma Network Working Group. Soft tissue and visceral sarcomas: ESMO clinical practice guidelines for diagnosis, treatment and follow-up. Ann Oncol. 2014;25(Suppl 3):iii102–12.
2. Stiller CA, Trama A, Serraino D, Rossi S, Navarro C, Chirlaque MD, Casali PG, RARECARE Working Group. Descriptive epidemiology of sarcomas in Europe: report from the RARECARE project. Eur J Cancer. 2013;49(3):684–95.
3. Toro JR, Travis LB, Wu HJ, Zhu K, Fletcher CD, Devesa SS. Incidence patterns of soft tissue sarcomas, regardless of primary site, in the surveillance, epidemiology and end results program, 1978–2001: an analysis of 26,758 cases. Int J Cancer. 2006;119(12):2922–30.
4. Cormier JN, Pollock RE. Soft tissue sarcomas. CA Cancer J Clin. 2004;54(2):94–109.

5. Swallow CJ, Catton CN. Local management of adult soft tissue sarcomas. J Am Acad Orthop Surg. 2009;17(1):40–7.
6. Gilbert NF, Cannon CP, Lin PP, Lewis VO. Soft-tissue sarcoma. J Am Acad Orthop Surg. 2009;17(1):40–7.
7. Fletcher CD, Hogendoorn P, Mertens F, Bridge J. WHO classification of tumours of soft tissue and bone. 4th ed. Lyon: IARC Press; 2013.
8. Spiguel A. Soft tissue sarcomas. Cancer Treat Res. 2014;162:203–23.
9. Garcia-Ortega DY, Villa-Zepeda O, Martinez-Said H, Cuellar-Hübbe M, Luna-Ortiz K. Oncology outcomes in retroperitoneal sarcomas: prognostic factors in a retrospective cohort study. Int J Surg. 2016;32:45–9.
10. Callegaro D, Fiore M, Gronchi A. Personalizing surgical margins in retroperitoneal sarcomas. Expert Rev Anticancer Ther. 2015;15(5):553–67.
11. Raut CP, Miceli R, Strauss DC, Swallow CJ, Hohenberger P, van Coevorden F, Rutkowski P, Fiore M, Callegaro D, Casali PG, Haas RL, Hayes AJ, Honore C, Cannell AJ, Jakob J, M S, Fairweather M, Pollock RE, Bonvalot S, Gronchi A. External validation of a multi-institutional retroperitoneal sarcoma nomogram. Cancer. 2016;122(9):1417–24.
12. Sanfilippo R, Miceli R, Grosso F, Fiore M, Puma E, Pennacchioli E, Barisella M, Sangalli C, Mariani L, Casali PG, Gronchi A. Myxofibrosarcoma: prognostic factors and survival in a series of patients treated at a single institution. Ann Surg Oncol. 2011;18(3):720–5.
13. Sambri A, Bianchi G, Righi A, Ferrari C, Donati D. Surgical margins do not affect prognosis in high grade myxofibrosarcoma. Eur J Surg Oncol. 2016;42(7):1042–8.
14. Fugløø HM, Maretty-Nielsen K, Hovgaard D, Keller JØ, Safwat AA, Petersen MM. Metastatic pattern, local relapse, and survival of patients with myxoid liposarcoma: a retrospective study of 45 patients. Sarcoma. 2013;2013:548628.
15. Haniball J, Sumathi VP, Kindblom LG, Abudu A, Carter SR, Tillman RM, Jeys L, Spooner D, Peake D, Grimer RJ. Prognostic factors and metastatic patterns in primary myxoid/round-cell liposarcoma. Sarcoma. 2011;2011:538085.
16. Vlenterie M, Jones RL, van der Graaf WT. Synovial sarcoma diagnosis and management in the era of targeted therapies. Curr Opin Oncol. 2015;27(4):316–22.
17. Eilber FC, Dry SM. Diagnosis and management of synovial sarcoma. J Surg Oncol. 2008;97(4):314–20.
18. James AW, Shurell E, Singh A, Dry SM, Eilber FC. Malignant peripheral nerve sheath tumor. Surg Oncol Clin N Am. 2016;25(4):789–802.
19. Farid M, Demicco EG, Garcia R, Ahn L, Merola PR, Cioffi A, Maki RG. Malignant peripheral nerve sheath tumors. Oncologist. 2014;19(2):193–201.
20. Guillou L, Wadden C, Coindre JM, Krausz T, Fletcher CD. "Proximal-type" epithelioid sarcoma, a distinctive aggressive neoplasm showing rhabdoid features. Clinicopathologic, immunohistochemical, and ultrastructural study of a series. Am J Surg Pathol. 1997;21(2):130–46.
21. Asano N, Yoshida A, Ogura K, Kobayashi E, Susa M, Morioka H, Iwata S, Ishii T, Hiruma T, Chuman H, Kawai A. Prognostic value of relevant clinicopathologic variables in epithelioid sarcoma: a multi-institutional retrospective study of 44 patients. Ann Surg Oncol. 2015;22(8):2624–32.
22. Baratti D, Pennacchioli E, Casali PG, Bertulli R, Lozza L, Olmi P, Collini P, Radaelli S, Fiore M, Gronchi A. Epithelioid sarcoma: prognostic factors and survival in a series of patients treated at a single institution. Ann Surg Oncol. 2007;14(12):3542–51.
23. Tejwani A, Kobayashi W, Chen YL, Rosenberg AE, Yoon S, Raskin KA, Rosenthal DI, Nielsen GP, Hornicek FJ, Delaney TF. Management of acral myxoinflammatory fibroblastic sarcoma. Cancer. 2010;116(24):5733–9.
24. Hwang S, Kelliher E, Hameed M. Imaging features of low-grade fibromyxoid sarcoma (Evans tumor). Skelet Radiol. 2012;41(10):1263–72.
25. Billings SD, Giblen G, Fanburg-Smith JC. Superficial low-grade fibromyxoid sarcoma (Evans tumor): a clinicopathologic analysis of 19 cases with a unique observation in the pediatric population. Am J Surg Pathol. 2005;29(2):204–10.

26. Cornillie J, van Cann T, Wozniak A, Hompes D, Schöffski P. Biology and management of clear cell sarcoma: state of the art and future perspectives. Expert Rev Anticancer Ther. 2016;16(8):839–45.
27. Malchau SS, Hayden J, Hornicek F, Mankin HJ. Clear cell sarcoma of soft tissues. J Surg Oncol. 2007;95(6):519–22.
28. Bianchi G, Charoenlap C, Cocchi S, Rani N, Campagnoni S, Righi A, Frisoni T, Donati DM. Clear cell sarcoma of soft tissue: a retrospective review and analysis of 31 cases treated at Istituto Ortopedico Rizzoli. Eur J Surg Oncol. 2014;40(5):505–10.
29. Pennacchioli E, Fiore M, Collini P, Radaelli S, Dileo P, Stacchiotti S, Casali PG, Gronchi A. Alveolar soft part sarcoma: clinical presentation, treatment, and outcome in a series of 33 patients at a single institution. Ann Surg Oncol. 2010;17(12):3229–33. doi:10.1245/s10434-010-1186-x.
30. Portera CA Jr, Ho V, Patel SR, Hunt KK, Feig BW, Respondek PM, Yasko AW, Benjamin RS, Pollock RE, Pisters PW. Alveolar soft part sarcoma: clinical course and patterns of metastasis in 70 patients treated at a single institution. Cancer. 2001;91(3):585–91.
31. Paolino G, Lora V, Cota C, Panetta C, Muscardin LM, Donati P. Early angiosarcoma of the scalp: a clinicopathological pitfall. Am J Dermatopathol. 2016;38(9):690–4.
32. Choi JH, Ahn KC, Chang H, Minn KW, Jin US, Kim BJ. Surgical treatment and prognosis of angiosarcoma of the scalp: a retrospective analysis of 14 patients in a single institution. Biomed Res Int. 2015;2015:321896.
33. Loss L, Zeitouni NC. Management of scalp dermatofibrosarcoma protuberans. Dermatol Surg. 2005;31(11 Pt 1):1428–33.
34. Gupta B, Verma N, Khurana N, Kumar JS. Dermatofibrosarcoma protuberans of the scalp, with fibrosarcomatous areas masquerading as epidermal inclusion cyst. BMJ Case Rep. 2016;2016:pii:bcr2016215427.
35. Machado I, Llombart B, Calabuig-Fariñas S, Llombart-Bosch A. Superficial Ewing's sarcoma family of tumors: a clinicopathological study with differential diagnoses. J Cutan Pathol. 2011;38(8):636–43.
36. Ulusan S, Koc Z, Canpolat ET, Colakoglu T. Radiological findings of primary retroperitoneal Ewing sarcoma. Acta Radiol. 2007;48(7):814–8.
37. Gronchi A, Colombo C, Raut CP. Surgical management of localized soft tissue tumors. Cancer. 2014;120(17):2638–48.
38. Gilbert NF, Cannon CP, Lin PP, Lewis VO. Soft-tissue sarcoma. J Am Acad Orthop Surg. 2009;17(1):40–7.
39. Gronchi A, Guadagnolo BA, Erinjeri JP. Local ablative therapies to metastatic soft tissue sarcoma. Am Soc Clin Oncol Educ Book. 2016;35:e566–75.
40. Fong Y, Coit DG, Woodruff JM, Brennan MF. Lymph node metastasis from soft tissue sarcoma in adults. Analysis of data from a prospective database of 1772 sarcoma patients. Ann Surg. 1993;217(1):72–7.
41. Riad S, Griffin AM, Liberman B, Blackstein ME, Catton CN, Kandel RA, O'Sullivan B, White LM, Bell RS, Ferguson PC, Wunder JS. Lymph node metastasis in soft tissue sarcoma in an extremity. Clin Orthop Relat Res. 2004;426:129–34.
42. Gronchi A, Raut CP. Treatment of localized sarcomas. Hematol Oncol Clin North Am. 2013;27(5):921–38.
43. Rosenberg SA, Tepper JE, Glatstein EJ, et al. The treatment of softtissue sarcomas of the extremities: prospective randomized evaluations of (1) limb-sparing surgery plus radiation therapy compared with amputation and (2) the role of adjuvant chemotherapy. Ann Surg. 1982;196:305–15.
44. Yang JC, Chang AE, Baker AR, et al. A randomized prospective study of the benefit of adjuvant radiation therapy in the treatment of soft tissue sarcomas of the extremity. J Clin Oncol. 1998;16:197–203.
45. Lee J, Park YJ, Yang DS, et al. Treatment outcome of conservative surgery plus postoperative radiotherapy for extremity soft tissue sarcoma. Radiat Oncol J. 2012;30(2):62–9.

46. Davis AM, O'Sullivan B, Bell RS, et al. Function and health status outcomes in a randomized trial comparing preoperative and postoperative radiotherapy in extremity soft tissue sarcoma. J Clin Oncol. 2002;20(22):4472–7.
47. Pervaiz N, Colterjohn N, Farrokhyar F, et al. A systematic meta-analysis of randomized controlled trials of adjuvant chemotherapy for localized resectable soft-tissue sarcoma. Cancer. 2008;113(3):573–81.
48. Woll PJ, Reichardt P, Le Cesne A, et al. Adjuvant chemotherapy with doxorubicin, ifosfamide, and lenograstim for resected soft-tissue sarcoma (EORTC 62931): a multicentre randomised controlled trial. Lancet Oncol. 2012;13(10):1045–54.
49. Frustaci S, Gherlinzoni F, De Paoli A, et al. Adjuvant chemotherapy for adult soft tissue sarcomas of the extremities and girdles: results of the Italian randomized cooperative trial. J Clin Oncol. 2001;19(5):1238–47.
50. Gronchi A, Stacchiotti S, Verderio P, Ferrari S, Martin Broto J, Lopez-Pousa A, Llombart-Bosch A, Dei Tos AP, Collini P, Jurado JC, De Paoli A, Donati DM, Poveda A, Quagliuolo V, Comandone A, Grignani G, Morosi C, Messina A, De Sanctis R, Bottelli S, Palassini E, Casali PG, Picci P. Short, full-dose adjuvant chemotherapy (CT) in high-risk adult soft tissue sarcomas (STS): long-term follow-up of a randomized clinical trial from the Italian Sarcoma Group and the Spanish Sarcoma Group. Ann Oncol. 2016;27(12):2283–8.
51. Edmonson JH, Ryan LM, Blum RH, et al. Randomized comparison of doxorubicin alone versus ifosfamide plus doxorubicin or mitomycin, doxorubicin, and cisplatin against advanced soft tissue sarcomas. J Clin Oncol. 1993;11:1269–75.
52. Gronchi A, Ferrari S, Quagliuolo V, Martin Broto J, Lopez-Pousa A, Grignani G, Ferraresi V, Blay J-Y, Rutkowski P, Merlo FD, Marchesi E, Ledesma P, Dei Tos AP, Bague Rosell S, Coindre J-M, Morosi C, Stacchiotti S, Picci P, Bruzzi P, Casali PG. Full-dose neoadjuvant anthracycline + ifosfamide chemotherapy is associated with a relapse free survival (RFS) and overall survival (OS) benefit in localized high-risk adult soft tissue sarcomas (STS) of the extremities and trunk wall: Interim analysis of a prospective randomized trial. Ann Oncol. 2016;27:LBA6_PR.
53. Vagnoni V, Brunocilla E, Schiavina R, Borghesi M, Passaretti G, Gentile G, Fiorentino M, Martorana G. Inguinal canal tumors of adulthood. Anticancer Res. 2013;33(6):2361–8.
54. Coleman J, Brennan M, Alektiar K, et al. Adult spermatic cord sarcomas: management and results. Ann Surg Oncol. 2003;10:669–75.
55. Brooks AD, Bowne WB, Delgado R, et al. Soft tissue sarcoma of the groin: diagnosis, management and prognosis. J Am Coll Surg. 2001;193:130–6.
56. Faguendes MA, Zietman AL, Althausen AF, et al. The management of spermatic cord sarcoma. Cancer. 1996;77:1873–6.
57. Ballo MT, Zagars GK, Pisters PW, et al. Spermatic cord sarcoma: outcome, patterns of failure and management. J Urol. 2011;166:1306–10.
58. Longbotham JH, Joyce RP. Retroperitoneal liposarcoma presentino as spermatic cord tumor. Urology. 1987;30:276.
59. Baldassarre E, Santacroce C, Barone M, Torino G, Siani A, Valenti GG, Chir G. Retroperitoneal well-differentiated liposarcoma presenting as an incarcerated inguinal. Hernia. 2007;28(8–9):315–7.
60. Forciea B. An eText of human anatomy and physiology, pp 232–6.
61. Gray H. Anatomy of the human body, pp 279–84.
62. Saladin KS. Human anatomy, McGraw Hill, pp 310–7.
63. Croce E, Olmi S. Chirurgia della parete addominale, Masson, pp 21–39.
64. George A, Grimer R. Early symptoms of bone and soft tissue sarcomas: could they be diagnosed earlier? Ann R Coll Surg Engl. 2012;94(4):261–6.
65. Ilaslan H, Schils J, Nageotte W, Lietman SA, Sundaram M. Clinical presentation and imaging of bone and soft-tissue sarcomas. Cleve Clin J Med. 2010;77(Suppl 1):S2–7.
66. Johnson CJ, Pynsent PB, Grimer RJ. Clinical features of soft tissue sarcomas. Ann R Coll Surg Engl. 2001;83(3):203–5.

67. Potter BK, Adams SC, Qadir R, Pitcher JD, Temple HT. Fungating soft-tissue sarcomas. Treatment implications and prognostic importance of malignant ulceration. J Bone Joint Surg Am. 2009;91(3):567–74.
68. Fernández EL, Plasencia LD, Palma JP, Pallares AC. Giant ulcerated pleomorphic liposarcoma of the chest wall. J Thorac Oncol. 2007;2(12):1126–7.
69. Valverde J, Vinagreiro M, Gouveia P, Koch P, Soares V, Gomes T. Sarcoma the great "masquerader" hematoma/deep vein thrombosis manifestation. Int J Surg Case Rep. 2016;28:348–51.
70. Niimi R, Matsumine A, Kusuzaki K, Okamura A, Matsubara T, Uchida A, Fukutome K. Soft-tissue sarcoma mimicking large haematoma: a report of two cases and review of the literature. J Orthop Surg (Hong Kong). 2006;14(1):90–5.
71. Collin T, Blackburn AV, Milner RH, Gerrand C, Ragbir M. Sarcomas in the groin and inguinal canal—often missed and difficult to manage. Ann R Coll Surg Engl. 2010;92(4):326–9.
72. Bell RS, O'Sullivan B, Mahoney JL, Nguyen C, Langer F, Catton C. The inguinal sarcoma: a review of five cases. Can J Surg. 1990;33(4):309–12.
73. Bertelsen CA, Eilber FR. Paraneoplastic syndromes with soft-tissue sarcoma: a report of two unusual cases. J Surg Oncol. 1983;24(3):170–2.
74. Bisagni G, Boni C, Fornaciari G, Gherlinzoni F. Soft tissues sarcoma with acute paraneoplastic hepatic dysfunction: case report. Tumori. 1996;82(5):499–501.
75. Buvarp Dyrop H, Vedsted P, Rædkjær M, Safwat A, Keller J. Routes to diagnosis for suspected sarcoma: the impact of symptoms and clinical findings on the diagnostic process. Sarcoma. 2016;2016:8639272.
76. Siegel GW, Biermann JS, Chugh R, Jacobson JA, Lucas D, Feng M, Chang AC, Smith SR, Wong SL, Hasen J. The multidisciplinary management of bone and soft tissue sarcoma: an essential organizational framework. J Multidiscip Healthc. 2015;8:109–15.
77. Amini B, Jessop AC, Ganeshan DM, Tseng WW, Madewell JE. Contemporary imaging of soft tissue sarcomas. J Surg Oncol. 2015;111(5):496–503.
78. Fenstermacher MJ. Imaging evaluation of patients with soft tissue sarcoma. Surg Oncol Clin N Am. 2003;12(2):305–32.
79. Knapp EL, Kransdorf MJ, Letson GD. Diagnostic imaging update: soft tissue sarcomas. Cancer Control. 2005;12(1):22–6.
80. Aga P, Singh R, Parihar A, Parashari U. Imaging spectrum in soft tissue sarcomas. Indian J Surg Oncol. 2011;2(4):271–9.
81. Caracciolo JT, Letson GD. Radiologic approach to bone and soft tissue sarcomas. Surg Clin North Am. 2016;96(5):963–76.
82. Mayerson JL, Scharschmidt TJ, Lewis VO, Morris CD. Diagnosis and management of soft-tissue masses. Instr Course Lect. 2015;64:95–103.
83. Fadul D, Fayad LM. Advanced modalities for the imaging of sarcoma. Surg Clin North Am. 2008;88(3):521–37, vi.
84. Chou SS, Hippe DS, Lee AY, Scherer K, Porrino JA, Davidson DJ, Chew FS, Ha AS. Gadolinium contrast enhancement improves confidence in diagnosing recurrent soft tissue sarcoma by MRI. Acad Radiol. 2017;pii:S1076-6332(17)30012-0.
85. Chan WP. Magnetic resonance imaging of soft-tissue tumors of the extremities: a practical approach. World J Radiol. 2013;5(12):455–9.
86. Kig DM, Hackbarth DA, Kilian CM, Carrera GF. Soft-tissue sarcoma metastases identified on abdomen and pelvis CT imaging. Clin Orthop Relat Res. 2009;467(11):2838–44.
87. Trieu J, Schlicht SM, Choong PF. Diagnosing musculoskeletal tumours: how accurate is CT-guided core needle biopsy? Eur J Surg Oncol. 2016;42(7):1049–56.
88. Benz MR, Tchekmedyian N, Eilber FC, Federman N, Czernin J, Tap WD. Utilization of positron emission tomography in the management of patients with sarcoma. Curr Opin Oncol. 2009;21(4):345–51.
89. Kubo T, Furuta T, Johan MP, Ochi M. Prognostic significance of (18)F-FDG PET at diagnosis in patients with soft tissue sarcoma and bone sarcoma; systematic review and meta-analysis. Eur J Cancer. 2016;58:104–11.

90. Strauss DC, Qureshi YA, Hayes AJ, Thway K, Fisher C, Thomas JM. The role of core needle biopsy in the diagnosis of suspected soft tissue tumours. J Surg Oncol. 2010;102(5):523–9.
91. Traina F, Errani C, Toscano A, Pungetti C, Fabbri D, Mazzotti A, Donati D, Faldini C. Current concepts in the biopsy of musculoskeletal tumors: AAOS exhibit selection. J Bone Joint Surg Am. 97(2):e7.
92. Mankin HJ, Mankin CJ, Simon MA. Members of the musculoskeletal tumor society: the hazards of the biopsy, revisited. J Bone Joint Surg Am. 1996;78(5):656–63.
93. Liu PT, Valadez SD, Chivers FS, Roberts CC, Beauchamp CP. Anatomically based guidelines for core needle biopsy of bone tumors: Implications for limb-sparing surgery. Radiographics. 2007;27(1):189–205, discussion 206.
94. Pohlig F, Kirchhoff C, Lenze U, Schauwecker J, Burgkart R, Rechl H, von Eisenhart-Rothe R. Percutaneous core needle biopsy versus open biopsy in diagnostics of bone and soft tissue sarcoma: a retrospective study. Eur J Med Res. 2012;17:29.
95. C S, Png W, Tan MH. Accuracy of core needle biopsy for musculoskeletal tumours. J Orthop Surg (Hong Kong). 2013;21(1):92–5.
96. Colletti SM, Tranesh GA, Whetsell CR, Chambers LN, Nassar A. High diagnostic accuracy of core needle biopsy of soft tissue tumors: an institutional experience. Diagn Cytopathol. 2016;44(4):291–8.
97. Trieu J, Schlicht SM, Oncol CPFEJS. Diagnosing musculoskeletal tumours: how accurate is CT-guided core needle biopsy? Eur J Surg Oncol. 2016;42(7):1049–56.
98. Coran A, Di Maggio A, Rastrelli M, Alberioli E, Attar S, Ortolan P, Bortolanza C, Tosi A, Montesco MC, Bezzon E, Rossi CR, Stramare R. Core needle biopsy of soft tissue tumors, CEUS vs US guided: a pilot study. J Ultrasound. 2015;18(4):335–42.
99. Shives TC. Biopsy of soft-tissue tumors. Clin Orthop Relat Res. 1993;289:32–5.
100. Dangoor A, Seddon B, Gerrand C, Grimer R, Whelan J, Judson I. UK guidelines for the management of soft tissue sarcomas. Clin Sarcoma Res. 2016;6:20. eCollection 2016
101. Randall RL, Bruckner JD, Papenhausen MD, Thurman T, Conrad EU III. Errors in diagnosis and margin determination of soft-tissue sarcomas initially treated at non-tertiary centers. Orthopedics. 2004;27(2):209–12.
102. Anastasiou I, Levis PK, Katafigiotis I, Karaolanis G, Palla VV, Felekouras E, Athanasiou A, Perdiki M, Mitropoulos D, Constantinides CA. High grade myofibroblastic sarcoma of paratesticular soft tissues. Case Rep Oncol Med. 2014;2014:Article ID 768379.
103. Radaelli S, Desai A, Hodson J, Colombo C, Roberts K, Gourevitch D, Gronchi A. Prognostic factors and outcome of spermatic cord sarcoma. Ann Surg Oncol. 2014;21:3557–63.
104. Pisters PW, Leung DH, Woodruff J, Shi W, Brennan MF. Analysis of prognostic factors in 1,041 patients with localized soft tissue sarcomas of the extremities. J Clin Oncol. 1996;14:1679–89.
105. Ballo MT, Zagars GK, Pisters PW, Feig BW, Patel SR, von Eschenbach AC. Spermatic cord sarcoma: outcome, patterns of failure and management. J Urol. 2001;166:1306–10.
106. Rodríguez D, Aria F. Olumi Management of spermatic cord tumors: a rare urologic malignancy. Ther Adv Urol. 2012;4(6):325–34.
107. Enoch S, Wharton SM, Murray DS. Management of leiomyosarcomas of the spermatic cord: the role of reconstructive surgery. World J Surg Oncol. 2005;3:23.
108. Dangle P, Basavaraj DR, Bhattarai S, Paul AB, Biyani CS. Leiomyosarcoma of the spermatic cord: case report and literature review. Can Urol Assoc J. 2007;1:55–8.
109. Balducci M, Di Rito A, Mantini G, Manfrida S, Chiesa S, Frascino V, D'Agostino GR, Valentini V, De Bari B. Role of radiotherapy in the treatment of fibrosarcoma of the spermatic cord: a case report and review of the literature. Tumori. 2011;97:e36–8.
110. Moroni M, Nesi G, Travaglini F, Rizzo M, Amorosi A, Dominici A. Rhabdomyosarcoma of the spermatic cord. A case report with review of the literature. Urol Int. 2003;71:114–7.
111. Coleman J, Brennan M, Alektiar K, et al. Adult spermatic cord sarcomas: management and results. Ann Surg Oncol. 2003;10:669–75.

112. Brooks AD, Bowne WB, Delgado R, et al. Soft tissue sarcoma of the groin: diagnosis, management and prognosis. J Am Coll Surg. 2001;193:130–6.
113. Fagundes MA, Zietman LA, Althausen AF. The management of spermatic cord sarcoma. Cancer. 1996;77:1873–6.
114. Kanso C, Roussel H, Zerbib M, Flam T, Debré B, Vieillefond A. Adult spermatic cord sarcoma: diagnosis and management. Prog Urol. 2011;21:53–8.
115. Stewart RJ, Martelli H, Oberlin O, Rey A, Bouvet N, Spicer RD, et al. Treatment of children with nonmetastatic paratesticular rhabdomyosarcoma: results of the malignant mesenchymal tumors studies (MMT 84 and MMT 89) of the International Society of Pediatric Oncology. J Clin Oncol. 2003;21:793–8.
116. Mizuno Y, Sumi Y, Nachi S, Ito Y, Marui T, Saji S, Matsutomo H. A case of a large retroperitoneal liposarcoma presenting as an incarcerated inguinal hernia. Hernia. 2006;10:439–42.
117. Bhandarwar AH, Bakhshi GD, Borisa AD, Shenoy SS, Kori CG, Vora S. Giant recurrent retroperitoneal liposarcoma presentino as a recurrent inguinal hernia. Clin Pract. 2011;1:e130.
118. Hirofumi N, Yoshio N, Minoru H, et al. Retroperitoneal liposarcoma presenting as an indirect inguinal hernia. Acta Med Okayama. 2001;55:51–4.
119. Bognár G, Bereczky B, Barabás L, Diczházi C, Ondrejka P. Recurrent retroperitoneal liposarcoma presenting as an incarcerated femoral hernia. Magy Seb. 2009;62(1):30–3.
120. Longbotham JH, Joyce RP. Retroperitoneal liposarcoma presenting as spermatic cord tumor. Urology. 1987;30:276–80.
121. Baldassarre E, Santacroce C, Barone M, Torino G, Siani A, Valenti G. Retroperitoneal well-differentiated liposarcoma presentino as an incarcerated inguinal hernia. G Chir. 2007;28(8/9):315–7.
122. Karakousis CP. Abdominoinguinal incision and other incisions in the resection of pelvic tumors. Surg Oncol. 2000;9(2):83–90.
123. Karakousis CP. The abdominoinguinal incision in limb salvage and resection of pelvic tumors. Cancer. 1984;54(11):2543–8.
124. Miyamoto S, Fujiki M, Nakatani F, Kobayashi E, Sakisaka M, Sakuraba M. Reconstruction of complex groin defects after sarcoma resection. Ann Plast Surg. 2017;78(4):443–7.
125. Radaelli S, Fiore M, Colombo C, Ford S, Palassini E, Sanfilippo R, Stacchiotti S, Sangalli C, Morosi C, Casali PG, Gronchi A. Vascular resection en-bloc with tumor removal and graft reconstruction is safe and effective in soft tissue sarcoma (STS) of the extremities and retroperitoneum. Surg Oncol. 2016;25:125e131.
126. Schwarzbach MH, Hormann Y, Hinz U, Bernd L, Willeke F, Mechtersheimer G, Böckler D, Schumacher H, Herfarth C, Büchler MW, Allenberg JR. Results of limb-sparing surgery with vascular replacement for soft tissue sarcoma in the lower extremity. J Vasc Surg. 2005;42(1):88–97.
127. Bonardelli S, Nadari F, Maffeis R, Ippolito V, Saccalani M, Lussardi L, et al. Limb salvage in lower-extremity sarcomas and technical details about vascular reconstruction. J Orthop Sci. 2000;5:555–60.
128. McGoldrick NP, Butler JS, Lavelle M, Sheehan S, Dudeney S, O'Toole GC. Resection and reconstruction of pelvic and extremity soft tissue sarcomas with major vascular involvement: Current concepts. World J Orthop. 2016;7(5):293–300.
129. Hohenberger P, Allenberg JR, Schlag PM, Reichardt P. Results of surgery and multimodal therapy for patients with soft tissue sarcoma invading to vascular structures. Cancer. 1999;85(2):396e408.
130. Ghert MA, Davis AM, Griffin AM, Alyami AH, White L, Kandel RA, Ferguson P, O'Sullivan B, Catton CN, Lindsay T, Rubin B, Bell RS, Wunder JS. The surgical and functional outcome of limb-salvage surgery with vascular reconstruction for soft tissue sarcoma of the extremity. Ann Surg Oncol. 12(12):1102–10.
131. Yfadopoulos D, Nikolopoulos D, Novi E, Psaroudakis A. Primary superficial femoral vein leiomyosarcoma: report of a case. Surg Today. 2011;41:1649–54.

132. Roland CL, Boland GM, Demicco EG, Lusby K, Ingram D, May CD, Kivlin CM, Watson K, Al Sannaa GA, Wang W-L, Ravi V, Pollock RE, Lev D, Cormier JN, Hunt KK, Feig BW, Lazar AJ, Torres KE. Primary vascular leiomyosarcoma: clinical observations and molecular variables. JAMA Surg. 2016;151(4):347–54.
133. Schwarzbach MHM, Hormann Y, Hinz U, Leowardi C, Böckler D, Mechtersheimer G, d Helmut F, Büchler MW, Allenberg J-R. Clinical results of surgery for retroperitoneal sarcoma with major blood vessel involvement. J Vasc Surg. 2006;44(1):46–55.
134. Emori M, Hamada K, Omori S, Joyama S, Tomita Y, Hashimoto N, Takami H, Naka N, Yoshikawa H, Araki N. Surgery with vascular reconstruction for soft-tissue sarcomas in the inguinal region: oncologic and functional outcomes. Ann Vasc Surg. 2012;26(5):693–9.
135. Covarelli P, Boselli C, Burini G, Parente B, Loschi D, Lenti M, Rulli A. Sarcoma of the thigh: radical excision with arterial reconstruction using the contralateral saphenous vein. Ann Surg Oncol. 2015;22:S334.
136. Poppler LH, Davidge K, Lu JC, Armstrong J, Fox IK, Mackinnon SE. Alternatives to sural nerve grafts in the upper extremity. Hand. 2015;10:68–75.
137. Melendez M, Brandt K, Evans GR. Sciatic nerve reconstruction: limb preservation after sarcoma resection. Ann Plast Surg. 2001;46(4):375–81.
138. Mackinnon SE. Surgical management of the peripheral nerve gap. Clin Plast Surg. 1989;16(3):587–603.
139. Kang S, Yoo HJ, Kim HS, Han I. Soft tissue sarcoma misdiagnosed as benign peripheral neurogenic tumor. J Orthop Sci. 2015;20(1):180–5.
140. Tsuchihara T, Nemoto K, Arino H, Amako M, Murakami H, Yoshizumi Y. Sural nerve grafting for long defects of the femoral nerve after resection of a retroperitoneal tumour. J Bone Joint Surg Br. 2008;90(8):1097–100.
141. Das Gupta TK, Brasfield RD. Solitary malignan-t Schwannoma. Ann Surg. 1970;171(3):419–28.
142. LoGiudice JA, Haberman K, Sanger JR. The anterolateral thigh flap for groin and lower abdominal defects: a better alternative to the rectus abdominis flap. Plast Reconstr Surg. 2014;133:162–8.
143. Kayano S, Sakuraba M, Miyamoto S, et al. Comparison of pedicled and free anterolateral thigh flaps for reconstruction of complex defects of the abdominal wall: review of 20 consecutive cases. J Plast Reconstr Aesthet Surg. 2012;65:1525–9.
144. Ninkovic M, Kronberger P, Harpf C, et al. Free innervated latissimus dorsi muscle flap for reconstruction of full-thickness abdominal wall defects. Plast Reconstr Surg. 1998;101:971–8.
145. Johannesmeyer D, Smith V, Cole DJ, Esnaola NF, Camp ER. The impact of lymph node disease in extremity soft-tissue sarcomas: a population-based analysis. Am J Surg. 2013;206:289–95.
146. Can Atalay Æ, Altinok M, Besim Seref Æ. The impact of lymph node metastases on survival in extremity. Soft tissue sarcomas. World J Surg. 2007;31:1433–7.
147. Behranwala KA, A'Hern R, Omar AM, Thomas JM. Prognosis of lymph node metastasis in soft tissue sarcoma. Ann Surg Oncol. 2004;11(7):714–9.
148. Sherman KL, Kinnier CV, Farina DA, Wayne JD, Laskin WB, Agulnik M, Attar S, Hayes JP, Peabody T, Peabody T. Examination of national lymph node evaluation practices for adult extremity soft tissue sarcoma. J Surg Oncol. 2014;110(6):682–8.
149. Mentzel T, Calonje E, Wadden C, Camplejohn RS, Beham A, Smith MA, Fletcher CD. Myxofibrosarcoma. Clinicopathologic analysis of 75 cases with emphasis on the low-grade variant. Am J Surg Pathol. 1996;20(4):391–405.
150. Wagner LM, Kremer N, Gelfand MJ, Sharp SE, Turpin BK, Nagarajan R, Tiao GM, Pressey JG, Yin J, Dasgupta R. Detection of lymph node metastases in pediatric and adolescent/young adult sarcoma: sentinel lymph node biopsy versus fludeoxyglucose positron emission tomography imaging—a prospective trial. Cancer. 2017;123(1):155–60.
151. Andreou D. Sentinel node biopsy in soft tissue sarcome. In: Andreou D, Tunn PU, editors. Treatment of bone and soft tissue sarcomas. New York, NY: Springer; 2009. p. 25–36.

152. Maduekwe UN, Hornicek FJ, Springfield DS, Raskin KA, Harmon DC, Choy E, Rosenberg AE, Nielsen GP, TF DL, Chen YL, Ott MJ, Yoon SS. Role of sentinel lymph node biopsy in the staging of synovial, epithelioid, and clear cell sarcomas. Ann Surg Oncol. 2009;16:1356–63.
153. Blazer DG III, Sabel MS, Sondak VK. Is there a role for sentinel lymph node biopsy in the management of sarcoma? Surg Oncol. 2003;12:201–6.
154. Ruka W, Emrich LJ, Driscoll DL, Karakousis CP. Prognostic significance of lymph node metastasis and bone, major vessel, or nerve involvement in adults with high-grade soft tissue sarcomas. Cancer. 1988;62:999–1006.
155. Ecker BL, Peters MG, McMillan MT, Sinnamon AJ, Zhang PJ, Kelz RR, Roses RE, Drebin JA, Fraker DL, Karakousis GC. Implications of lymph node evaluation in the management of resectable soft tissue sarcoma. Ann Surg Oncol. 2017;24(2):425–33.
156. Gaakeer HA, Albus-Lutter CE, Gortzak E, Zoetmulder FA. Regional lymph node metastases in patients with soft tissue sarcomas of the extremities, what are the therapeutic consequences? Eur J Surg Oncol. 1988;14(2):151–6.
157. Mazeron JJ, Suit HD. Lymph nodes as sites of metastases from sarcomas of soft tissue. Cancer. 1987;60:1800–8.
158. Al-Refaie WB, Andtbacka RH, Ensor J, et al. Lymphadenectomy for isolated lymph node metastasis from extremity soft-tissue sarcomas. Cancer. 2008;112:1821–6.

Chapter 11
Malignant Inguinal Adenopathy: Considerations for the Radiation Oncologist

Martin Korzeniowski and Juanita Crook

Abbreviations

3D-CRT	3D conformal radiotherapy
AP	Anterior-posterior
AP-PA	Parallel-opposed anteroposterior
Chemo-RT	Combined modality treatment
CR	Complete response
CT	Computed tomography
CTCAE	Common terminology criteria for adverse events
CTV	Clinical target volume
EORTC	European Organization for Research and Treatment of Cancer
FDG-PET	Fluorodeoxyglucose-positron emission tomography
GTV	Gross tumor volume
GOG	Gynecologic Oncology Group
GOG 88	Gynecologic Oncology Group protocol 88 trial
ILND	Inguinal lymph node dissection
IMRT	Intensity-modulated radiotherapy
InPACT	International Penile Advanced Cancer Trial (NCT 02305654)
LVI	Lymphovascular invasion
MRI	Magnetic resonance imaging
NCCN	National Comprehensive Cancer Network
PA	Posterior-anterior
PET-CT	Positron emission tomography–computed tomography
PLND	Pelvic lymph node dissection

M. Korzeniowski, M.D., F.R.C.P.C. • J. Crook, M.D., F.R.C.P.C. (✉)
Division of Radiotherapy and Developmental Radiotherapeutics, Department of Radiation Oncology, University of British Columbia, BC Cancer Agency, 399 Royal Avenue, Kelowna, V1Y5L3 British Columbia, Canada
e-mail: martin.korzeniowski@medportal.ca; JCrook@bccancer.bc.ca

K.A. Delman, V.A. Master (eds.), *Malignancies of the Groin*,
https://doi.org/10.1007/978-3-319-60858-7_11

PTV	Planning target volume
RT	Radiotherapy
RTOG	Radiation Therapy Oncology Group
SCC	Squamous cell carcinoma
SLNB	Sentinel lymph node biopsy
TROG	Trans Tasman Radiation Oncology Group
UKCCCR	UK Coordinating Committee on Cancer Research
VMAT	Volumetric arc therapy

Introduction

Primary malignancies involving structures of the lower pelvis and perineum, including the vulva, penis, and anal canal, frequently spread to the groin. For each of these malignancies, radiotherapy (RT) may play a role in definitive management, adjuvant treatment, or palliation. This has important implications in clinical decision-making for the radiation oncologist who must appreciate the prognostic factors affecting the risk of metastases, the indications for treating the inguinal region, the associated considerations in planning treatment, and the acute and late effects of treatment.

While there are differences in the natural history of vulvar, penile, anal, and other perineal malignancies, the most common histology for these sites is squamous cell carcinoma (SCC), which tends to follow a stepwise spread from the primary site to the groin nodal basin. First echelon nodes include the superficial inguino-femoral, second echelon the deep inguino-femoral, and third echelon the external iliac nodes. This chapter will discuss the clinical prognostic factors for metastatic spread to the groin, technical aspects of treatment planning, and radiation toxicities.

Risk of Groin Metastases and Indications for Radiotherapy

Anatomical Considerations

The Superficial and Deep Femoral Lymph Nodes

Many lower pelvic and skin malignancies spread to the groin. Defining this region is important for the radiation oncologist because of implications for field design, choice of technique, and sparing of neighboring organs at risk. The groin nodes include both the superficial inguinal and deeper femoral nodes. These nodal regions lie in the subcutaneous tissue that overlies the femoral triangle defined by the inguinal ligament superiorly, sartorius muscle inferolaterally, and the adductor longus medially. These nodes are divided into superficial and deep based on their location relative to the cribriform fascia [1].

Pathways of primary lymphatic spread to the groin begin with the superficial inguinal nodes around the fossa ovalis and the saphenous vein [1]. Secondary drainage occurs through the lymphatic channels traversing the cribriform fascia, culminating in lymph nodes surrounding the femoral vessels (especially the medial aspect of the femoral artery) at the level of the greater saphenous vein inferiorly [2]. Beyond this, lymphatic drainage occurs primarily to the external iliac lymph nodes within the pelvis.

Appreciating the principles of sequential lymphatic spread is paramount for understanding treatment paradigms. Any treatment strategies utilizing RT as part of adjuvant or primary treatment must encompass both the superficial and deep inguino-femoral lymph nodes.

Vulvar Cancer

The lymphatics of the vulva consist of a network spanning superiorly from the prepuce, through the labia minora, and inferiorly to the vulvar fourchette. Lymph node scintigraphy has demonstrated that radiolabeled tracers or colloid dyes injected into the vulva follow well-lateralized flow when the primary is >2 cm from midline. For lesions originating in the perineum, clitoris, or anterior labia minora, flow is often bilateral [3]. The deep femoral lymph nodes should be included in the radiation volume, as isolated metastases can occur in the absence of superficial inguinal involvement [4].

As primary vulvar cancer invades through the layers of the dermis into the underlying tissues, it gains access to a rich dermal lymphatic plexus. Tumors with a depth of invasion <1 mm have a very small risk of groin node involvement, 1% or less [5]. In the absence of lymphovascular invasion (LVI) or high-grade disease, such superficial tumors do not require formal inguinal staging or prophylactic radiation. The risk of inguinal metastases rises exponentially as invasion increases from 1–3, 3–5, and >5 mm, with metastases present in 8%, 27%, and 34%, respectively [6, 7]. Other factors prognostic for lymph node spread include poor differentiation, fixed or clinically suspicious nodes, and LVI [7]. Clinical exam is not reliable since occult metastatic disease is seen in up to 25% of patients. Therefore, any patient presenting with invasion >1 mm, high-grade tumor, LVI, or clinically positive nodes requires surgical staging to rule out metastatic involvement.

Surgical series have demonstrated that a well-lateralized primary will rarely spread to contralateral lymph nodes. In a series of patients treated with unilateral lymphadenectomy for stage I disease, contralateral groin failure occurred in <3% [8]. However, midline structures (such as the clitoris or medial labia) have redundant and bilateral drainage in up to two-thirds of cases [3]. Thus, surgical staging of bilateral inguinal regions is advisable. Likewise, patients with ipsilateral groin node involvement are at risk of contralateral metastases and may require further staging/treatment of the contralateral groin. Similarly, unilateral irradiation of the hemipelvis is not recommended.

In cases where surgical staging is not possible, diagnostic computed tomography (CT), positron emission tomography-computed tomography (PET-CT), or magnetic resonance imaging (MRI) may help guide clinical decision-making. Multidisciplinary consultation is recommended to weigh the choice between observation with radiation therapy reserved for salvage and up-front treatment, subject to the individual patient scenario.

Sentinel lymph node biopsy (SLNB) is increasingly available and is recommended as a first-line surgical staging for high-risk patients in the most recent National Comprehensive Cancer Network (NCCN) guidelines [9]. However, not all patients who are sentinel lymph node positive may undergo a subsequent therapeutic inguinal lymph node dissection (ILND) due to reasons of comorbidities or refusal. These patients may still be treated with definitive RT, preferably combined with chemotherapy, given the poor prognosis of untreated malignant adenopathy. The results of the GROINSS-V II study, a nonrandomized observational study of SLNB-positive patients according to management with adjuvant RT or chemo-RT, will be helpful.

Vulvar Cancer: The Pelvic Lymph Nodes

The next echelon for regional lymphatic spread is the deep pelvic lymph nodes, particularly the external iliac lymph nodes. The rate of pelvic lymph node involvement may be as high as 7.5–16% for those with positive inguinal lymph nodes [10]. Historically, pelvic lymph node dissection (PLND) was performed as the standard of care for patients with positive inguinal nodes, for those with midline lesions, or for those with invasion of deep structures [10]. However, patients with de novo pelvic lymph node involvement or who develop later pelvic recurrence all have more extensive groin node involvement (≥3 inguino-femoral nodes) [11]. Furthermore, the morbidity of PLND and the high risk of subsequent distant relapse (up to 66%) have changed clinical practice for gynecological surgeons such that PLND is now offered more selectively.

The Gynecologic Oncology Group (GOG protocol 37) randomly assigned 114 patients with positive groin nodes to PLND or RT to the pelvis and groin to 45–50.4 Gy [7]. A 2-year survival advantage was demonstrated for RT over surgery (68% vs. 54% $p = 0.02$). The survival benefit was attributed to a decrease in groin relapses (5.1% vs. 23.6%) in the radiotherapy arm. If more than one LN was involved, the difference in 2-year survival was even greater, being 63% versus 37%. For the 53 patients who underwent PLND, 15 had pelvic lymph node metastases, but only one patient experienced pelvic failure in the PLND arm, while four patients had pelvic recurrences in the radiotherapy arm (1.8% vs. 6.8%). This trend did not reach statistical significance due to the small numbers but suggested that surgical resection was better at preventing pelvic relapses. This study supports the routine use of adjuvant RT to the pelvis and groin for all patients with either clinically fixed inguinal nodes or ≥2 involved nodes. An update has demonstrated a persistent survival benefit at 6 years favoring the RT arm (HR, 0.61) [12].

Controversy remains regarding the treatment of the pelvis for patients with one positive groin node. The sample size of GOG 37 ($n = 114$) was too small to answer this question definitively. Other institutional series have reported equivalent outcomes, but were also subject to small sample size and short follow-up [13–15]. A larger and more contemporary retrospective series [16] has suggested that survival is improved with adjuvant RT, although the advantage may be limited to those with extracapsular extension [17]. The role of adjuvant RT to the pelvis and groins for patients with only a single positive groin node remains controversial.

Primary Anal Cancers

Like the dermal lymphatics of the vulva, the lymphatic drainage of the anal canal is rich with multiple pathways of drainage between the different levels of the anal canal and the rectum [9, 18, 19]. For true anal canal malignancies (distal to the dentate line), the primary lymph drainage is to the inguinal and femoral lymph nodes [20]. Above the dentate line, drainage occurs along the hemorrhoidal vessels to perirectal and internal iliac lymph nodes. Similar to gynecological malignancies of the lower pelvis, cancers of the anal canal are also characterized by early lymphatic spread with pelvic lymph node metastases in 25–35% of cases treated with surgical resection alone [21–23].

Involvement of the inguinal lymph nodes is dependent on the T-stage of the primary and varies between 20 and 60% depending on tumor size and extent of local invasion [22, 23]. Occult metastatic involvement is seen in 13% of clinically node-negative patients and reaches up to 30% for patients with large tumors (T3) or those invading local structures (T4) [21, 24]. Nodal status is an important prognostic factor for patients treated with RT. As was shown in the large randomized trial by the European Organization for Research and Treatment of Cancer (EORTC), patients who were node negative had higher rates of both overall survival and local control ($p = 0.0017$) [25]. On multivariate analysis, the number and size of lymph nodes did not impact either outcome.

Because many metastatic lymph nodes are below the threshold of clinical detection, CT and MR imaging are limited in their ability to rule out metastatic involvement. A recent systematic review and meta-analysis has suggested that the addition of fluorodeoxyglucose-positron emission tomography (FDG-PET) imaging resulted in nodal upstaging in 21% of patients (95% CI 13–30), altering the TNM stage in 41% [26]. Staging with FDG-PET scans is an important addition for radiation treatment planning since it alters clinical target volumes.

Penile Cancers

Penile cancer is an uncommon malignancy, representing less than 1% of male malignancies in the western world. However, lymph node status remains the most important prognostic factor for overall survival [27]. Like vulvar cancer, the lymphatic drainage from the penis follows a stepwise pattern, first involving the superficial and deep inguinal lymph nodes, and then the pelvis. Inguinal drainage is typically bilateral for penile cancers and may involve either side [28]. Single-photon emission computed tomography (SPECT) in penile cancer has shown primary drainage from the penis to occur in the superior and central inguinal zones [29]. Clinically palpable nodes are frequently encountered in penile cancer, but due to the prevalence of concurrent infection, only approximately 50% are malignant [30]. The former practice of a trial of antibiotics to assess regression is no longer recommended since this delays management of those with metastatic involvement.

Several factors predict the risk of lymph node involvement, including histology, T-stage, grade, and the presence of LVI. Lymph node involvement is very rare in verrucous carcinoma but increases to 30% in SCC and is most common in basaloid variants of SCC [27]. Histologic grade, as described by Broder, is also very important in predicting lymph node metastases, being 10–15% for grade 1 lesions, while grade 2 and 3 lesions demonstrate rates of metastases between 60 and 70% [31]. Likewise, this series has demonstrated the importance of T-stage for predicting lymph node risk with 5–10% of T1 patients having involved lymph nodes, while T2 and T3 tumors had 60–70% involvement. Finally, LVI is associated with a 60–80% risk of inguinal metastatic disease [27].

Historically, only about 20% of patients with clinically negative nodes had micrometastatic disease on ILND [32]. Nonetheless, patients at high risk of micrometastases, especially >T1B or >grade 2 tumors, should undergo surgical evaluation prior to definitive therapy (either SLNB or ILND). The challenges with predicting lymph node involvement de novo are that the pathological features necessary for decision-making are based on information typically attained after primary surgical resection; grade, depth of invasion, and LVI are less reliably assessed on biopsy material. This is particularly challenging for penile cancers where organ preservation using radiotherapy may be selected.

Guidelines developed by the European Association of Urology suggest that only Tis, TaG1, and T1G1 tumors without high-risk features should be monitored with surveillance of the inguinal lymph nodes. For patients presenting with clinically negative nodes, but high-risk features, systematic staging with CT, PET-CT, and fine-needle aspiration is recommended. In these high-risk patients, up-front prophylactic ILND or SLNB should be considered. Delayed "therapeutic" resection is associated with markedly inferior 10-year disease-free survival compared to patients who undergo immediate lymphadenectomy (30% vs. 71%, $p = 0.002$) [33]. In those high-risk patients who are not surgical candidates, treatment with RT to the inguinal regions and pelvis could be considered [34]. However, in the era of SLNB, the necessary pathologic assessment of lymph nodes can usually be obtained without the same risks of ILND.

Malignant Melanoma

Melanoma of the lower extremity or perineal region can present with inguinal adenopathy. Cutaneous melanoma is characterized by an aggressive phenotype, with early spread to regional lymph nodes and/or distant metastases. Overall, 15%

present with nodal metastases, and the risk of regional metastatic spread is related to tumor thickness, ulceration, and mitotic index. The risk is greater for tumors with >1 mm of invasion (T2) (25% vs. 5%). Therefore, all patients presenting with >1 mm invasion should have radiographic staging with CT and consideration of SLNB. The role of routine elective nodal dissection remains controversial as, unlike penile cancer, there has been no demonstrated overall survival benefit when compared to delayed therapeutic lymphadenectomy.

The risk of regional recurrence is dependent on the number of positive nodes, presence of extracapsular extension, and the number/site of involvement [35, 36]. The risk of regional recurrence within a nodal basin is moderate (9%) for a single positive lymph node but increases to 15% for 2–4 nodes and 17% for 5–10 nodes ($p < 0.001$). Other reviews have shown even higher failure rates, with recurrences ranging from 25 to 60%, depending on the number of lymph nodes involved [37]. As with vulvar cancers, recurrences almost double in the presence of extracapsular extension (28% vs. 15%). Similarly, nodal recurrences also depend on lymph node size (24% for <3 cm, 42% for 3–6 cm, and 80% for >6 cm).

Bibault et al. retrospectively reviewed 86 consecutive patients treated for locally advanced melanoma [38]. After lymph node dissection, 69% of the patients received adjuvant RT. The presence of extracapsular extension was a significant prognostic factor for regional relapses ($p = 0.019$). For patients with extracapsular extension, doses of radiation ≥50 Gy improved 5-year regional control (80% vs. 35% $p = 0.004$). This is an important factor in decision-making for the radiation oncologist, as not only is extracapsular extension an important adverse prognostic factor that requires adjuvant treatment, but it also mandates a dose of at least 50 Gy to achieve reasonable regional control.

Clinical practice guidelines have been established for adjuvant radiation therapy [39]. Extranodal extension is a strong indication for adjuvant treatment, and adjuvant radiation should be considered in the inguinal area if there are three or more involved lymph nodes or if any one lymph node exceeds 4 cm in size.

While adjuvant radiation therapy improves local and regional control for melanoma, it has not affected cause-specific survival or overall survival [40]. In the ANZMTG 01.02/TROG 02.01 trial, 123 patients were randomized to adjuvant RT (48 Gy in 20 fractions); relapses were reduced with adjuvant RT: 21% versus 36%, but there was no overall or cause-specific survival benefit. While this may be used as an argument against the use of adjuvant RT for melanoma, groin recurrences carry significant morbidity. With a multitude of targeted agents entering clinical practice, patients may be living longer with metastatic melanoma, suggesting an even greater role for locoregional control.

In a scenario where a patient with a high-risk primary (as indicated above) declines surgical evaluation of the inguinal lymph nodes or is deemed not to be a surgical candidate, staging of the groin with CT, and preferably PET-CT, is recommended. In the absence of clinical disease, elective nodal irradiation is not recommended. While two small studies have been published on the subject, the majority of evidence is limited to head and neck melanomas and should not be extrapolated to areas of the groin [41, 41].

Technical Aspects and Treatment Planning

Patient Setup and CT Simulation

The patient should be in the supine position for CT simulation for inguino-pelvic radiotherapy. As the required dose is generally greater than small bowel tolerance, treatment with a full bladder to displace small bowel out of the pelvis is recommended. Custom immobilization, such as a Vac-Lok bag (TM Civco Medical Solutions, Iowa), improves day-to-day reproducibility of setup. The use of delayed oral contrast will help delineate loops of small bowel. Simulation with an empty rectum also helps to ensure day-to-day reproducibility.

Palpable nodes and surgical scars from ILND can be marked clinically with radiopaque wire since the postsurgical bed may be underestimated using CT alone. Depending on the choice of treatment technique and patient body habitus, the use of bolus may be required to ensure appropriate coverage of the lymph node target volumes, especially the superficial inguinal lymph nodes in a thin patient.

Radiation Therapy Techniques

External Beam Radiotherapy

Pelvic external beam RT can irradiate multiple clinical targets (i.e., groins, pelvis, primary site) in one treatment plan. Coverage of the primary disease is individualized and based on the location. The focus of this section will be the coverage of the inguino-pelvic nodal drainage regions.

Treatment fields/techniques have evolved over time due to advances in imaging, image-guidance, and treatment delivery. Irradiation of the groin for patients with carcinoma of the vulva, anus, distal vagina, or other perineal sites presents certain technical challenges. In the Gynecologic Oncology Group protocol 88 trial (GOG 88), prophylactic radiation was compared to a bilateral inguinal lymph node dissection for clinically node-negative patients. The prescription point for radiotherapy patients was fixed to a depth of 3 cm for all patients and resulted in inadequate coverage of the inguinal node regions in many patients, especially those with higher body mass index [42]. Lessons learned from clinical trials like GOG 88 underline the importance of clearly delineated clinical target volumes and the selection of appropriate treatment techniques in order to avoid geographical misses and/or inadequate dose coverage.

Historically, treatment fields (and thus target delineation) were designed using fluoroscopic simulation based on bony landmarks. Treatment was based on simple parallel-opposed anteroposterior (AP-PA) beams or four-field techniques (AP-PA combined with right and left opposed laterals). Field sizes were chosen to cover the external iliac, the deep inguino femoral, and the internal and external iliac lymph nodes, sometimes including as high as the common iliac lymph nodes. Modifications on these basic approaches can include a wider anterior field to cover the inguinal region matched with a narrower posterior field limited to the pelvis and the addition of a photon or electron boost. As techniques have become more complex and delivery more accurate, using approaches such as 3D conformal radiotherapy [3D-CRT] or intensity-modulated radiotherapy [IMRT], delineation of clinical target volumes has become dependent on diagnostic imaging with CT and MRI [43–46]. Treatment planning techniques will be discussed here, starting with the most simple and increasing in complexity.

Wide AP-PA Technique

The simplest technique, from a treatment planning perspective, is that of a wide parallel-opposed beam arrangement using photons. First described by Perez et al., this technique uses equally sized anterior and posterior (opposed) fields [47]. The lateral border is placed just lateral to the greater trochanter to ensure full coverage of the groin [43]. The dose is preferentially weighted to the anterior fields by using lower energy photons (i.e., 6 MV) and by changing the depth of prescription normalization to ensure adequate coverage of the posterior aspect of the femoral vessels [43]. While simple and effective, this technique carries an increased risk of femoral head fractures [48] and has given way to the improved conformality and organ-sparing techniques such as 3D conformal radiotherapy or intensity-modulated radiotherapy.

Wide AP and Narrow PA Techniques

The simplest way to reduce dose to the femoral heads is to use an asymmetric field distribution, such that a wide anterior (AP) field is set to cover the pelvis and inguinal nodes, while a narrower posterior (PA) field includes just the pelvis, excluding the femoral heads [49]. The consequence is that the only contribution to the inguino-femoral target is that of the anterior field, which creates a dose gradient, risking underdose to the inguino-femoral vessels. To compensate for this, the dose to the inguino-femoral regions can be supplemented with a photon or electron field boost.

Photon Boost Techniques

The photon boost technique was first described in the pre-CT era and relied on clinical markup [50]. With modern planning techniques, the photon boost field may be placed either by matching at the anterior skin surface with the divergent edge of the posterior field or may be matched at the depth of the femoral vessels at the level of the mid-obturator foramen [43]. A *modified segmental boost technique* is favored using partially split right and left anterior oblique fields such that the medial surface of the partially/half beam-blocked groin fields matches the divergence of the posterior photon field. This is considered ideal for centers capable of treatment with a mono-isocentric technique [43].

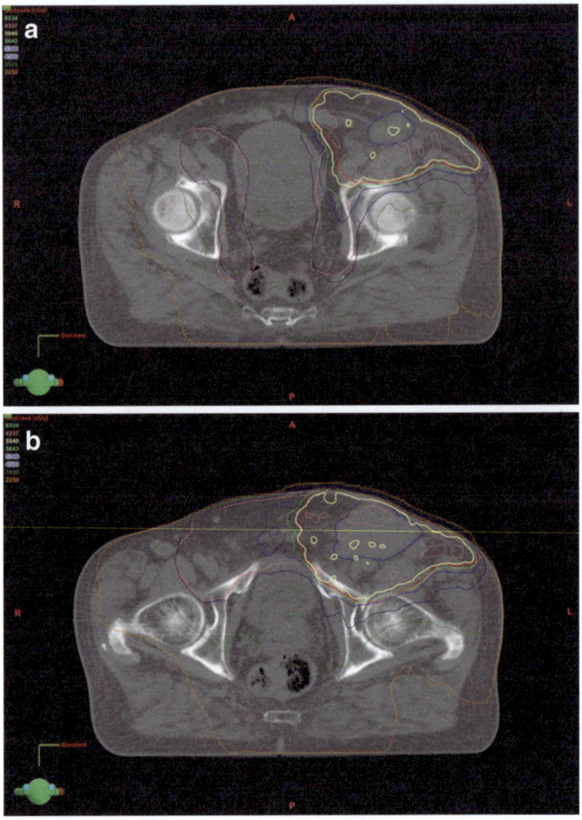

Fig. 11.1 (**a**, **b**) 4-field 3D-conformal pelvic plan with electron boost (20MeV) for a patient with penile cancer and involved inguinal node. Elective nodal PTV highlighted in *pink*, nodal boost PTV is highlighted in *red*. Prescription is 45 Gy to elective nodal PTV and 14.4 Gy to the nodal boost PTV. The 95% isodose (*green*) shows deficient coverage at a depth

Electron Field Boost

Supplementing the inguinal target volume with an electron field boost offers several advantages over photons. The rapid dose falloff of electrons allows adequate dose to the inguinal regions while sparing the femoral heads. Additionally, the bell-shaped isodose distribution of electrons deep to the skin surface may help with improving coverage at depth near the femoral vessels. While the femoral-sparing benefits of electron boosts are highly desirable, the correct choice of prescription point and beam energy is required in order to avoid missing part of the intended target at a depth. Furthermore, choosing a match point for electron fields can be difficult given the bell-shaped isodose distribution at depth. See Fig. 11.1a, b for illustrative dosimetry seen for a patient treated with electron boost. Please note the physical limitations of electrons for covering volumes that extend deep into the inguino-femoral triangle.

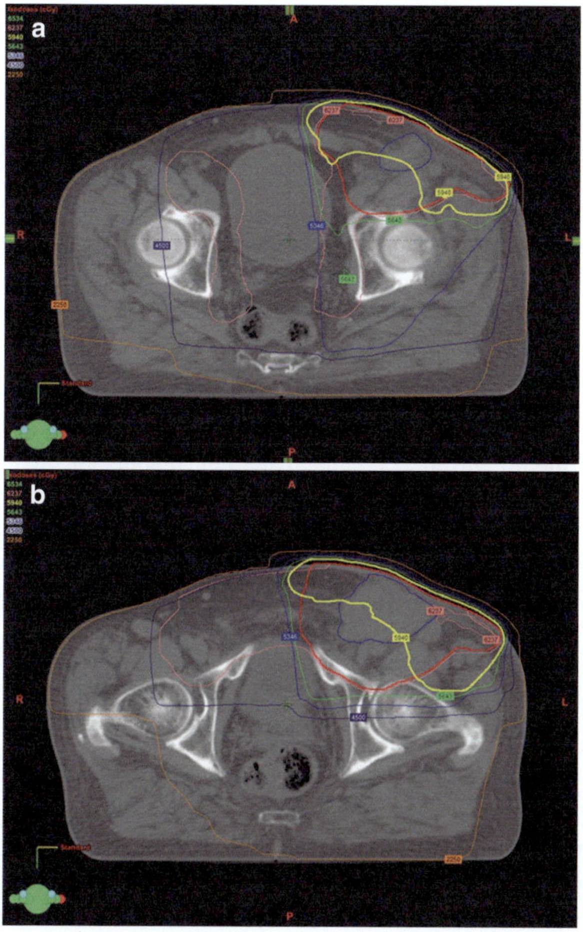

Fig. 11.2 (**a**, **b**) 4-field 3D-conformal plan for penile cancer with involved inguinal node. Elective nodal PTV shown in *pink* and boost volume PTV shown in *red*. Prescription is 45 Gy/25 fractions with a 14.4 Gy/8 nodal boost. Although the 100% isodose cuts through the middle of the boost target volume, it is now adequately covered by the 95% isodose. The disadvantage of this plan is that the entire pelvic contents receive 45 Gy

Four-Field Box and 3D Conformal Therapy

An alternative technique uses a 4-field or a 3D conformal plan. Here, opposing right and left lateral fields are added to create a box-like photon distribution. The shape of the pelvis and location of the inguino-femoral lymph node region lends itself well to this technique, as 4-field distributions offer a more homogenous distribution compared to AP-PA techniques. See Fig. 11.2a, b for illustrative dosimetry seen in 4-field 3D conformal distributions. Please note the improved coverage of the nodal planning target volume (PTV) boost.

Although the 100% prescription isodose cuts through the middle of the boost target volume, it is now adequately covered by the 95% isodose. Note that entire pelvic contents receiving 45 Gy.

Historical bony landmarks (superiorly at the L4–L5 interspace, inferiorly at the greater trochanter, laterally 1.5–2 cm beyond the pelvic brim, anteriorly at the pubic symphysis, and posteriorly at the S2–3 interspace) [51] have been replaced with 3D planning adapted to target volumes delineated using CT-based planning systems. While effective in covering the areas at risk, 4-field distributions still expose central pelvic organs to unnecessary radiation. In order to improve the therapeutic window by reducing toxicity, advanced techniques using IMRT have been developed.

Intensity-Modulated Radiotherapy (IMRT)

IMRT allows more conformal treatment through modulation of radiation beam intensity, providing increased degrees of freedom in spatial and temporal dimensions [52]. Using predefined organ and target volume constraints, an inverse-planning process creates a highly conformal treatment plan, reducing the dose to important organs at risk, such as the bowel, bladder, and rectum, without compromising the coverage of target volumes [46].

For IMRT (and 3D-CRT) clinical target definitions are used to guide the treatment planning process. Gross tumor volume (GTV) is defined clinically or radiographically and delineates residual gross disease. The clinical target volume (CTV) defines the nodal regions according to their accompanying vessels, the internal and external iliac, as well as common femoral vessels, with a 0.7–1 cm expansion (cropped for anatomical boundaries), to allow for microscopic disease. The anteromedial margin may need to be larger to encompass all lymph nodes, especially if using highly conformal techniques. The planning target volume (PTV) includes a 1 cm isotropic expansion on the CTV such that volumes extend superiorly to the L5–S1 interspace [45]. The additional margin is to accommodate treatment setup variation and beam penumbra. Please see Fig. 11.3a, b for illustrative dosimetry of a volumetric arc therapy (VMAT) plan. Note the highly conformal distribution and organ-at-risk sparing properties. Given these advantages, these techniques have become preferred at our center.

Dosimetric comparisons between plans generated with IMRT or 3D-CRT show a significant decrease in mean dose to the rectum, bladder, and small bowel by 41%, 26%, and 27%, respectively [45]. While there are some concerns about highly conformal treatments missing the intended target volumes by nature of tight margins, early clinical outcomes in management of vulvar cancers with IMRT appear favorable to historical comparisons [44]. If done well, highly conformal treatments like IMRT allow the safe delivery of higher doses of radiotherapy, without compromising tumor control.

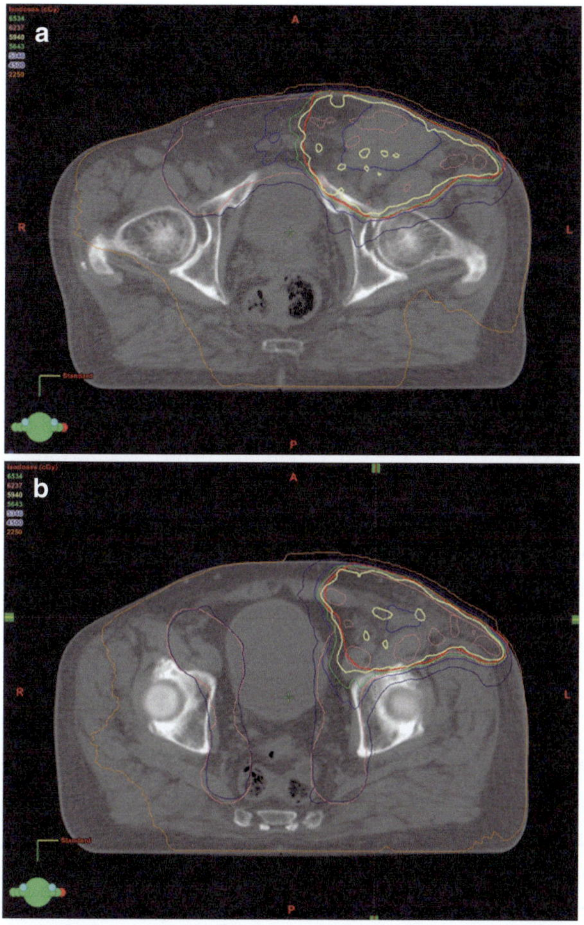

Fig. 11.3 (**a**, **b**) VMAT pelvic and nodal boost plan for penile cancer with involved inguinal node. Elective nodal PTV shown in *pink* and nodal boost PTV shown in *red*. Prescription is 45 Gy/25 to elective nodal PTV and 14.4 Gy/8 to the nodal boost PTV. The 100% isodose (*yellow*) now covers the boost volume perfectly. The bladder dose is reduced (<45 Gy)

Treatment Strategies and Applications of Radiotherapy

Radiation plays a variety of roles for the management of vulvar, anal, and penile squamous cell carcinomas and malignant melanoma of the skin that have spread to the groin lymph nodes. Definitive radiotherapy, either alone or combined with concurrent chemotherapy, can be an effective strategy for treating malignancies involving the inguinal nodes. Additionally, it may be used in a neoadjuvant manner to improve resectability, and finally, radiation may be used in an adjuvant setting, to help mitigate the risk of micrometastatic disease and decrease locoregional relapses. We will review the various treatment strategies in each of the discussed sites and the evidence supporting their use.

Vulvar Carcinoma

Vulvar Cancer: Definitive or Neoadjuvant Radiation Therapy

Vulvar cancer is treated primarily with surgical resection due to the significant morbidity and toxicity of radiation, which can be magnified with the use of radiosensitizing chemotherapy. However, patients with locally advanced lesions or unresectable lymph nodes may be converted to surgical candidates with neoadjuvant RT or chemo-RT. Furthermore, vulvar cancer afflicts older women who may never be surgical candidates because of age or comorbidities. Definitive chemo-RT may be used in these situations. Multiple studies have supported the role of chemo-RT as an alternative to surgery for the nonsurgical candidate or to help facilitate surgical resection [44, 53, 54].

While most of the evidence comes from phase II trials, the principles are supported by the multinational guidelines published by the NCCN [9]. In the GOG trial 101, 96 women with unresectable vulvar cancer were treated with split-course radiation to 47.6 Gy (two courses of 23.8 Gy) combined with cisplatin and 5-FU chemotherapy. This was followed by resection of the residual tumor and bilateral ILND. For the subset of women with N2 or N3 lymphadenopathy ($n = 41$), 95% became resectable, and of these, 41% achieved a complete pathological response [54].

The subsequent GOG trial 205 study eliminated the split course and adopted daily fractionated RT to 57.6 Gy (in 1.8 Gy daily) combined with weekly cisplatin [53]. Surgical resection was planned for residual disease, and clinical response was confirmed by biopsy. Interim results report that 64% of the 58 patients ($n = 37$) had a complete clinical response, of which 78% ($n = 29$) were confirmed histologically. The study continues in a second stage of accrual.

The pelvis and standard areas at risk require RT to a dose of 45–50.4 Gy in standard 1.8 Gy fractions. Gross primary disease and positive lymph nodes (or scenarios where significant extracapsular extension is present) require higher doses ranging from 59.4 to 64.8 Gy, depending on the extent of disease and whether concurrent chemotherapy is used. Concurrent weekly cisplatin at 40 mg/m^2, as per the approach for SCC cervix, has become the standard of care.

Vulvar Cancer: Adjuvant Radiation Therapy

The main indications for adjuvant radiation therapy in vulvar cancer are the presence of multiple involved inguinal nodes or any extracapsular extension. As discussed above, in GOG 37, 114 patients were randomly assigned to adjuvant pelvic and groin radiation (45–50 Gy, $n = 59$) or PLND ($n = 55$) after radical vulvectomy and ILND. At a median follow-up of 74 months, adjuvant RT reduced the incidence of inguinal recurrence from 24 to 5% and was associated with an improvement in the 6-year overall survival from 41 to 51%. This study has paved the way for modern practice in adjuvant inguinal and pelvic radiation for vulvar cancer [12].

The current recommendation is to treat the pelvis to 45–50.4 Gy in 1.8 Gy per fraction using either 3D-CRT or IMRT. Extracapsular extension or gross lymphadenopathy requires a higher dose, graduated according to the bulk of disease to maximize locoregional control. Consistent with recent guidelines, adjuvant radiotherapy to the inguinal and pelvic lymph node region is omitted for patients who are node negative based on lymphadenectomy or SLNB, even in the presence of high-risk features (LVI, deep invasion, or close/positive margins) [9].

Anal Canal Cancers

Definitive Chemoradiotherapy

As discussed previously, surgical series in the 1980s established the role of abdominoperineal resection for the definitive management of early and advanced anal SCC. While limited wide local excision can still be considered for early T1N0 cancers [21], surgical resection for locally advanced primary cancers requires abdominoperineal resection and permanent colostomy. Primary chemoradiotherapy was established in the 1980s and 1990s as definitive management in order to avoid the morbidity associated with permanent colostomy [55, 56]. Local control and complete response (CR) were reported from 71 to 93%. Abdominoperineal resection is now typically reserved for patients who fail primary chemo-RT.

Subsequently, two major European trials have demonstrated the superiority of combined modality treatment (chemo-RT) compared to RT alone. The UK Coordinating Committee on Cancer Research (UKCCCR) ACT 1 trial randomized 585 patients of any stage to RT alone (45 Gy to the pelvis with a 15–35 Gy boost to the primary) or to chemo-RT with 5-FU and mitomycin-C. Both complete response (39% vs. 30%, $p = 0.08$) and locoregional control (53.7% vs. 29.5%) were improved (HR 0.46, $p < 0.001$) [57, 58]. A similar local control benefit was seen in the EORTC trial where 110 patients were randomized between radiotherapy alone (total dose of 60–65 Gy) and radiotherapy combined with infusional 5-FU and bolus mitomycin-C [25]. At 5 years, locoregional control was improved favoring the combined modality treatment arm (68% vs. 50%, $p = 0.02$).

Recently the results of a phase II trial, Radiation Therapy Oncology Group (RTOG) 05-29, have further defined the treatment of anorectal cancers. Using concurrent 5-FU and a dose-painting IMRT technique, the primary tumor and inguinal and pelvic lymph node regions are treated to differential dose levels depending on their associated risk. Elective nodal regions are treated to 42–45 Gy in 28–30 fractions (for T2 N0 and T3-T4 N0 patients, respectively), while involved metastatic nodes are boosted to 50.4–54 Gy in 30 fractions depending on their size (<3 cm or >3 cm, respectively)

[59]. The NCCN guidelines recommend and support the use of multi-field treatment techniques and recommend using PET-CT for identifying pathologically involved lymph nodes [60].

Penile Cancer

The role of radiation in the management of penile cancer varies depending on the stage, indications for treatment, and patient-specific factors. For localized presentations, organ preservation should be considered and discussed with patients. In this capacity, patients may undergo radiation therapy (external or brachytherapy) to the primary tumor and, for high-risk primaries, undergo surgical staging of the inguinal groin nodes [34].

Penile Cancer: Definitive/Preoperative Management

Given the low incidence of penile cancer in western societies, large multicenter randomized controlled trials are currently lacking. InPACT (International Penile Advanced Cancer Trial (NCT 02305654)) will randomize node-positive patients to standard ILND or neoadjuvant chemotherapy or chemo-RT. Fundamental in this trial design was the extrapolation of treatment strategies from published literature on other SCCs with HPV etiology, particularly vulvar cancer.

Downstaging penile cancer with neoadjuvant chemotherapy using triple regimens consisting of paclitaxel, ifosfamide, and cisplatin yields response rates reaching 50% in patients with locally advanced presentations [61]. However, only 10% achieve a complete pathological response, and in another phase II series, only 38.5% of patients demonstrated an objective response with docetaxel, cisplatin, and 5-FU chemotherapy [62]. Lessons learned from gynecological malignancies and head and neck cancers suggest that SCC responds well to combined modality strategies with chemo-RT and should be further explored for squamous histology cancers of the penis.

For patients who are surgically fit but have unresectable disease at diagnosis, neoadjuvant chemo-RT strategies for downstaging have been described [63, 64]. In practice, select patients with locally advanced unresectable disease are considered for neoadjuvant chemo-RT strategies. In one of the largest series for chemo-RT for penile cancer, 26 patients were treated with cisplatin-based chemotherapy combined with a median radiation dose 49 Gy (range 18–70 Gy) [65]. Progression-free survival was only 6 months in the absence of surgical intervention. The small sample size, low radiation dose, and heterogeneity of the patient population make conclusions difficult in this study. However, the small number of patients per institution suggests that chemoradiation strategies remain underutilized and that further studies, such as InPACT, looking at the use of chemo-RT strategies in a systematic fashion, are warranted.

Melanoma

Malignant melanoma is primarily treated surgically. While some small institutional series advocate definitive radiation, this should only be considered in the context of medically inoperable patients [66, 67]. For the vast majority of situations, radiation therapy will be used in an adjuvant setting.

Adjuvant Radiation Therapy

The cornerstone trial by the Trans Tasman Radiation Oncology Group (TROG) defined high-risk melanoma (as applicable to the groin metastases) as ≥3 inguinal nodes, extranodal extension, or lymph node size ≥4 cm. The planned treatment volume included the dissected lymph node field and lymphadenectomy scar [40]. In this study, adjuvant RT improved locoregional control but did not affect overall survival or relapse-free survival. At 3 years, the cumulative incidence of lymph node relapse was 19% in the radiation group versus 31%. Extranodal spread was the only independent risk factor for infield relapse [HR 1.77; $p = 0.001$]. Patients were treated with 48 Gy in 20 fractions, which is now considered the standard dose and fractionation for adjuvant treatment.

While the TROG study has had the most impact on adjuvant radiotherapy, it is supported by other retrospective studies. Corry et al. reviewed 113 patients with regional node involvement. Forty-two patients had complete surgical resection of macroscopic disease and were treated with adjuvant radiotherapy to a median dose of 50 Gy. For these patients, ten were alive and failure-free (26%), while eight failed with nodal relapse (including three with in-transit metastases). Furthermore, in patients experiencing failure, more than half (52%) had distant relapse as their first site. The authors recommended adjuvant postoperative RT for proven nodal metastases at high risk of regional recurrence (multiple nodes, extracapsular extension, or recurrent nodal disease) [68].

Historically, melanoma has been considered radio resistant, and although hypofractionation is an accepted means of overcoming radioresistance, randomized studies have not confirmed an advantage to this approach for melanoma. In RTOG 83-05, 137 patients with measureable lesions were randomized to 32 Gy in four fractions or 50 Gy in 20 fractions, with no difference in the clinical response rate (23–24%, respectively) [69]. Nonetheless, a fraction size of 2.5 Gy or greater has been adopted as the standard of care.

Treatment Toxicity

The acute and late toxicities of radiation therapy to the groin and pelvis are ultimately related to the ability to spare organs at risk. The choice of treatment technique plays a major role, even in the face of unfavorable patient anatomy and tumor location. The National Cancer Institute has standardized reporting of adverse events in the Common Terminology Criteria for Adverse Events (CTCAE), currently in its fourth version, and is a commonly accepted means of grading treatment toxicity on a scale of 1–5. Grade 1 toxicities (mild) are asymptomatic and based on clinical observations that do not require intervention. Grade 2 (moderate) toxicity necessitates local or noninvasive interventions. Grade 3 (severe) toxicity is medically significant and requires hospitalization, while Grade 4 toxicity is life-threatening and Grade 5 toxicity is fatal [70, 71]. The typical structures that contribute to the risk of acute and late effects of inguinal/pelvic radiotherapy include the small bowel, rectum, bladder, urethra, vagina, skin, femoral heads, and bone marrow. Constraints have been developed within many protocols but vary depending on the definitions used and techniques chosen. Given the heterogeneity and dependence on technique, organ-at-risk tolerance should be individualized for the situation.

Lower-Extremity Lymphedema

Lymphedema of the lower extremity is a common complication that increases with multimodality treatment including surgical resection, chemotherapy, and radiation therapy. Surgical resection disrupts lymphatic pathways, while radiation induces fibrosis of smaller lymphatic channels, resulting in chronic swelling of the lower limb. This can be quite debilitating, reflected in changes in quality-of-life domain scores [71]. It is important to counsel patients on this risk and refer early for symptom management.

The Bladder and Urethra

Acutely, radiation therapy will cause denudation of the bladder mucosa, resulting in symptoms of urinary frequency, urgency, dysuria, but very rarely hematuria. Long-term complications are related to microvascular damage and the ensuing fibrotic changes that happen over months to years following treatment. These late changes are attributed to collagen deposition in the bladder wall [72, 73]. Late hematuria should be assessed by cystoscopy and treated with laser photocoagulation. More global dysfunction can include decreased contractility of the muscular wall or fibrosis causing decreased bladder capacity [73]. Obstructive symptoms include hesitancy, decreased stream, and increased urinary frequency secondary to incomplete emptying. To reduce the risk of toxicity, published constraints limit the dose to the bladder such that <50% of the bladder receives 35 Gy, <35% receives 40 Gy, and <5% receives 50 Gy [74]. Dose tolerance for the urethra is harder to define. Generally, the dose tolerance for a 5% risk of severe toxicity is accepted to be between 65 and 74 Gy but may be difficult to achieve when the primary site of malignancy is the penis or vulva [75]. High doses to the cauda equina may cause neurovascular damage to the afferent and efferent pathways leading to the bladder [76], but as the dose tolerance of peripheral nerves exceeds 60 Gy, this is not a common complication.

Small Bowel and the Rectum

The small bowel and rectum may be implicated in radiation to the groins and lower pelvis. The small bowel tolerance (D2 cc of 50 Gy) is easily exceeded given the high doses required for treating gross disease (up to 59.4–64 Gy), causing significant complications throughout treatment. Mucosal inflammation/denudation leads to malabsorption of fats, sugars, and vitamin B1 [77]. The acute symptoms of radiation enteritis include abdominal cramping, tenesmus, and significant watery diarrhea. Severe long-term complications from radiotherapy are rare and include bowel adhesions, stricture, obstruction, and necrosis, but chronic changes to bowel habits, including diarrhea, are much more common.

The proximity of the rectum to important nodal drainage areas within the pelvis places it at least partially within the high-dose region [78]. Given the high-dose region will vary in volume and location depending on the primary malignancy, dose constraints vary depending on the site of origin. For endometrial cancers, the rectal dose should be limited to ≤30 Gy for 60% of the rectal volume, but the dose limit is higher for cervical cancers, such that 85% of the rectal volume may receive ≤40 Gy. This is a pragmatic decision related to the prescription; higher doses are required for curative treatment of cervical cancer than for adjuvant treatment of endometrial cancer. In every case, the dose should be limited to as low as reasonably achievable. The symptoms of radiation proctitis include diarrhea, tenesmus, hemorrhoid flare-up, and bleeding. Long-term complications include hematochezia, decreased rectal compliance, and chronic diarrhea and, at the severe end of the spectrum, ulceration and fistula formation.

The risk of gastrointestinal side effects during and after treatment depends on many factors including preexisting comorbidities including diabetes, hypertension and a history of tobacco abuse, prior pelvic surgery, and coexisting inflammatory bowel disease. Treatment factors include the volume and length of bowel irradiated, dose and fractionation, and use of radiosensitizing chemotherapy.

Reproductive Organs

Radiation treatment has many effects on reproductive organs for males and females undergoing radiation treatment, both acutely and in the long-term. Although reproductive function is in the past for many of these patients, sexual function may be important. Acutely, radiation to the penis causes dysuria from urethral mucositis, penile edema, and possible secondary infection in necrotic tumor. Late effects may include superficial necrosis, soft tissue ulceration, urethral strictures, and hypopigmentation/dyschromia. If a significant length of the penile shaft is treated, fibrosis of the vascular channels in erectile tissue frequently results in impotence. Between the deleterious effects of the tumor, surgical intervention, and radiation treatment, many men experience feelings of loss of masculinity and identity [79].

Complications of treating the vulva and inguinal lymph nodes have similar side effects to treatment of the penis. Acutely, patients experience epilation of pubic hair and moist desquamation. This is worse with simple parallel-opposed beams than for 3D-CRT or IMRT techniques. Long-term complications include telangiectatic changes of the skin, atrophy, altered pigmentation, and dyschromia. Changes in the vagina including mucosal atrophy, vaginal shortening and/or narrowing, and loss of lubrication all result in dyspareunia. Many women will benefit from referral to an oncology sexual health clinic.

Younger males and females are at significant risk of developing infertility after radiotherapy. Both the sperm and ovaries are extremely sensitive to radiation, and small doses (2–3 and 10 Gy) can lead to permanent infertility. Fertility preservation should be discussed if appropriate. Preventative measures such as testicular shielding and ovarian transposition may also be considered prior to treatment.

Conclusions

Regardless of the site of origin, malignant inguinal adenopathy shares many common features. Features of the primary tumor that predict for inguinal node involvement are grade, depth of invasion, tumor size, and the presence of LVI. These features demand that the inguinal region be assessed surgically if possible or treated in an adjuvant fashion. The high-risk groin is defined as having multiple involved nodes, extranodal extension of disease, or lymph node diameter >4 cm. These features have implications for local inguinal recurrence, as well as signaling coexistent pelvic nodal metastases. Consequently, not only does the inguinal region require adjuvant treatment, if the pathologic status of the pelvic nodes is unknown, the pelvis should be included in the clinical target volume.

The techniques for inguinal radiotherapy have evolved such that treatment can now be provided while minimizing morbidity to the central pelvic organs; however, patients should still be counselled about the risks of leg edema.

As the majority of malignancies that metastasize to the inguinal region are of squamous origin, combination with weekly cisplatin chemotherapy should be considered, whether for neoadjuvant, adjuvant, or definitive management (Table 11.1).

Table 11.1 Predictive factors for malignant groin node involvement and recurrence

Site	Predictive factors for groin node involvement	Predictive factors for groin node recurrence/pelvic node involvement
Vulvar cancer	Depth of invasion >1 mm Poorly differentiated histology Lymphovascular invasion Clinically palpable nodes	>1 lymph node involved Extracapsular extension Clinically fixed lymph nodes
Anal cancer	T-stage Size and extent of local invasion	Groin node involvement
Penile cancer	Histology (SCC or basaloid variants) Histologic grade, lymphovascular invasion T-stage	>1 lymph node involved Extracapsular extension Clinically fixed lymph nodes
Malignant melanoma	Depth of invasion (tumor thickness) Mitotic index Ulceration	>4 lymph node involvement Extracapsular extension Lymph node size >3 cm, recurrent nodal disease

References

1. Coleman R. Anatomy and natural history. In: Patricia E, Levenback C, editors. Cancer of the female lower gential tract. 1st ed. Hamilton, ON: BC Decker Inc; 2001. p. 47–5.
2. Hacker, N.F., Eifel, P., McGuire, W., Wilkinson E.J. Vulva. In: Hoskins WJ, Perez CA, Young RC, editors. Principles and practice of gynecologic oncology Philadelphia, PA: J. B. Lippincott Co; 1992. p. 537–566.
3. Iversen T, Aas M. Lymph drainage from the vulva. Gynecol Oncol. 1983;16(2):179–89.
4. Chu J, Tamimi HK, Figge DC. Femoral node metastases with negative superficial inguinal nodes in early vulvar cancer. Am J Obstet Gynecol. 1981;140(3):337–9.
5. Kelley JL III, Burke TW, Tornos C, Morris M, Gershenson DM, Silva EG, et al. Minimally invasive vulvar carcinoma: an indication for conservative surgical therapy. Gynecol Oncol. 1992;44(3):240–4.
6. Hacker NF, Van der Velden J. Conservative management of early vulvar cancer. Cancer. 1993;71(4 Suppl):1673–7.
7. Homesley HD, Bundy BN, Sedlis A, Yordan E, Berek JS, Jahshan A, et al. Prognostic factors for groin node metastasis in squamous cell carcinoma of the vulva (a Gynecologic Oncology Group study). Gynecol Oncol. 1993;49(3):279–83.
8. Stehman FB, Bundy BN, Dvoretsky PM, Creasman WT. Early stage I carcinoma of the vulva treated with ipsilateral superficial inguinal lymphadenectomy and modified radical hemivulvectomy: a prospective study of the Gynecologic Oncology Group. Obstet Gynecol. 1992;79(4):490–7.
9. Greer BE, Koh WJ. New NCCN guidelines for vulvar cancer. J Natl Compr Cancer Netw. 2016;14(5 Suppl):656–8.
10. Di Saia PJ, Morrow CP, Townsend DE. Cancer of the vulva. Calif Med. 1973;118(6):13–8.
11. Hacker NF, Berek JS, Lagasse LD, Leuchter RS, Moore JG. Management of regional lymph nodes and their prognostic influence in vulvar cancer. Obstet Gynecol. 1983;61(4):408–12.
12. Kunos C, Simpkins F, Gibbons H, Tian C, Homesley H. Radiation therapy compared with pelvic node resection for node-positive vulvar cancer: a randomized controlled trial. Obstet Gynecol. 2009;114(3):537–46.

13. Origoni M, Sideri M, Garsia S, Carinelli SG, Ferrari AG. Prognostic value of pathological patterns of lymph node positivity in squamous cell carcinoma of the vulva stage III and IVA FIGO. Gynecol Oncol. 1992;45(3):313–6.
14. Burke TW, Levenback C, Coleman RL, Morris M, Silva EG, Gershenson DM. Surgical therapy of T1 and T2 vulvar carcinoma: further experience with radical wide excision and selective inguinal lymphadenectomy. Gynecol Oncol. 1995;57(2):215–20.
15. Mariani L, Lombardi A, Atlante M, Atlante G. Radiotherapy for vulvar carcinoma with positive inguinal nodes. Adjunctive treatment. J Reprod Med. 1993;38(6):429–36.
16. Parthasarathy A, Cheung MK, Osann K, Husain A, Teng NN, Berek JS, et al. The benefit of adjuvant radiation therapy in single-node-positive squamous cell vulvar carcinoma. Gynecol Oncol. 2006;103(3):1095–9.
17. Fons G, Groenen SM, Oonk MH, Ansink AC, van der Zee AG, Burger MP, et al. Adjuvant radiotherapy in patients with vulvar cancer and one intra capsular lymph node metastasis is not beneficial. Gynecol Oncol. 2009;114(2):343–5.
18. Frost DB, Richards PC, Montague ED, Giacco GG, Martin RG. Epidermoid cancer of the anorectum. Cancer. 1984;53(6):1285–93.
19. Pintor MP, Northover JM, Nicholls RJ. Squamous cell carcinoma of the anus at one hospital from 1948 to 1984. Br J Surg. 1989;76(8):806–10.
20. Paradis P, Douglass HO Jr, Holyoke ED. The clinical implications of a staging system for carcinoma of the anus. Surg Gynecol Obstet. 1975;141(3):411–6.
21. Boman BM, Moertel CG, O'Connell MJ, Scott M, Weiland LH, Beart RW, et al. Carcinoma of the anal canal. A clinical and pathologic study of 188 cases. Cancer. 1984;54(1):114–25.
22. Doggett SW, Green JP, Cantril ST. Efficacy of radiation therapy alone for limited squamous cell carcinoma of the anal canal. Int J Radiat Oncol Biol Phys. 1988;15(5):1069–72.
23. Tanum G. Treatment of relapsing anal carcinoma. Acta Oncol. 1993;32(1):33–5.
24. Ortholan C, Resbeut M, Hannoun-Levi JM, Teissier E, Gerard JP, Ronchin P, et al. Anal canal cancer: management of inguinal nodes and benefit of prophylactic inguinal irradiation (CORS-03 Study). Int J Radiat Oncol Biol Phys. 2012;82(5):1988–95.
25. Bartelink H, Roelofsen F, Eschwege F, Rougier P, Bosset JF, Gonzalez DG, et al. Concomitant radiotherapy and chemotherapy is superior to radiotherapy alone in the treatment of locally advanced anal cancer: results of a phase III randomized trial of the European Organization for Research and Treatment of Cancer Radiotherapy and Gastrointestinal Cooperative Groups. J Clin Oncol. 1997;15(5):2040–9.
26. Jones M, Hruby G, Solomon M, Rutherford N, Martin J. The role of FDG-PET in the initial staging and response assessment of anal cancer: a systematic review and meta-analysis. Ann Surg Oncol. 2015;22(11):3574–81.
27. Ficarra V, Akduman B, Bouchot O, Palou J, Tobias-Machado M. Prognostic factors in penile cancer. Urology. 2010;76(2 Suppl 1):S66–73.
28. Pow-Sang MR, Ferreira U, Pow-Sang JM, Nardi AC, Destefano V. Epidemiology and natural history of penile cancer. Urology. 2010;76(2 Suppl 1):S2–6.
29. Leijte JA, Valdes Olmos RA, Nieweg OE, Horenblas S. Anatomical mapping of lymphatic drainage in penile carcinoma with SPECT-CT: implications for the extent of inguinal lymph node dissection. Eur Urol. 2008;54(4):885–90.
30. Heyns CF, Fleshner N, Sangar V, Schlenker B, Yuvaraja TB, van Poppel H. Management of the lymph nodes in penile cancer. Urology. 2010;76(2 Suppl 1):S43–57.
31. Solsona E, Iborra I, Rubio J, Casanova JL, Ricos JV, Calabuig C. Prospective validation of the association of local tumor stage and grade as a predictive factor for occult lymph node micrometastasis in patients with penile carcinoma and clinically negative inguinal lymph nodes. J Urol. 2001;165(5):1506–9.
32. Parra RO. Accurate staging of carcinoma of the penis in men with nonpalpable inguinal lymph nodes by modified inguinal lymphadenectomy. J Urol. 1996;155(2):560–3.

33. Ornellas AA, Kinchin EW, Nobrega BL, Wisnescky A, Koifman N, Quirino R. Surgical treatment of invasive squamous cell carcinoma of the penis: Brazilian National Cancer Institute long-term experience. J Surg Oncol. 2008;97(6):487–95.
34. Crook J. The role of radiotherapy in the management of penile cancer. Curr Probl Cancer. 2016;39(3):158–65.
35. Oxenberg J, Kane JM III. The role of radiation therapy in melanoma. Surg Clin North Am. 2014;94(5):1031–47.
36. Shi W. Role for radiation therapy in melanoma. Surg Oncol Clin N Am. 2015;24(2):323–35.
37. Calabro A, Singletary SE, Balch CM. Patterns of relapse in 1001 consecutive patients with melanoma nodal metastases. Arch Surg. 1989;124(9):1051–5.
38. Bibault JE, Dewas S, Mirabel X, Mortier L, Penel N, Vanseymortier L, et al. Adjuvant radiation therapy in metastatic lymph nodes from melanoma. Radiat Oncol. 2011;6:12-717X-6-12.
39. Coit DG, Thompson JA, Algazi A, Andtbacka R, Bichakjian CK, Carson WE III, et al. Melanoma, version 2.2016, NCCN clinical practice guidelines in oncology. J Natl Compr Cancer Netw. 2016;14(4):450–73.
40. Henderson MA, Burmeister BH, Ainslie J, Fisher R, Di Iulio J, Smithers BM, et al. Adjuvant lymph-node field radiotherapy versus observation only in patients with melanoma at high risk of further lymph-node field relapse after lymphadenectomy (ANZMTG 01.02/TROG 02.01): 6-year follow-up of a phase 3, randomised controlled trial. Lancet Oncol. 2015;16(9):1049–60.
41. Bonnen MD, Ballo MT, Myers JN, Garden AS, Diaz EM Jr, Gershenwald JE, et al. Elective radiotherapy provides regional control for patients with cutaneous melanoma of the head and neck. Cancer. 2004;100(2):383–9.
42. Stehman FB, Bundy BN, Thomas G, Varia M, Okagaki T, Roberts J, et al. Groin dissection versus groin radiation in carcinoma of the vulva: a Gynecologic Oncology Group study. Int J Radiat Oncol Biol Phys. 1992;24(2):389–96.
43. Gilroy JS, Amdur RJ, Louis DA, Li JG, Mendenhall WM. Irradiating the groin nodes without breaking a leg: a comparison of techniques for groin node irradiation. Med Dosim. 2004;29(4):258–64.
44. Beriwal S, Shukla G, Shinde A, Heron DE, Kelley JL, Edwards RP, et al. Preoperative intensity modulated radiation therapy and chemotherapy for locally advanced vulvar carcinoma: analysis of pattern of relapse. Int J Radiat Oncol Biol Phys. 2013;85(5):1269–74.
45. Beriwal S, Heron DE, Kim H, King G, Shogan J, Bahri S, et al. Intensity-modulated radiotherapy for the treatment of vulvar carcinoma: a comparative dosimetric study with early clinical outcome. Int J Radiat Oncol Biol Phys. 2006;64(5):1395–400.
46. Heron DE, Gerszten K, Selvaraj RN, King GC, Sonnik D, Gallion H, et al. Conventional 3D conformal versus intensity-modulated radiotherapy for the adjuvant treatment of gynecologic malignancies: a comparative dosimetric study of dose-volume histograms. Gynecol Oncol. 2003;91(1):39–45.
47. Perez CA, Grigsby PW, Galakatos A, Swanson R, Camel HM, Kao MS, et al. Radiation therapy in management of carcinoma of the vulva with emphasis on conservation therapy. Cancer 1993;7148.:3707–3716.
48. Ma L, Chang W, Lau-Chin M, Tate EM, Boyer AL. Using static MLC fields to replace partial transmission cerrobend blocks in treatment planning of rectal carcinoma cases. Med Dosim. 1998;23(4):264–6.
49. Wang CJ, Chin YY, Leung SW, Chen HC, Sun LM, Fang FM. Topographic distribution of inguinal lymph nodes metastasis: significance in determination of treatment margin for elective inguinal lymph nodes irradiation of low pelvic tumors. Int J Radiat Oncol Biol Phys. 1996;35(1):133–6.
50. Petereit DG, Mehta MP, Buchler DA, Kinsella TJ. A retrospective review of nodal treatment for vulvar cancer. Am J Clin Oncol. 1993;16(1):38–42.
51. Chao KS, Williamson JF, Grigsby PW, Perez CA. Uterosacral space involvement in locally advanced carcinoma of the uterine cervix. Int J Radiat Oncol Biol Phys. 1998;40(2):397–403.

52. Leibel SA, Fuks Z, Zelefsky MJ, Wolden SL, Rosenzweig KE, Alektiar KM, et al. Intensity-modulated radiotherapy. Cancer J. 2002;8(2):164–76.
53. Moore DH, Ali S, Koh WJ, Michael H, Barnes MN, McCourt CK, et al. A phase II trial of radiation therapy and weekly cisplatin chemotherapy for the treatment of locally-advanced squamous cell carcinoma of the vulva: a gynecologic oncology group study. Gynecol Oncol. 2012;124(3):529–33.
54. Montana GS, Thomas GM, Moore DH, Saxer A, Mangan CE, Lentz SS, et al. Preoperative chemo-radiation for carcinoma of the vulva with N2/N3 nodes: a gynecologic oncology group study. Int J Radiat Oncol Biol Phys. 2000;48(4):1007–13.
55. Cummings BJ, Keane TJ, O'Sullivan B, Wong CS, Catton CN. Epidermoid anal cancer: treatment by radiation alone or by radiation and 5-fluorouracil with and without mitomycin C. Int J Radiat Oncol Biol Phys. 1991;21(5):1115–25.
56. Nigro ND, Seydel HG, Considine B, Vaitkevicius VK, Leichman L, Kinzie JJ. Combined preoperative radiation and chemotherapy for squamous cell carcinoma of the anal canal. Cancer 1983;5157.:1826–1829.
57. Epidermoid anal cancer: results from the UKCCCR randomised trial of radiotherapy alone versus radiotherapy, 5-fluorouracil, and mitomycin. UKCCCR Anal Cancer Trial Working Party. UK Co-ordinating Committee on Cancer Research. Lancet 1996;348(9034):1049–54.
58. Northover J, Glynne-Jones R, Sebag-Montefiore D, James R, Meadows H, Wan S, et al. Chemoradiation for the treatment of epidermoid anal cancer: 13-year follow-up of the first randomised UKCCCR Anal Cancer Trial (ACT I). Br J Cancer. 2010;102(7):1123–8.
59. Kachnic LA, Winter K, Myerson RJ, Goodyear MD, Willins J, Esthappan J, et al. RTOG 0529: a phase 2 evaluation of dose-painted intensity modulated radiation therapy in combination with 5-fluorouracil and mitomycin-C for the reduction of acute morbidity in carcinoma of the anal canal. Int J Radiat Oncol Biol Phys. 2013;86(1):27–33.
60. Benson AB III, Arnoletti JP, Bekaii-Saab T, Chan E, Chen YJ, Choti MA, et al. Anal carcinoma, version 2.2012: featured updates to the NCCN guidelines. J Natl Compr Cancer Netw. 2012;10(4):449–54.
61. Pagliaro LC, Williams DL, Daliani D, Williams MB, Osai W, Kincaid M, et al. Neoadjuvant paclitaxel, ifosfamide, and cisplatin chemotherapy for metastatic penile cancer: a phase II study. J Clin Oncol. 2010;28(24):3851–7.
62. Nicholson S, Hall E, Harland SJ, Chester JD, Pickering L, Barber J, et al. Phase II trial of docetaxel, cisplatin and 5FU chemotherapy in locally advanced and metastatic penis cancer (CRUK/09/001). Br J Cancer. 2013;109(10):2554–9.
63. Eliason M, Bowen G, Bowen A, Hazard L, Samlowski W. Primary treatment of verrucous carcinoma of the penis with fluorouracil, cis-diamino-dichloro-platinum, and radiation therapy. Arch Dermatol. 2009;145(8):950–2.
64. Pedrick TJ, Wheeler W, Riemenschneider H. Combined modality therapy for locally advanced penile squamous cell carcinoma. Am J Clin Oncol. 1993;16(6):501–5.
65. Pond GR, Milowsky MI, Kolinsky MP, Eigl BJ, Necchi A, Harshman LC, et al. Concurrent chemoradiotherapy for men with locally advanced penile squamous cell carcinoma. Clin Genitourin Cancer. 2014;12(6):440–6.
66. Farshad A, Burg G, Panizzon R, Dummer R. A retrospective study of 150 patients with lentigo maligna and lentigo maligna melanoma and the efficacy of radiotherapy using Grenz or soft X-rays. Br J Dermatol. 2002;146(6):1042–6.
67. Harwood AR. Conventional fractionated radiotherapy for 51 patients with lentigo maligna and lentigo maligna melanoma. Int J Radiat Oncol Biol Phys. 1983;9(7):1019–21.
68. Corry J, Smith JG, Bishop M, Ainslie J. Nodal radiation therapy for metastatic melanoma. Int J Radiat Oncol Biol Phys. 1999;44(5):1065–9.
69. Sause WT, Cooper JS, Rush S, Ago CT, Cosmatos D, Coughlin CT, et al. Fraction size in external beam radiation therapy in the treatment of melanoma. Int J Radiat Oncol Biol Phys. 1991;20(3):429–32.

70. National Cancer Institute. Common terminology criteria for adverse events (CTCAE) version 4.0. 2009.
71. Yost KJ, Cheville AL, Al-Hilli MM, Mariani A, Barrette BA, McGree ME, et al. Lymphedema after surgery for endometrial cancer: prevalence, risk factors, and quality of life. Obstet Gynecol. 2014;124(2 Pt 1):307–15.
72. Mangar SA, Foo K, Norman A, Khoo V, Shahidi M, Dearnaley DP, et al. Evaluating the effect of reducing the high-dose volume on the toxicity of radiotherapy in the treatment of bladder cancer. Clin Oncol (R Coll Radiol). 2006;18(6):466–73.
73. Ritchey M, Ferrer F, Shearer P, Spunt SL. Late effects on the urinary bladder in patients treated for cancer in childhood: a report from the Children's Oncology Group. Pediatr Blood Cancer. 2009;52(4):439–46.
74. Radiation Therapy Oncology Group. RTOG 0529: a phase II evaluation of dose-painted IMRT in combination with 5-fluorouracil and mitomycin-C for reduction of acute morbidity in carcinoma of the anal canal;s 2011. https://www.rtog.org/ClinicalTrials/ProtocolTable/StudyDetails.aspx?action=openFile&FileID=4641. Accessed 13 Aug 2016.
75. Vainshtein J, Abu-Isa E, Olson KB, Ray ME, Sandler HM, Normolle D, et al. Randomized phase II trial of urethral sparing intensity modulated radiation therapy in low-risk prostate cancer: implications for focal therapy. Radiat Oncol. 2012;7:82-717X-7-82.
76. Pieters RS, Niemierko A, Fullerton BC, Munzenrider JE. Cauda equina tolerance to high-dose fractionated irradiation. Int J Radiat Oncol Biol Phys. 2006;64(1):251–7.
77. PDQ Supportive and Palliative Care Editorial Board. Gastrointestinal complications (PDQ(R)): health professional version. PDQ cancer information summaries. Bethesda, MD: PDQ Supportive and Palliative Care Editorial Board; 2002.
78. von der Maase H, Overgaard J. Microcolony survival assay for jejunal crypt cells exposed to radiation alone and combined with cancer chemotherapeutic agents—methodological problems. Int J Radiat Biol Relat Stud Phys Chem Med. 1983;43(1):45–56.
79. Bullen K, Edwards S, Marke V, Matthews S. Looking past the obvious: experiences of altered masculinity in penile cancer. Psychooncology. 2010;19(9):933–40.

Chapter 12
Surgical Technique for Open Inguinal Lymphadenectomy

Andrew J. Spillane and John F. Thompson

Anatomy

The femoral triangle is defined by the sartorius muscle laterally and the adductor longus muscle medially. These two muscles meet at the apex of the femoral triangle inferiorly. The inguinal ligament defines the superior extent of the femoral triangle. In the operation of open inguinal lymphadenectomy, the fatty and lymphatic contents of the femoral triangle are cleared, commencing superiorly on the lower abdominal musculature 5–7 cm above the inguinal ligament and extending inferiorly to the apex of the femoral triangle. The major vascular structures in the femoral triangle are all preserved during the operation, including the common femoral artery (CFA), superficial femoral artery (SFA), and profunda femoral artery (PFA) and the common femoral vein (CFV), superficial femoral vein (SFV), and profunda femoris vein (PFV). However, the long saphenous vein (LSV) is removed as part of the operative specimen after ligation at the saphenofemoral junction (SFJ) and again at the inferior extent of the dissection.

Further structures that act as landmarks in the dissection include:

1. The lateral cutaneous nerve of the thigh as it emerges from under the inguinal ligament laterally and courses obliquely across the upper sartorius muscle
2. The superficial external pudendal artery, which is usually located just inferior to the saphenofemoral venous junction (not in the above figure)
3. The cutaneous branches of the superficial branch of the femoral nerve as they course inferolaterally across the sartorius muscle further down the thigh and the saphenous nerve that courses toward the saphenous vein in the lower medial thigh
4. The femoral canal with its contents including the lymph node of Cloquet medial to the femoral vein as it passes behind the inguinal ligament to become the external iliac vein

A.J. Spillane, M.D., F.R.A.C.S. • J.F. Thompson, M.D., F.R.A.C.S., F.A.C.S. (✉)
Melanoma Institute Australia, The University of Sydney, Sydney, NSW, Australia
e-mail: John.Thompson@melanoma.org.au

K.A. Delman, V.A. Master (eds.), *Malignancies of the Groin*,
https://doi.org/10.1007/978-3-319-60858-7_12

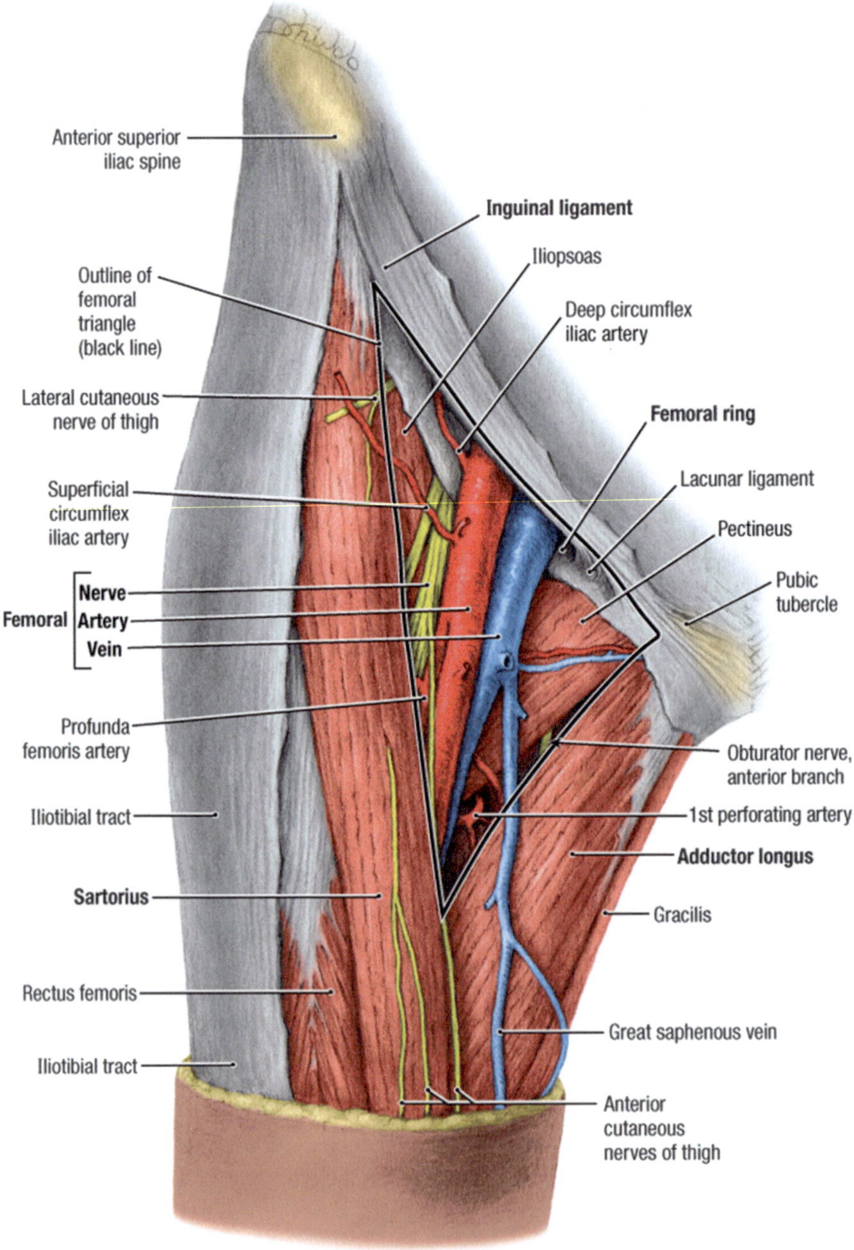

Indications

The most common indication for open inguinal lymphadenectomy is for the regional control of malignant solid tumors when metastatic spread to groin lymph nodes has occurred. In Australia, metastatic melanoma is the most common indication, followed by metastatic squamous cell carcinoma. Less common indications include metastatic Merkel cell carcinoma, metastatic sarcoma, and metastasis from rare skin adnexal tumors. The relative frequency of primary tumor pathologies differs in countries that have a lower incidence of melanoma and other sun exposure-related skin cancers.

In the past, the usual method of detection of metastatic melanoma in inguinal lymph nodes was by clinical or radiological assessment, with confirmation by fine-needle cytology or core biopsy. Some centers performed elective lymph node dissection for intermediate- and high-risk melanoma patients, although the great majority of patients did not have metastatic nodal disease when the operative specimens were examined. Nowadays the diagnosis is most commonly made after detection of micrometastatic disease by sentinel lymph node biopsy (SLNB).

Extent of Lymphadenectomy

In cases with apparently isolated malignancy in the inguinofemoral region, there is ongoing controversy as to whether the appropriate extent of surgery is inguinal lymphadenectomy alone or ilioinguinal lymphadenectomy, with clearance of iliac and obturator lymph nodes. It is unclear whether there are significant advantages to adding a pelvic dissection to the superficial inguinal lymphadenectomy or for that matter whether adding pelvic dissection adds any significant morbidity [1–3]. In an international survey, when 191 melanoma surgeons were asked what operation they would do for a positive inguinal SLNB, the replies were almost evenly split: 1/3 inguinal dissection, 1/3 ilioinguinal dissection, and 1/3 operation dependent on the specific circumstances [4]. For melanoma patients, there is currently a randomized controlled trial pilot study evaluating the question of extent of surgery [5].

Preoperative Staging

From the cancer perspective, it is common to perform full preoperative metastatic staging with whole-body PET/CT or CT chest, abdomen, and pelvis as well as CT or MRI of the brain.

There are circumstances where inguinal lymphadenectomy may still be recommended even when low-volume distant metastatic disease is identified. These include situations where the multidisciplinary team is worried about the likelihood of disease progressing in the inguinal area, with loss of regional control. In melanoma patients with low-volume distant metastatic disease, this is less likely now, with the availability of effective systemic therapies, which are often used first, and inguinal or ilioinguinal lymphadenectomy reserved for regional salvage if it is necessary.

Consideration may also be given to suitability for enrollment in any neoadjuvant or adjuvant therapy trials that are open at the institution.

Preoperative Assessment

General anesthesia is required. The preoperative workup is tailored to be appropriate for the patient's preexisting medical comorbidities. Open inguinal lymphadenectomy is not deep body cavity surgery, and there is a very low risk of major bleeding. Postoperative pain is usually not at a high level. The major early complications are poor wound healing and the development of a seroma. As far as feasible, it is desirable to mitigate risks by stopping smoking and optimizing the management of diabetes and any other systemic disease.

Surgical Technique

Positioning

The patient is positioned supine on the operating table. Most surgeons externally rotate the hip and flex the knee, placing the foot at the level of the opposite mid-calf area to improve access to the groin. However, this is not essential and the operation can also be done without this maneuver. Preoperatively surface marking with an operative marking pen of the location of the known disease and extent of dissection as well as the planned incision is logical.

Incisions

There are several suitable surgical incisions. These include single long incisions (straight or curvilinear or sigmoid), but it is useful to remove an ellipse of skin to reduce the risk of skin edge necrosis and reduce the laxity of the tissues after closure (Fig. 12.1, Panel a). Separate minimal access incisions sited above and below the inguinal ligament may also be used [6].

Procedure

Standard "thin" skin flaps are raised. They are usually around 5 mm thick but thinner if there is bulky disease. Superiorly, the flaps are raised for at least 5 cm above the inguinal ligament, laterally to the line of the anterior superior iliac spine and medially to the line of the pubic tubercle.

Below the inguinal ligament, thin skin flaps are similarly raised for the extent of the femoral triangle described above.

Superiorly at the upper limit of the dissection on the lower abdominal wall, the fatty tissue is then incised down to the abdominal musculature. The fatty flap is mobilized inferiorly, exposing the external oblique aponeurosis down to the inguinal ligament. The medial extent of this dissection is in line with the pubic tubercle and laterally with the anterior superior iliac spine.

The tissue plane of dissection changes here. As you dissect onto the investing fascia and femoral sheath coverings at the inferior extent of the inguinal ligament in the upper thigh, care has to be taken to identify and protect the common femoral vessels (Fig. 12.1, Panel b). Laterally, just below the inguinal ligament, the deep fascia over the sartorius muscle is incised along the line of its longitudinal fibers to reveal the underlying muscle. The lateral cutaneous nerve of the thigh has to be protected when dividing this fascia.

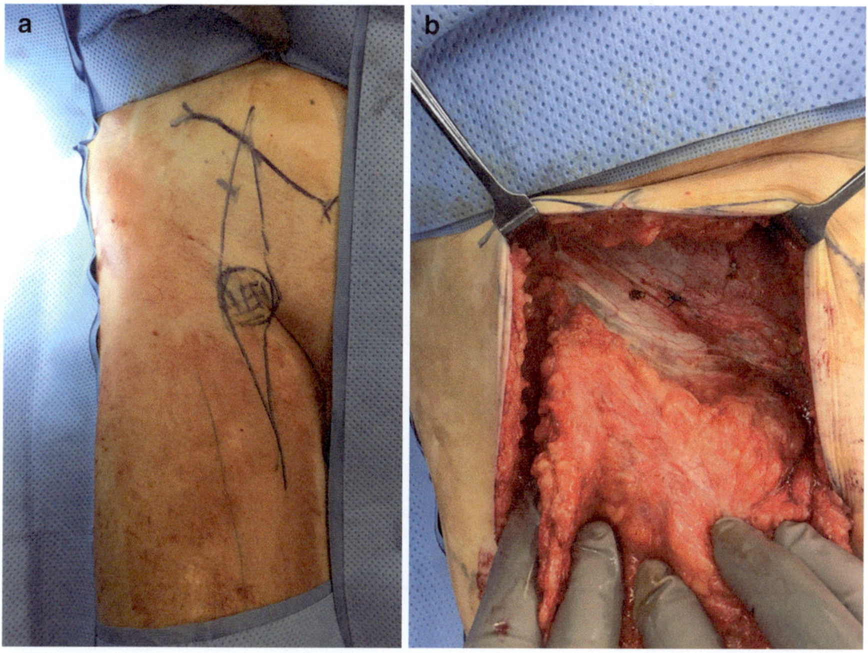

Fig. 12.1 Operative photos for open inguinal lymphadenectomy. Panel (**a**): skin markings for incision. Inguinal ligament and site of known metastatic lymph node marked. Panel (**b**): fatty tissue mobilized off lower abdominal wall to level of inguinal ligament. Care should be taken here when dissecting onto the femoral vessels. Panel (**c**): dissection defined laterally along the line of the sartorius muscle. The tributaries of the long saphenous vein and cutaneous branches of the femoral nerve are shown crossing the muscle. Panel (**d**): once the lateral extent of dissection is defined (see Panel **c**), the en bloc dissection starts at the level of the inguinal ligament dissecting in the fascial plane from the medial edge of the sartorius muscle across the common femoral artery avoiding the femoral nerve in the deeper plane. Panel (**e**): dissection exposes the common and superficial femoral arteries and then starts to expose the common femoral vein. The cutaneous branches of the femoral nerve that are not near the tumor can be preserved and are seen crossing the sartorius muscle inferolaterally. Panel (**f**): a vessel loop sling is around the saphenofemoral junction. The tissue up to the neck of the femoral canal is dissected in front of the pectineus muscle, but the femoral canal was ablated by earlier hernia surgery. Panel (**g**): the dissection proceeds caudally along the front of the superficial femoral artery. The fascia over the adductor longus muscle medially can be removed to give a clean plane of dissection. Panel (**h**): the caudal extent of the dissection is the apex of the femoral triangle where the long saphenous vein is divided again. Panel (**i**): because the single long ellipse incision leaves the femoral vessels exposed in the base of the dissection, the sartorius muscle is reflected after dividing its origin from the anterior superior iliac spine. The segmental neurovascular supply can be seen entering the muscle. It will be sutured to the inguinal ligament to cover the vessels

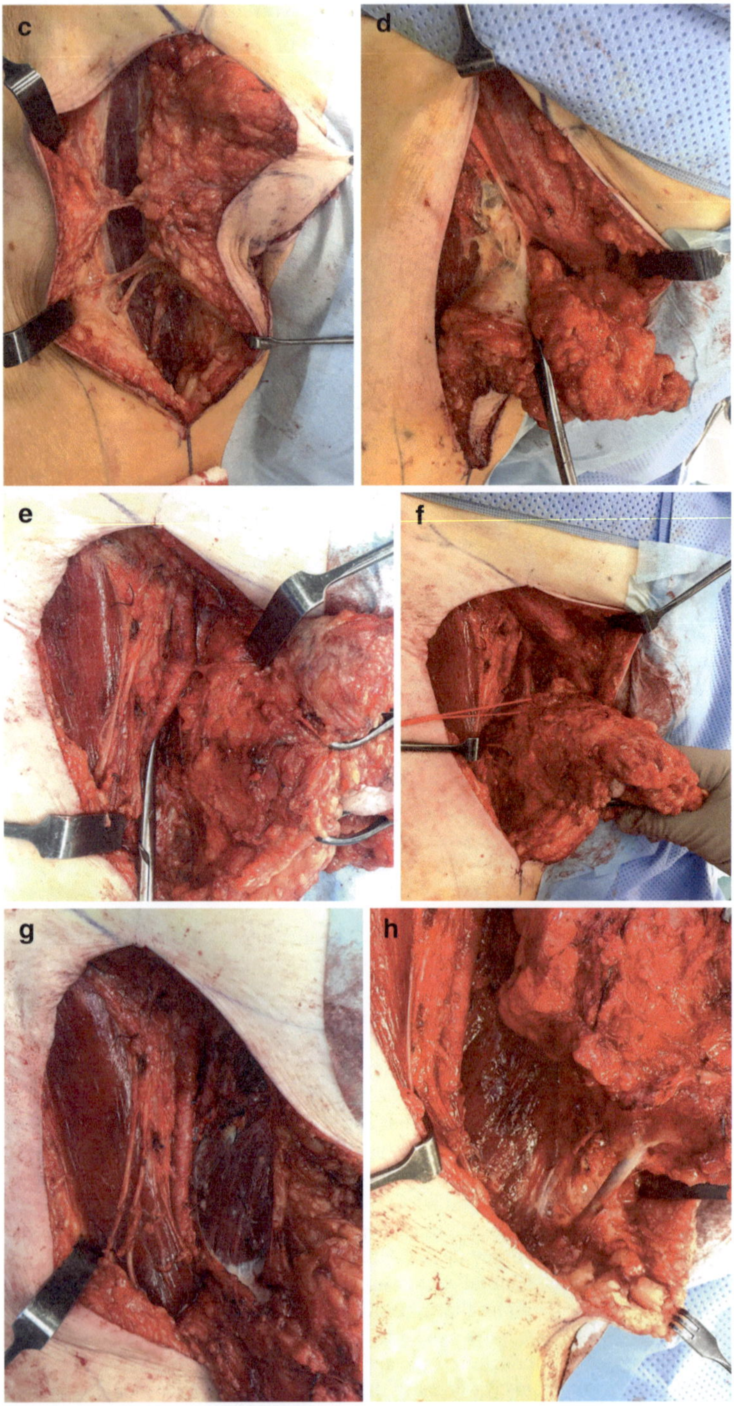

Fig. 12.1 (continued)

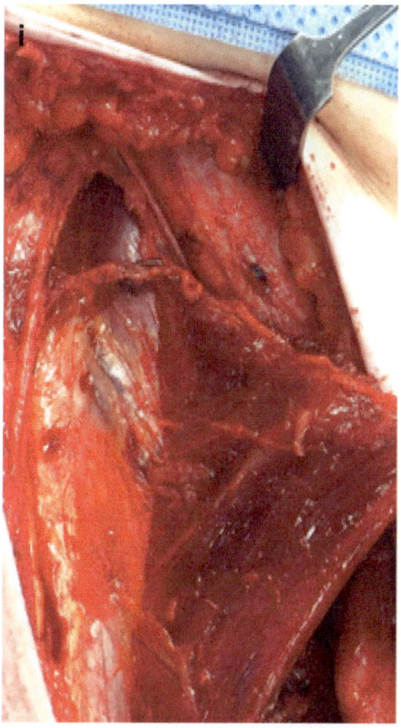

Fig. 12.1 (continued)

The lateral extent of the dissection is then extended inferiorly so that the deep fascia over the sartorius muscle is divided along its length from the anterior superior iliac spine along the lateral edge of the sartorius muscle to the apex of the femoral triangle. Several cutaneous branches of the femoral nerve are encountered passing inferolaterally across or sometimes through the sartorius muscle. These can either be preserved or sacrificed. In addition, there are usually several tributaries of the LSV that must be secured and divided (Fig. 12.1, Panel c).

Having freed up the lateral extent of the dissection, as above, the surgeon can then start from the medial edge of the sartorius muscle at the level of the inguinal ligament, bringing the en bloc specimen in a medial direction (Fig. 12.1, Panel d). From the medial edge of the sartorius muscle, the plane of dissection is direct to the lateral edge of the CFA and SFA. As the femoral nerve is in a deeper plane, there is no need to display the nerve, and the only encounter with the nerve is with its cutaneous branches described above (Fig. 12.1, Panel e). The fascial covering over the femoral vessels formed by the femoral sheath is incised to get onto the adventitial plane of the vessels proper (Fig. 12.1, Panel e).

The dissection can then proceed, starting superiorly just below the inguinal ligament from the lateral edge of the CFA and SFA, then onto their anterior, and then the anteromedial surfaces of these vessels. This exposes the CFA and SFA and in the process mobilizes the contents of the femoral triangle off these vessels in a medial direction (Fig. 12.1, Panel f). The superficial external pudendal artery and a further medial branch of the SFA slightly more distally are both usually seen coming from the medial aspect of the artery (Fig. 12.1, Panel e). The superficial external pudendal artery is usually sited just below the SFJ. The profunda femoral artery is not usually exposed in the dissection. As the CFA and SFA are exposed and the dissection extends more medially, the SFV, CFV, and the LSV and its saphenofemoral junction are encountered (Fig. 12.1, Panel f). Again, taking the fascial plane over the veins and dissecting onto the adventitial wall of the veins allows the surgeon to in turn expose the anterolateral, anterior, and then medial aspects of the CFV and SFV. When the SFJ is clearly delineated, with display of the CFV above and SFV below, the SFJ can be doubly ligated and divided. Alternatively, a stitch-tie may be used. At the medial edge of the CFV at its superior extent, as it passes under the inguinal ligament, the location of the femoral canal and slightly more inferomedially the adductor longus tendon becomes apparent. At this stage, the femoral canal can be fully explored to remove the lymph node of Cloquet, or if there is no disease in this area, then this can be done separately, later. If that is the case, the tissues entering this canal can be divided flush and the remaining dissection continued in a more inferior direction (Fig. 12.1, Panel f).

The medial extent of the dissection is the tendon of the adductor longus muscle superiorly and more inferiorly the line of the adductor longus muscle and part of the adductor brevis muscle until they cross deep to the sartorius muscle. Having defined the medial extent of dissection, the en bloc contents of the femoral triangle can be mobilized inferiorly along the superficial femoral vessels exposing them fully while bringing the specimen inferiorly (Fig. 12.1, Panel g). Note that the CFV and SFV are medial to the SFA just below the inguinal ligament but the SFV winds posteriorly in the thigh and is not displayed when it becomes posterior to the SFA. The fascia over the adductor longus and adductor brevis muscles, along with a few minor related vessels, is often removed with the specimen as it provides a clean plane of dissection at the lower medial extent (Fig. 12.1, Panel g).

The inferior limit of the dissection is the apex of the femoral triangle. This is defined by the crossing of the adductor longus behind the sartorius muscle. It is also the point where the lower aspect of the dissection identifies the LSV again. It is then doubly ligated and divided (Fig. 12.1, Panel h).

If not already done, the femoral canal is explored separately, removing its contents, usually including the lymph node of Cloquet. The femoral canal is then obliterated with nylon or PDS sutures.

Two large suction drains are commonly used. One can be inserted into the cavity laterally, extending to its upper limit, and one from distally beyond the lowest extent of the incision with the drain extending up to the level of the inguinal ligament. Deep tissues are closed with absorbable sutures, and the skin is closed with staples or sutures.

Technical Variations

1. If the wound has the femoral vessels in the base, as always occurs with a single long incision, the upper portion of the sartorius muscle can be freed from its origin from the anterior superior iliac spine and reflected from its lateral aspect

to pivot on its medial edge where the segmental neurovascular supply of the muscle enters. The upper edge of the muscle is then sutured to the lower aspect of the inguinal ligament. This sartorius flap provides coverage of the femoral vessels, which will protect them if skin wound breakdown occurs (Fig. 12.1, Panel i).

2. It is possible to spare the long saphenous vein. This requires that it be dissected out of the en bloc specimen and is not advised in cases with macroscopic residual disease in the area around the LSV. The suggested benefit is that it assists with lowering the rate of lymphedema, but this is substantiated by only low-level evidence [7].

Postoperative Care

In the past, patients have been kept on bed rest for several days or even for up to a week. However, a recent non-randomized assessment of various time periods to mobilization postoperatively did not indicate a significant difference in complication rates across a range of 1–≥10 days bed rest postoperatively [8].

In the authors' practices, earlier mobilization is used, and patients are discharged from the hospital with one or two drains still in situ after 2–7 days. The time of discharge depends on their physical capability, residential situation, residential distance from medical support, and the availability of home nursing support. The duration of hospital stay varies widely from country to country and is often determined by hospital bed availability and cost considerations.

The patient may require some ambulatory aid initially. There may be benefit for review and advice from a physiotherapist and lymphatic therapist.

Antithrombotic prevention treatment with TED Stockings™ and daily subcutaneous enoxaparin 20–40 mg subcutaneous daily or heparin 5000 u subcutaneous twice daily are recommended. This commences several hours preoperatively and continues until discharge.

Usually there is perioperative coverage with intravenous antibiotics such as ceftriaxone or cephalothin. There is no proven role for using antibiotics beyond the immediate perioperative period.

The drains are usually removed when the daily volume is <30–40 mL for 2 days in a row.

Quality Assurance

The surgeon knows the extent of the procedure that has been performed, but being able to validate this to multidisciplinary colleagues and for the purposes of clinical trials is important.

Options include:

1. Lymph node retrieval numbers. Recommended node retrieval numbers for inguinal lymphadenectomy ranges are 5 [9] or 6 [10]. In an analysis of a large prospective database documenting 105 inguinal lymphadenectomies, the median retrieval was 11 nodes with an interquartile range of 10–14, and 90% of the time there were 8 or more lymph nodes retrieved [11]. Melanoma Institute Australia sets its quality threshold at 8, and if fewer than eight lymph nodes are retrieved, then the pathologist is questioned about the thoroughness of the specimen cut up. If there are repeatedly low counts from the same surgeon, the quality of the surgery must be examined [12].
2. Operation report analysis.
3. Photographic evidence of the dissection.

Complications and Management of Complications

The complications can be categorized as short term, medium term, and long term.

Short-term complications: These are mainly those related to cardiorespiratory complications of general anesthesia, bleeding and thrombotic complications, wound healing issues, and issues with drain malfunction.

Medium-term complications: These typically relate to wound healing issues with or without seroma, infection, and wound edge necrosis. The most fragile part of the wound is that part crossing the groin crease. Incisions which avoid this area can moderate the severity of wound complications [6].

V.A.C.® (Acelity, USA) dressings can be useful if there is wound breakdown. Leaving the drains in place until lymphatic drainage is minimal is the best way to avoid many of these issues.

Late complications: Lymphedema, scarring and fibrosis in the groin area, and neural symptoms are the main long-term issues. These can lead to chronic mobility restriction or pain. Scrotal or vulval edema can be an issue when bilateral inguinal dissection is performed, but this problem often improves with time.

References

1. Spillane AJ, Haydu L, et al. Quality assurance parameters and predictors of outcome for ilio-inguinal and inguinal dissection in a contemporary melanoma patient population. Ann Surg Oncol. 2011;18(9):2521–8.
2. Allan CP, Hayes AJ, et al. Ilioinguinal lymph node dissection for palpable metastatic melanoma to the groin. ANZ J Surg. 2008;78(11):982–6.
3. Faries MB, Thompson JF, et al. The impact on morbidity and length of stay of early versus delayed complete lymphadenectomy in melanoma: results of the multicenter selective lymphadenectomy trial (I). Ann Surg Oncol. 2010;17(12):3324–9.

4. Pasquali S, Spillane AJ, de Wilt JH, McCaffery K, Rossi CR, Quinn MJ, Saw RP, Shannon KF, Stretch JR, Thompson JF. Surgeons' opinions on lymphadenectomy in melanoma patients with positive sentinel nodes: a worldwide web-based survey. Ann Surg Oncol. 2012;19:4322–9.
5. https://clinicaltrials.gov/ct2/results?term=EAGLE+FM&Search=Search
6. Spillane AJ, Tucker M, Pasquali S. A pilot study reporting outcomes for melanoma patients of a minimal access ilio-inguinal dissection technique based on two incisions. Ann Surg Oncol. 2011;18:970.
7. Zhang SH, Sood AK, Sorosky JI, Anderson B, Buller RE, Preservation of the saphenous vein during inguinal lymphadenectomy decreases morbidity in patients with carcinoma of the vulva. Cancer. 2000;89(7):1520–5.
8. Faut M, Heidema RM, Hoekstra HJ, van Ginkel RJ, SLB B, Kruijff S, van Leeuwen BL. Morbidity after inguinal lymph node dissections: it is time for a change. Ann Surg Oncol. doi:10.1245/s10434-016-5461-3.
9. Balch CM, Durant JR, Bartolucci AA, The Southeastern Cancer Study Group. The impact of surgical quality control in multi-institutional group trials involving adjuvant cancer treatments. Ann Surg. 1983;198(2):164–7.
10. https://www.clinicaltrials.gov/ct2/show/NCT00297895. Accessed 14 Sept 2016.
11. Spillane AJ, Haydu L, McMillan W, Stretch J, Thompson JF. Quality assurance parameters and predictors of outcome for ilio-inguinal and inguinal dissection in a contemporary melanoma patient population. Ann Surg Oncol. 2011 Sep;18(9):2521–8.
12. Read R, Pasquali S, Haydu L, Thompson JF, Stretch JR, Saw R, Quinn MJ, Shannon K, Spillane AJ. Quality assurance in melanoma surgery: the evolving experience at a large tertiary referral centre. EJSO. 2015;41(7):830–6. http://dx.doi.org/10.1016/j.ejso.2014.12.003

Chapter 13
Surgical Technique for Minimally Invasive Inguinal Lymphadenectomy

Clara Farley, Keith A. Delman, and Viraj A. Master

Abbreviation

VIL Videoscopic inguinal lymphadenectomy

Introduction

Complete inguinal lymphadenectomy remains the standard of care for patients with nodal metastases to the groin from most malignancies [1]. While surgical resection by means of traditional open inguinal lymphadenectomy, detailed elsewhere in this book, has been associated with improvements in survival [2], complication rates may be as high as 50% with most of the morbidity related to the large incision required for adequate exposure (Table 13.1) [3, 8, 10, 12, 14, 15]. These complications can include dehiscence (Fig. 13.1), infection, seroma formation, and/or skin flap necrosis. This high complication rate prevented many surgeons and medical oncologists from recommending the procedure [16, 17]. As a result, alternative means of resection

C. Farley, M.D. (✉)
Department of Surgery, Emory University, Suite H100, 1364 Clifton Road NE, Atlanta, GA 30322, USA
e-mail: crfarle@emory.edu

K.A. Delman, M.D., F.A.C.S.
Division of Surgical Oncology, Department of Surgery, Emory University School of Medicine, 1364 Clifton Road NE, Room H127, Atlanta, GA 30322, USA
e-mail: kdelman@emory.edu

V.A. Master, M.D., Ph.D., F.A.C.S.
Department of Urology, Emory University School of Medicine, 1364 Clifton Road NE, Atlanta, GA 30322, USA
e-mail: vmaster@emory.edu

K.A. Delman, V.A. Master (eds.), *Malignancies of the Groin*,
https://doi.org/10.1007/978-3-319-60858-7_13

Table 13.1 Selected studies of wound complications following open inguinal lymphadenectomy for melanoma

Study	No. of patients	Overall wound complications (%)
Shaw [3]	58	43
Coit [4]	42	64
Beitsch [5]	168	51
Karakousis [6]	205	52
Serpell [7]	27	71
De Vries [8]	14	35
Van Akkooi [9]	129	29
Sabel [10]	212	19
Guggenheim [11]	43	48
Poos [12]	129	21
Chang [13]	53	77

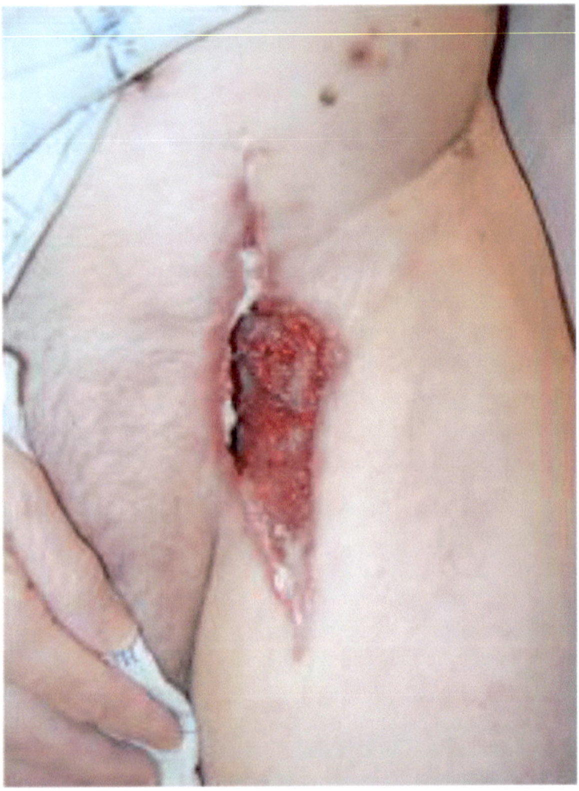

Fig. 13.1 Wound dehiscence following open inguinal lymphadenectomy

were explored in hopes of reducing perioperative morbidity. Unfortunately, several technical modifications to reduce morbidity such as relocating the skin incision, creating thicker skin flaps, preserving the saphenous vein, and omitting sartorius transposition have been attempted, but have not substantially decreased complication rates [18]. On the contrary, videoscopic inguinal lymphadenectomy (VIL), a minimally invasive technique designed to minimize wound complications, demonstrates promise as an alternative to traditional open surgery while achieving comparable oncological control in both genitourinary tumors and melanoma [19–21]. This chapter will outline the steps required to complete a successful resection using VIL.

Videoscopic Inguinal Lymphadenectomy (VIL)

Bishoff et al. first reported use of endoscopic technologies to perform groin dissection in 2003 [22]. This technique was described in two cadavers and one living patient in whom he converted the case to the standard open approach due to failure to mobilize the nodal mass superiorly. Sotelo et al. subsequently reported a series of 14 minimally invasive lymphadenectomies for penile cancer in which no wound-related complications were noted [23]. In 2009, Delman et al. modified the approach to allow for a dissection that would be anatomically appropriate for melanoma [19, 20]. The following is a description of the standard VIL technique.

Preparation and Position

Upon entering the operating room, patients are positioned on a split-leg table, with the legs externally rotated and abducted, and the boundaries of the femoral triangle are mapped out with a surgical pen (Fig. 13.2). Accurate marking is necessary for both correct trocar placement and to aid in determining the extent of the dissection during the case. Clipping and prepping are performed via standard techniques. The suprapubic skin should be included in the sterile field so that development of crepitus can be monitored. Prior to starting the case, appropriate antibiotic prophylaxis is given. The surgeon is positioned between the patient's leg, and the assistant stands to the outside of the operative limb (Fig. 13.3). The laparoscopic tower should be positioned on the side of the operative limb with the monitors placed at the patient's shoulders.

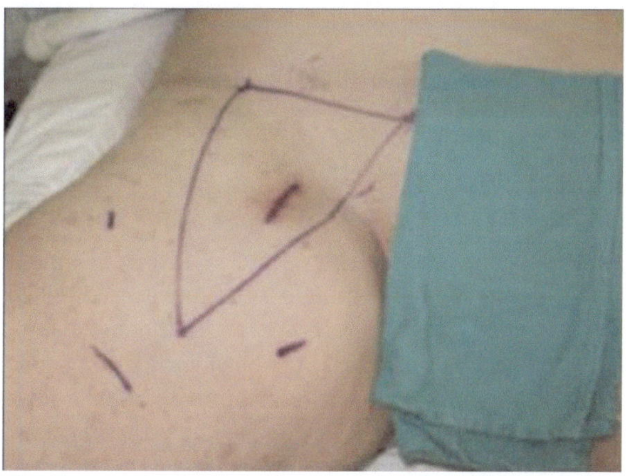

Fig. 13.2 Patient positioned on a split-leg table with the femoral triangle, sentinel lymph node biopsy site, and port locations marked

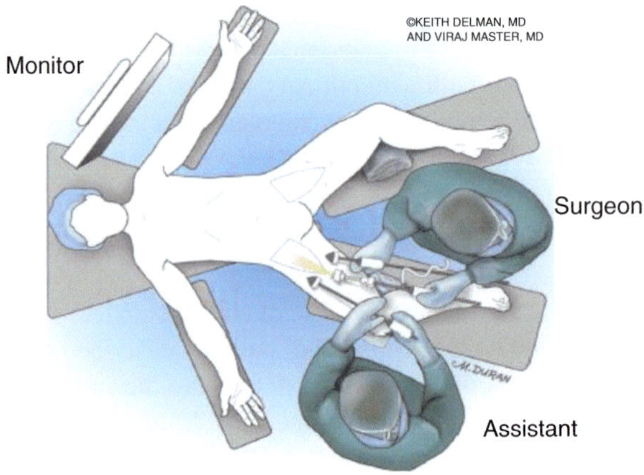

Fig. 13.3 Patient and surgeon positioning for video-assisted lymphadenectomy

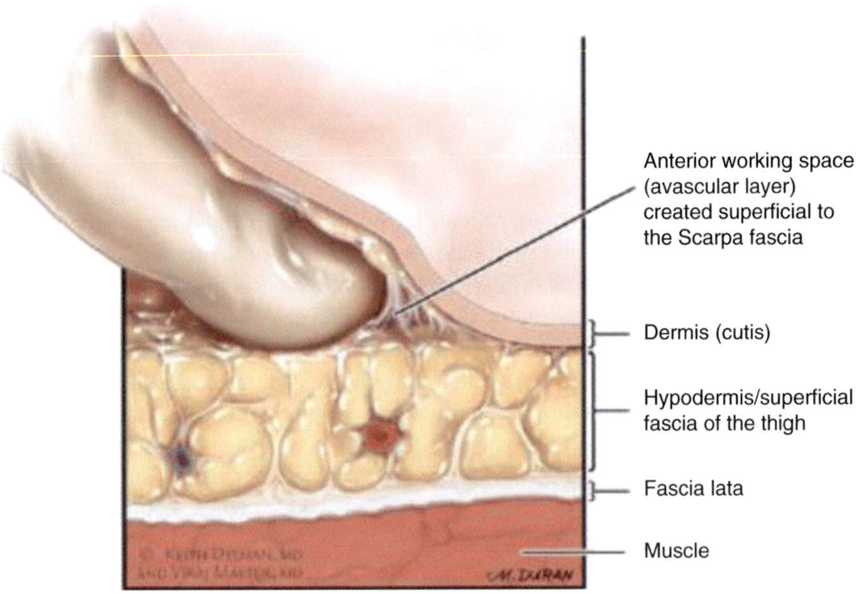

Fig. 13.4 Correct level for the development of the anterior plane of dissection

Trocar Placement

A three-incision technique is used. The first is a 12-mm incision placed 3 cm distal to the apex of the femoral triangle. A scalpel is used to incise the skin and dissect down from Camper's fascia to Scarpa's fascia, although the precise delineation at this point is not critical, as this incision is outside of the boundaries of the template. Scarpa's fascia, a glistening thin film, is then incised, and a finger is used to develop a space extending out 5 cm on each side of the incision (Fig. 13.4). This blunt finger dissection allows enough space to insert two additional 10-mm trocars to be placed under direct visualization. A 12-mm balloon port trocar is then placed in the original incision, and the dissected space is insufflated to 25 mmHg for 10 min. The pressure is then decreased to 15 mmHg to prevent end-tidal CO_2 elevation. Under direct visualization with a 0° 10-mm laparoscope, two 10-mm short bladeless trocars are inserted approximately a handsbreadth from the visualizing port. The trocars should be positioned 3 cm outside of the medial and lateral boundaries of the previously delineated femoral triangle.

Boundaries of Dissection

At this stage, it is critical to ensure that the anterior working space is completely developed before proceeding with additional dissection. This anterior working space is defined as the area created between the fibrofatty packet containing the lymph nodes and the "flaps" that are created when dissecting along Scarpa's fascia. The dissection should be extended superficial to Scarpa's fascia. If the glistening undersurface layer of Scarpa's fascia is identified, the flap is too thick and the plane must be changed. The correct tissue thickness is approximately 3–5 mm in most patients and allows the surgeon to see the cutaneous vessels when the skin flap is transilluminated with a camera. Loss of vessel visualization is often associated with flap necrosis of this area.

For melanoma patients with primary lesions of the trunk, dissection is routinely carried 5 cm above the inguinal ligament along the abdominal wall with an endoscopic dissecting stick in tandem with ultrasonic shears. Medial and lateral boundaries of the dissection consist of the adductor longus and sartorius muscle fascia, which should be correlated to the previous skin markings via transillumination. The fibrofatty packet may be rolled inward on both sides using an endoscopic sponge or Kittner. This maneuver is continued superiorly and inferiorly as much as possible to assist in defining the posterior tail of the node packet. Throughout the dissection, small perforating vessels are routinely encountered and should be controlled using an ultrasonic dissecting scalpel, LigaSure, or clips. Lymph vessels should be sealed with the ultrasonic dissecting scalpel. The deep thigh fascia (fascia lata) constitutes the posterior boundary, and its violation is readily apparent when reddish muscle fibers are encountered.

Saphenous Vein Division and Vascular Dissection

The saphenous vein is visualized within the apex of the femoral triangle and divided with the vascular load of an endoscopic linear cutting stapler. Careful dissection within the femoral triangle enables the identification of the femoral artery pulse. Laparoscopic ultrasound probes may be used to delineate the vessel anatomy if needed. The dissection is then carried from an inferior to superior direction on top of the artery. The femoral vein is identified using the artery as a landmark. Both vessels are then skeletonized, along with all of the tissue between the femoral vein and adductor longus. This dissection is more easily accomplished when the assistant elevates the packet, allowing the surgeon to work below it in the dissection plane.

Saphenofemoral Junction Dissection and Transection

Following completion of the vascular dissection, blunt dissection in the saphenofemoral junction is performed to identify the inferior edge of the saphenous vein. A right-angle dissector and a Hunter grasper are the preferred tools for this maneuver. An endoscopic linear cutting stapler with a vascular load is then used to transect the vein at this level (Fig. 13.5). While exposing the saphenofemoral junction, continued inferomedial dissection around the femoral vein will enable resection of the deep inguinal nodes. Dissection should be continued to the level of the femoral canal until the pectineus muscle is visible. This will ensure complete nodal retrieval and also provide exposure for a biopsy of Cloquet's node, although this element has been largely abandoned in patients with only sentinel node involvement. Some fascial attachments to the inguinal ligament may remain at this point (Fig. 13.6).

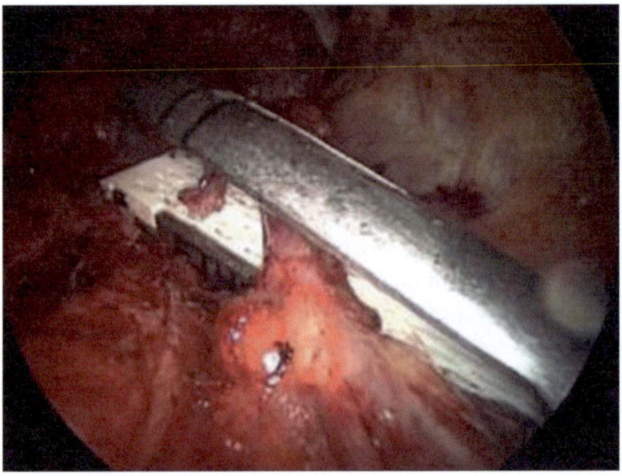

Fig. 13.5 Division of the saphenous vein as it crosses the adductor musculature using an endoscopic linear cutting stapler

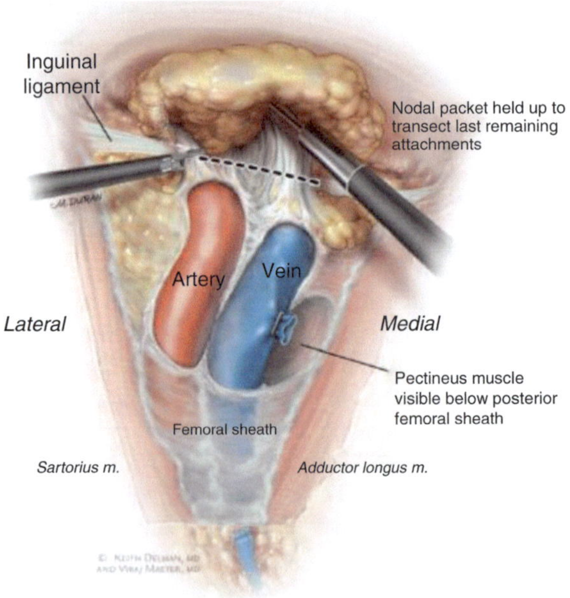

Fig. 13.6 Release of tissue at the superior border along the inguinal ligament

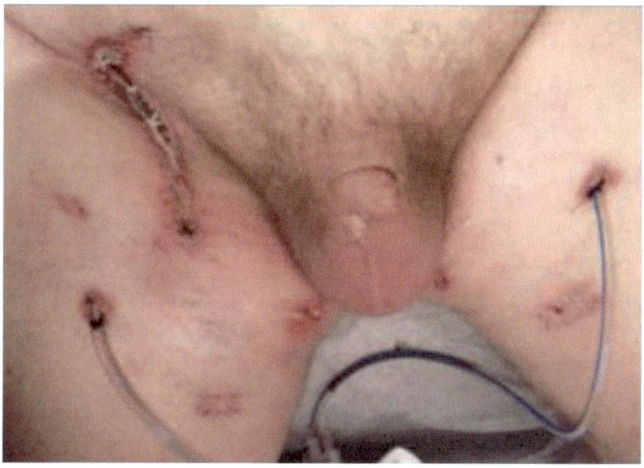

Fig. 13.7 Drain placement is shown on a patient who underwent a right-sided open inguinal lymphadenectomy and a left-sided VIL. Note the right-sided wound dehiscence following open lymphadenectomy

In order to completely separate the nodal packet from these attachments, the tissue must be dissected off the fascia by inferior retraction of the nodal packet. This technique will provide the appropriate visualization for blunt dissection or, in some cases, dissection of the tissue from the inguinal ligament using the ultrasonic dissecting scalpel.

Packet Removal, Drain Placement, and Postoperative Management

Once the nodal packet is free, it is withdrawn in a laparoscopic specimen bag through the apical port. If the packet is too large, the extraction site may need to be extended. Direct visualization is used to confirm complete dissection of all lymphatic tissue. To complete the procedure, a 19-French fully fluted drain is placed through the lateral port site, and the skin is closed (Fig. 13.7). The patient is encouraged to ambulate on the day of surgery and given a regular diet. Discharge is routinely planned for the same day, unless concomitant pelvic node or deep pelvic node dissection is performed. The drain remains in place until the output is <30 mL/day.

Pathology and Follow-Up

Based on the initial experience of Delman et al. in assessing 32 patients undergoing 45 procedures, the median nodal yield for VIL was 11 with a mean of 11 and a range of 4–24 [20]. A total of 8 procedures yielded node counts ≥15, while 33 (73%) of the 45 procedures had a nodal yield of 8 or greater, demonstrating a significantly higher lymph node yield than with open lymphadenectomy [24]. Of the 45 procedures that were performed, 18 dissections (40%) were for melanoma, 19 (42%) for penile carcinoma, 4 (9%) for scrotal/urethral carcinoma, and 4 (9%) for other pathologies. Median drain duration was 15 days with a range of 7–25 days.

Table 13.2 Published VIL data

Authors	Year	VIL patients	VIL procedures	Lymph node yield	Overall wound complications
Bishoff et al. [22]	2003	3[a]	3	–	–
Tobias-Machado et al. [25]	2008	15	20	10.8 (7–16)	3 (20%)
Sotelo et al. [23]	2009	8	14	9 (4–15)	0 (0%)
Delman et al. [20]	2011	32	45	11 (4–24)	8 (18%)
Xu et al. [26]	2011	17	17	16 (11–23)	2 (12%)
Sudhir et al. [27]	2012	22	39	n.r.	5 (13%)
Schwentner et al. [28]	2013	16	28	7.1 (4–13)	1 (3.57%)
Zhou et al. [29]	2013	7	11	12.3 (7–15)	2 (11%)
Abbott et al. [24]	2013	13	13	11 (9–15)	0 (0%)
Pahwa et al. [30]	2013	10	10	10.6 (7–14)	2 (20%)
Sommariva et al. [31]	2016	23	23	21 (15–25)	4 (17%)

n.r. not recorded
[a]Two fresh cadavers, one patient. Dissection was not possible in the patient due to the adherence of enlarged lymph nodes to the femoral vein

Eight (18%) of the Forty-five procedures had complications, including seroma, focal skin necrosis, and cellulitis, but there was no evidence of wound breakdown or dehiscence. These promising initial reports have been further corroborated by several recent publications (Table 13.2).

Summary

While the management of regional nodal metastases in patients with cutaneous and genitourinary tumors continues to evolve, completion lymph node dissection remains a significant component of the therapeutic algorithm. As the oncological application of endoscopic and laparoscopic techniques has expanded, VIL has emerged as a minimally invasive technique designed to reduce wound complications while achieving comparable oncological control in both genitourinary tumors and melanoma.

References

1. National Comprehensive Cancer Network [Internet]. NCCN clinical practice guidelines in oncology. Melanoma. v.3.2016. Updated 7 Jul 2016. http://www.nccn.org/professionals/physician_gls/pdf/melanoma.pdf. Accessed 12 Aug 2016.
2. Morton DL, Mozzillo N, Thompson JF, Kashani-Sabet M, Kelley MC, Gammon G. An international, randomized, double-blind, phase 3 study of the specific active immunotherapy agent,

Onamelatucel-L (Canvaxin compared to placebo as a post-surgical adjuvant in AJCC stage IV melanoma. Society of Surgical Oncology 59th Annual Cancer Symposium. San Diego, CA; 2006.
3. Shaw JH, Rumball EM. Complications and local recurrence following lymphadenectomy. Br J Surg. 1990;77:760–4.
4. Coit DG, Peters M, Brennan MF. A prospective randomized trial of perioperative cefazolin treatment in axillary and groin dissection. Arch Surg. 1991;126(11):1366–71.
5. Beitsch P, Balch C. Operative morbidity and risk factor assessment in melanoma patients undergoing inguinal lymph node dissection. Am J Surg. 1992;164(5):462–6.
6. Karakousis CP, Driscoll DL. Groin dissection in malignant melanoma. Br J Surg. 1994;81(12):1771–4.
7. Serpell JW, Carne PW, Bailey M. Radical lymph node dissection for melanoma. ANZ J Surg. 2003;73(5):294–9.
8. de Vries M, Vonkeman WG, van Ginkel RJ, Hoekstra HJ. Morbidity after inguinal sentinel lymph node biopsy and completion lymph node dissection in patients with cutaneous melanoma. Eur J Surg Oncol. 2006;32(7):785–9.
9. van Akkooi AC, Bouwhuis MG, van Geel AN, Hoedemaker R, Verhoef C, Grunhagen DJ, Schmitz PL, Eggermont AM, de Wilt JH. Morbidity and prognosis after therapeutic lymph node dissections for malignant melanoma. Eur J Surg Oncol. 2007;33(1):102–8.
10. Sabel MS, Griffith KA, Arora A, Shargorodsky J, Blazer DG, Rees R, Wong SL, Cimmino VM, Chang AE. Inguinal node dissection for melanoma in the era of sentinel lymph node biopsy. Surgery. 2007;141:728–35.
11. Guggenheim MM, Hug U, Jung FJ, Rousson V, Aust MC, Calcagni M, Kunzi W, Giovanoli P. Morbidity and recurrence after completion lymph node dissection following sentinel lymph node biopsy in cutaneous malignant melanoma. Ann Surg. 2008;247(4):687–93.
12. Poos HPAM, Kruijff S, Bastiaannet E, van Ginkel RJ, Hoekstra HJ. Therapeutic groin dissection for melanoma: risk factors for short-term morbidity. Eur J Surg Oncol. 2009;35:877–83.
13. Chang SB, Askew RL, Xing Y, Weaver S, Gershenwald JE, Lee JE, Royal R, Lucci A, Ross MI, Cormier JN. Prospective assessment of postoperative complications and associated costs following inguinal lymph node dissection (ILND) in melanoma patients. Ann Surg Oncol. 2010;17(10):2764–72.
14. Hughes TM, Thomas JM. Combined inguinal and pelvic lymph node dissection for stage III melanoma. Br J Surg. 1999;86:1493–8.
15. Mortenson M, Xing Y, Weaver S, Lee JE, Gershenwald JE, Lucci A, Mansfield PF, Ross MI, Cormier JN. Fibrin sealant does not decrease seroma output or time to drain removal following inguino-femoral lymph node dissection in melanoma patients: A randomized controlled trial (NCT00506311). World J Surg Oncol. 2008;6:63.
16. Bilimoria K, Balch C, Bentrem D, Talamonti M, Ko C, Lange J, Winchester D, Wayne J. Complete lymph node dissection for sentinel node-positive melanoma: assessment of practice patterns in the United States. Ann Surg Oncol. 2008;15:1566–76.
17. Johnson TV, Hsiao W, Delman KA, Jani AB, Brawley OW, Master VA. Extensive inguinal lymphadenectomy improves overall 5-year survival in penile cancer patients: results from the Surveillance, Epidemiology, and End Results program. Cancer. 2010;116(12):2960–6.
18. Sarnaik AA, Puleo CA, Zager JS, Sondak VK. Limiting the morbidity of inguinal lymphadenectomy for metastatic melanoma. Cancer Control. 2009;16(3):240–7.
19. Delman KA, Kooby DA, Ogan K, Hsiao W, Master V. Feasibility of a novel approach to inguinal lymphadenectomy: minimally invasive groin dissection for melanoma. Ann Surg Oncol. 2010;17(3):731–7.
20. Delman KA, Kooby DA, Rizzo M, Ogan K, Master V. Initial experience with videoscopic inguinal lymphadenectomy. Ann Surg Oncol. 2011;18(4):977–82.
21. Martin BM, Master VA, Delman KA. Videoscopic Inguinal lymphadenectomy for metastatic melanoma. Cancer Control. 2013;20(4):255–60.
22. Bishoff JT, Basler JW, Teichman JM, Thompson IM. Endoscopic subcutaneous modified inguinal lymph node dissection (ESMIL) for squamous cell carcinoma of the penis. J Urol. 2003;169:78.

23. Sotelo R, Sánchez-Salas R, Carmona O, Garcia A, Mariano M, Neiva G, Trujillo G, Novoa J, Cornejo F, Finelli A. Endoscopic lymphadenectomy for penile carcinoma. J Endourol. 2007;21(4):364–7.
24. Abbott AM, Grotz TE, Rueth NM, Hernandez Irizarry RC, Tuttle TM, Jakub JW. Minimally invasive inguinal lymph node dissection (MILND) for melanoma: experience from two academic centers. Ann Surg Oncol. 2013;20(1):340–5.
25. Tobias-Machado M, Tavares A, Silva MN, Molina WR, Forseto PH, Juliano RV, Wroclawski ER. Can video endoscopic inguinal lymphadenectomy achieve a lower morbidity than open lymph node dissection in penile cancer patients? J Endourol. 2008;22:1687–91.
26. Xu H, Wang D, Wang Y, Li Y, Chen Y, Liang Z. Endoscopic inguinal lymphadenectomy with a novel abdominal approach to vulvar cancer: description of technique and surgical outcome. J Minim Invasive Gynecol. 2011;18:644–50.
27. Sudhir R, Krishnappa R, Khanna S, Sekon R, Koul R. Video endoscopic inguinal lymphadenectomy (VEIL): minimally invasive radical inguinal lymphadenectomy technique Indian. J Surg Oncol. 2012;3:257–61.
28. Schwentner C, Todenhöfer T, Seibold J, Alloussi SH, Mischinger J, Aufderklamm S, Stenzl A, Gakis G. Endoscopic inguinofemoral lymphadenectomy—extended follow-up. J Endourol. 2013;27(4):497–503.
29. Zhou XL, Zhang JF, Zhou SJ, Yuan XQ. Endoscopic inguinal lymphadenectomy for penile carcinoma and genital malignancy: a preliminary report. J Endourol. 2013;27:657–61.
30. Pahwa HS, Misra S, Kumar A, Kumar V, Agarwal A, Srivastava R. Video endoscopic inguinal lymphadenectomy (VEIL)—a prospective critical perioperative assessment of feasibility and morbidity with points of technique in penile carcinoma. World J Surg Oncol. 2013;11:42. doi:10.1186/1477-7819-11-42.
31. Sommariva A, Pasquali S, Cona C, Ciccarese AA, Saadeh L, Campana LG, Meroni M, Rossi CR. Videoscopic ilioinguinal lymphadenectomy for groin lymph node metastases from melanoma. Br J Surg. 2016;103(8):1026–32.

Chapter 14
Other Approaches for Reducing Surgical Risk

Antonio Sommariva

Risk Factors

Surgery is currently the most effective treatment modality for patients with inguinal lymph node metastasis from cutaneous and genitourinary tumors. All surgeons dealing with groin metastases know that inguinal lymphadenectomy (IL) is burdened by a substantial morbidity including wound infection and dehiscence, seroma, leg lymphedema, and deep venous thromboembolism (DVT). In addressing the issues of how to decrease the risk of complications after inguinal lymphadenectomy, it is important to first define the recognized risk factors and for what types of complications such factors are important. It is also appropriate to bear in mind that often a complication in itself is a risk factor for another complication. For example, it is known that the onset of infection or hematoma in the groin favors the onset of lower limb lymphedema in the postoperative phase. The issue of reducing surgical risk is complex, and the level of evidence of the studies is not always adequate to allow definitive conclusions. In the evaluation of the available literature on morbidity after IL, we should bear in mind that the vast majority of the studies are retrospective and even in those where data collection is prospective in design, a wide range of variation in definition and grading of the complications as well as in the follow-up of the patients is found.

According to the Centers for Disease Control and Prevention (CDC), surgical wound classifications, wounds from superficial lymphadenectomy (neck, axilla, and groin) can be considered as a class I/clean. However, groin dissection is associated with an infection rate relatively higher than that reported after nodal dissections in other anatomic regions [1, 2] and is also higher than that expected for a typical "clean" operation, which ranges between 1 and 5% of cases. One of the potential

A. Sommariva, M.D.
Surgical Oncology Unit, Veneto Institute of Oncology, IOV-IRCCS,
Via Gattamelata 64, 35128 Padova, Italy
e-mail: antonio.sommariva@iov.veneto.it

K.A. Delman, V.A. Master (eds.), *Malignancies of the Groin*,
https://doi.org/10.1007/978-3-319-60858-7_14

Table 14.1 Recognized risk factors for postoperative morbidity after IL

Age
Male gender
Obesity
Diabetes
Smoking
Cardiovascular/pulmonary disease
Tumor burden
Surgeon case load
Radiotherapy
Patient mobilization

explanations is the bacterial load and pathogenicity of the bacterial flora of the groin, which is associated with the difficulty in maintaining adequate hygiene in the folds of the groin, particularly in overweight patients. Contributing factors are the relatively large area of dissection, the density of lymphatic vessels, and the critical vascular supply of the groin skin. The relatively high incidence of wound infection after groin dissection is also related to the higher risk in these patients of other postoperative complications, such as wound breakdown and seroma, which frequently lay the ground and contribute to bacterial contamination of the surgical field. Approximately one-third of seromas lead to infection requiring drainage or drain placement. The microorganisms isolated in the groin include gram-negative rods, *Staphylococci*, diphtheroids, and streptococci. In the same way, an infected wound frequently leads to dehiscence. For this reason, it is difficult to distinguish the process underneath the infection, which is always multifactorial, including several well-known factors (Table 14.1).

Age is an established risk factor for postoperative complications. The reason why older patients are at increased risk of postoperative complications is probably multifactorial, in part related to several associated morbidities affecting these patients. The higher risk of wound complications in older patients can be explained by the deterioration of wound healing with age. Comorbidities, including cardiovascular and/or pulmonary disease and diabetes, have an established association with complications after IL [3]. Diabetes itself, associated with wound problems after several surgical procedures, represents an independent risk factor for wound complications and seroma after IL [1, 4, 5]. Another reason why patients at a later age are more likely to develop complications can also be explained by factors related to postoperative management. It is possible that elderly patients present a later mobilization and that wound care is more difficult and less accurate than in young patients. Moreover, the significantly higher incidence of leg lymphedema observed in patients >50 years can be explained by delayed detection and referral for intervention, despite the knowledge that early diagnosis and treatment play a pivotal role in halting progression and preventing complications of lymphedema [6].

Another significant risk factor for morbidity after IL is obesity. Patients with an increased body mass index (BMI) are at significant risk for wound complications, as seen in several studies. In a multivariable analysis, a BMI of more than 25 was the only factor associated with a higher incidence of wound infection in two single-center studies [3, 4]. A prospective study estimated that a BMI >30 increased the risk of wound complications by more than 11-fold [1]. Moreover, obesity represents a significant risk factor for postoperative lymphedema after IL [6]. Obese patients are at higher risk for lymphedema because they have baseline impaired venous and lymphatic function. As shown in experimental models, the negative effect of obesity on lymphedema is increased after surgery as a result of an impaired lymphangiogenesis [7].

Another factor clearly related to postoperative morbidity is the indication and extent of dissection. In melanoma, IL for clinical disease is burdened by a higher postoperative complication rate (wound infection/dehiscence) and lymphedema compared to completion lymphadenectomy for positive sentinel biopsy [8, 9]. Also, in penile cancer, variables pertaining to the extent of disease burden (i.e., number of lymph nodes, AJCC stage) have been demonstrated to be significantly related to postoperative morbidity [10]. The result is thinner flaps or an increased tension on the wound that can favor skin necrosis and wound dehiscence. Surrogate risk factors for tumor burden are recognized in the length of surgery, the size of the largest lymph node, the transposition of the sartorius muscle, and the number of lymph nodes. This difference might be related to the surgeon's attitude or necessity to be more radical in patients with clinical disease. Regarding lymphedema, the presence of macroscopic disease seems to influence the onset of postoperative lymphedema as well. Patients with positive sentinel lymph node biopsy (SLNB) undergoing dissection (the so-called completion lymphadenectomy) showed a lower incidence of leg lymphedema with respect to those operated for clinically palpable disease [8, 11]. This observation finds two potential explanations. First, patients with clinical disease present an impaired lymphatic drainage due to the greater number of lymph nodes involved, which causes a more pronounced lymphatic obstruction. Second, surgery for clinical disease strives for complete clearance of the affected basin, leading to a greater thoroughness and disruption of lymphatic collaterals during dissection. Also, a more extensive surgery, including the iliac lymph nodes, has been significantly associated with a worse outcome, although this latter factor is still under discussion and is probably secondary to indication (more disease burden) rather than extensive surgery itself [5]. In melanoma, it is not clear whether the addition of deep dissection (i.e., obturator and iliac) could represent a significant risk factor for postoperative morbidity. Morbidity rates seem unaffected by a combined superficial and deep groin dissection, even though chronic lymphedema showed a trend in later onset in one study [12].

Other factors, such as smoking [9, 13], male gender [9], patient mobilization [3, 4], radiotherapy [14], and surgeon case load [15], have been evaluated and should also be taken into account when planning IL.

Preventing Bacterial Infection

Perioperative administration of antibiotics after groin surgery has been considered as a measure to reduce the rate of wound infection. A prospective randomized controlled trial on perioperative use of cefazolin in preventing wound complications after axillary and groin dissection did not show any significant benefit of antibiotic administration on wound complications in the inguinal region [13]. No guidelines are available for if and how long prophylactic antibiotics should be administrated, and the practice varies among centers. Prolonging antibiotics until drain removal or in the presence of undrained hematoma or seroma is not supported by any evidence and could not be recommended as standard of care in clinical practice. Shaving the surgical site, accurate sterilization of the groin before starting the procedure, and placement of drains laterally, as far away as possible from the bacteria-laden skin of the inner thigh, groin and genitals, and anus, all represent a pragmatic approach for limiting the risk of infections. During surgery, a diligent control of lymphatics and hemostasis prevents conditions such as seroma and hematoma that can favor infection. Wound irrigation and removal of any devitalized tissue should also be carried out. After surgery, the wound should be kept clean and dry. In obese patients, the abdomen skinfolds can make wound care problematic and favor excessive moisture of the skin, which becomes an ideal culture medium for pathogens. In this patient subgroup, the application of negative pressure wound therapy (NPWT) systems could be beneficial, although they have never been tested following groin dissection in a properly designed study. Epidermal vacuum dressing consists in a pump connected to a designed dressing which generates a negative pressure of 80 mm/Hg on the skin, allowing removal of fluids away from the wound through a combination of absorbency and evaporation [16].

The Choice of Skin Incision

One of the most effective ways to reduce wound-related morbidity after groin dissection would simply be by avoiding skin incision. Video-assisted groin dissection technique is the most promising and valuable approach towards this goal and is covered in a separate chapter [17].

The choice of the type and length of skin incision should be made with the main aim to permit full access and a direct view of the tumor limits of the inguinal and iliac dissection as well as to guarantee an effective clearance of the lymph nodes and a reliable control of bleeding and lymphatic leak. The type and length of skin incision play a pivotal role in wound morbidity. The ischemia of the skin flaps is the most important factor affecting wound morbidity after groin surgery (Fig. 14.1). Skin necrosis in a body area as the groin—moist and rich in cutaneous folds and bacterial colonization—is often complicated by infection, which can determine prolonged wound healing and eventually an increased risk of lymphedema due to hampered lymphatic regeneration. The blood supply of the groin is maintained by three

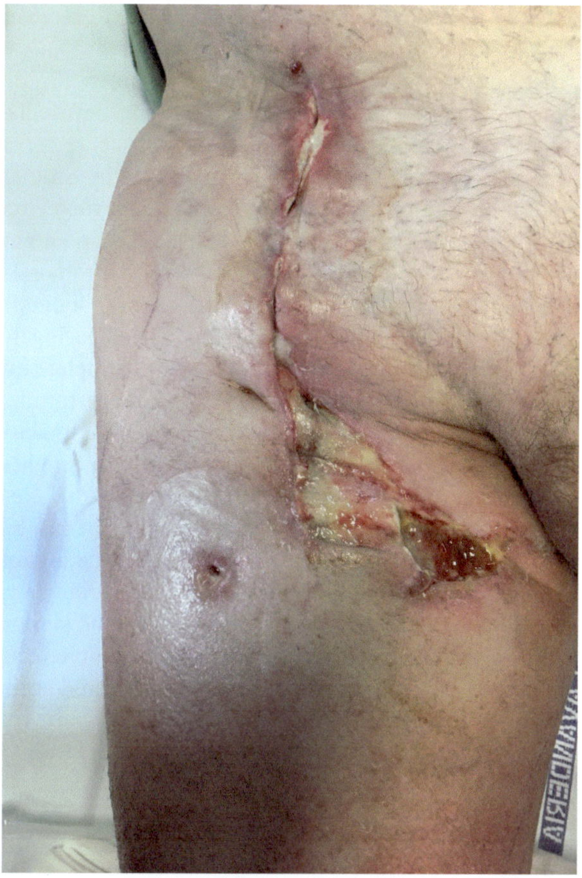

Fig. 14.1 Skin necrosis after S-shaped incision during IL

main collaterals: the epigastric artery, the circumflex iliac artery, and the external pudendal artery [18]. These arteries are generally transected by the classic vertical incision creating cutaneous areas at risk of ischemia. These small branches lie in the Camper fascia and tend to be parallel to the skin creases and the inguinal ligament. One surgical principle derived from these anatomical landmarks is that particular attention should be paid in preserving the Camper layer during flap preparation, avoiding lesions of the microvascular arterial plexus. Skin flaps should include at least 2–3 mm of subcutaneous fat and then become thicker as the base of the flap is reached. A careful skin flap preparation plays an important role in preventing wound edge ischemia, and particular attention should be paid in the case of obese patients with multiple redundant skinfolds in the groin [19]. At the end of dissection, the skin edge should be systematically checked and any ischemic area resected. Excision of at least 4 cm width of skin showed a significant lower rate of early complication with respect to excision of little or no skin [20]. New technology, such as intraoperative indocyanine green fluorescence angiography, is effective for visual assessment of tissue perfusion, and its application during IL seems a promising tool for preventing wound necrosis and dehiscence [21].

Regarding the type of incision, it is well recognized that the vertical or S-shaped skin incision leads to a greater risk of skin devascularization [22]. Oblique incisions, parallel to the inguinal ligament, transect fewer anastomotic vessels than vertical ones, preventing flap necrosis. Oblique incision allows good exposure for the iliac and obturator area, and, in case of radical vulvectomy or penectomy, the medial part of the incision can easily be extended if an en bloc resection is needed. With oblique incision, the access to the apex of the femoral triangle is sometimes problematic, and exposure with retractors (even if lighted) under the lower skin edge may cause damage of the microcirculation, increasing the risk of necrosis. Moreover, an oblique incision does not always allow complete exposure of the surgical field, and it is not uniformly adopted by surgeons performing groin dissection. A single incision below the inguinal ligament, more proximal to the apex of the femoral triangle, does not show a significant benefit over a single incision above [23]. In cutaneous tumors of the lower limb, where an optimal clearance of the distal inguinal nodes is mandatory, a double incision technique has been proposed. Adopting two separate oblique incisions, below and above the inguinal ligament, allows a better exposure of the distal portion of the femoral triangle and represents a good surrogate to single longitudinal incision. Although no significant advantage with respect to the vertical incision is demonstrated, the double incision technique can be useful in some cases where wound healing is considered at risk for previous surgery or in the presence of multiple risk factors [24].

Lymphatics and Vessel Control

Seroma formation (lymphocele) represents the most common complication after groin dissection. Meticulous control of lymphatic vessels during dissection is pivotal in preventing postoperative seroma. After sentinel lymph node biopsy (SLNB), lymphovascular control with Ligaclips is associated with a better postoperative outcome compared with diathermy use [25]. Although a longer operative time is expected, multiple small ligations with absorbable suture or clips are essential. Clip ligation carries minimal risk to surrounding structures; however, they may be dislodged during dissection and only offer a control of macroscopic vessels with minimal effect on the microscopic vascular and lymphatic network.

More recently, new devices have been tested for lymph node dissections, the most popular based on ultrasound or radiofrequency energy delivery. The hypothesis is that by reducing the thermal-induced injury and secondary inflammation on tissues and by complete sealing of vessels and lymphatics, postoperative morbidity could be reduced compared to the classic "electrocautery/clips" technique. Ultrasonic dissection devices are expected to seal vessels by denaturing hydrogen bonds and sealing the vessels with a coagulum. Radiofrequency devices use bipolar energy by denaturing the collagen and elastin in the vessel wall into a permanent seal. Ultrasonic scalpels (USS) and radiofrequency scalpels (RFS) are widely used in laparoscopic surgery, to minimize smoke and collateral damage during tissue dissection and to maintain adequate vascular control. These devices have been shown to produce less thermal injury in animal studies, and it is postulated that their use for lymphatic dissection might reduce bleeding, postoperative drainage, and seroma development. In small comparative studies of lymph node dissection in breast cancer lymphadenectomy, USS showed controversial results in terms of lymphatic fistula, lymphocele, and hematoma. In RCTs of axillary dissection for breast cancer, lymphadenectomy with USS was able to significantly reduce the serous drainage and hospitalization stay [26, 27]. In patients undergoing axillary or inguinal lymphadenectomy, a recent prospective randomized trial failed to show any significant reduction of complications (seroma, hematoma, and surgical site infection) between dissection with USS and ligation/monopolary electrocautery [28]. Also, operative time and length of hospital stay seem similar, although lymphedema was significantly higher after US dissection. The reasons for this should be further investigated, but a hypothesis could be that USS leads to a more efficient sealing of lymphatics with subsequent more evident lymphatic stasis in the limb. There is just one single study comparing USS, RFS, and electrocautery and clip application after SLNB for melanoma [29]. This study showed a significant reduction after RFS use on incidence of lymphocele compared with electrocautery and clip application or USS. The effectiveness of USS and RFS for IL is far from being definitely proven, and prospective comparative trials are necessary. These studies should be designed not only comparing the results in terms of morbidity but also considering the cost for healthcare systems of these new devices.

Fibrin sealants (FS) have been proposed as a potential method to reduce lymphatic leak after lymphadenectomies. Fibrin sealant or fibrin glues are hemostatic agents derived from plasma. They are composed of a solution of several molecules in different combinations (thrombin, fibrinogen, aprotin, fibronectin, and human factor XIII) that essentially replicate the final step of coagulation cascade, stop fibrinolysis, and reinforce the clot. A meta-analysis of six RCTs did not show any significant advantage of FS over standard closure in patients undergoing groin dissection [30]. This finding is in line with a similar analysis on FS use after breast and axillary surgery [31]. We should also consider the relatively high cost of FS and the potential risk (although never observed) of transmitting infective agents as they derive from pooled human plasma. Nevertheless, due to their simplicity of use and their low toxicity, they are still adopted in many centers. Further studies are needed with a larger sample size and better methodological quality before a definitive conclusion on their utility after IL can be made.

An interesting and innovative field is the application of microsurgical lymphatic-venous anastomoses (LVA) performed simultaneously with groin dissection in primary prevention of lymphedema [32]. In preliminary experiences, no lymphedema occurred after microsurgical primary lymphovascular anastomosis. The technique consists in direct anastomosis between lymphatics distal to the inguinal node and a collateral branch of the great saphenous vein. After blue dye injection, lymphatics are visualized and isolated cranially to the inguinal nodes, closed by titanium clips, cut from nodal capsule, and prepared for anastomosis. The main concern related to lymphovascular anastomosis in patients with groin lymph node metastases is the potential danger of diffusion of cancer cells between the lymphatic system of the leg and the trunk and the systemic blood circulation. Further research is needed to investigate this approach in terms of costs and operative times as well as in terms of oncology outcomes.

Saphenous Vein Preservation (SVP)

This technique was first described in 1988 [33] and consists in the isolation of the vein along its entire course in the apex of Scarpa's triangle up to the junction with the femoral vein, obtained through the meticulous ligation of all the tributary vessels. Preservation of the SV appears to reduce the cost and morbidity of IL [34].

In some comparative studies, both retrospective and prospective [35–38], the technique proved to consistently reduce the incidence of postsurgical lymphedema, especially long-term lymphedema (after 2 years). These data have been confirmed by a meta-analysis, which showed a significant reduction of lymphedema in the SVP group (odds ratio 0.24; 95% CI 0.11–0.53) [39]. In the same analysis, wound-related complication rates (infection and dehiscence) also seem to be lower (odds ratio 0.4; 95% CI 0.16–0.96 and 0.34; 95% CI 0.19–0.59), retrospectively. In one study, SVP also showed a lower occurrence of lymphocele [38]. However, the incidence of DVT is similar with the classic approach. Available data show that SVP is a relatively simple technique, which does not stretch operating time nor is associated with a greater blood loss. From an oncology point of view, SVP guarantees the excision of an equivalent number of lymph nodes, and the recurrence rate is similar to that of the vein ligation technique.

No clear explanations can be found on how the preservation of the saphenous vein may prevent lymphatic stasis after IL. Limb lymphedema is a morbid condition characterized by a difficult discharge of interstitial fluids. The preservation of the most important superficial vein of the leg can partly compensate for the accumulation of fluids that find an alternative way of drainage. Moreover, the better trophism of the skin ensured by a more adequate venous drainage and less edema can also explain better results in terms of infection and dehiscence. Moreover, a more meticulous dissection with multiple ligation can probably account for the lower incidence of seroma observed in one study.

The exact mechanism through which preservation of the saphenous vein can determine less lymphedema is not clear. In patients undergoing vascular bypass procedures in whom SV is entirely or partially disconnected, the simple loss of the SV rarely leads to significant lower extremity edema [40]. Chronic venous insufficiency can affect lymphatic function in the lower limb. The delay of lymph flow may correlate with the severity of clinical venous disease and/or the magnitude of venous reflux. Moreover, a dilated saphenous vein and/or varicose vein of their tributaries may directly obstruct flow through the lymph vessels. These phenomena are in general reversible with surgical treatment of venous incompetence making the hypothesis of SVL as direct cause of lymphedema after IL uncertain [41]. The mechanism underneath the venodynamics and lymphodynamics in the leg after groin dissection interacts as an unpredictable and mutually dependent outflow system. After saphenous vein ligation, the balance between the two systems is probably lost. Venous permeability due to capillary hyperpressure leads to an increased infiltration and edema. Under this situation, the impaired lymphatic system secondary to the lymphadenectomy cannot compensate the interstitial fluid overload, and clinical lymphedema may occur. The effects of disrupted groin lymphatic vessels can be overcome by preservation of the SV.

On the basis of these data, the preservation of the saphenous vein is capable of reducing the risk of lymphedema and other complications and can be recommended, especially in the presence of patients with established risk factors (obesity, previous irradiation of the groin). It must take into account, however, that the SVP is not feasible in patients with large tumor load due to the risk of an inadequate tumor dissection and, in any case, where an obvious infiltration of the vessels by a metastatic lymph node or scars from previous surgery (SLNB) is present.

Preservation of the Fascia

In the classic description of IL, the fascia overlying the sartorius, adductor longus, psoas, and external oblique muscles is excised en bloc with the fibro-fatty tissue of the groin. The rationale to remove the deep fascia performing dissection in the avascular plan outside the fascial layer is mainly oncological, aimed at reduces the risk of local recurrence in the groin. Preservation of the muscle fascia has been tested in a single study after axillary and groin dissection [42, 43]. The incidence of long-term leg lymphedema appears low (14%) without any evidence of higher risk of local recurrence with respect to similar published reports. The reason why the preservation of muscle fascia leads to a lower occurrence of postoperative leg edema is not fully understood; preservation of the lymphatics under the fascial layer is demonstrated and visualized after intraoperative injection of lymphazurin blue dye. Moreover, preserving the fascia probably causes less scarring in the area of muscle dissection, favoring better lymphatic flow and regeneration.

Another important anatomic structure in the groin is the fascia lata, which separates the deep inguinal lymph nodes (underneath) from the superficial inguinal nodes (above). Preservation of this fascial structure is possible and seems associated with a lower morbidity profile, including wound-related complications (infections, skin necrosis), seroma, and leg lymphedema. The technique has been tested in inguinal lymphadenectomy for vulvar and penile carcinoma and is associated with a limited clearance of distal and later lymph nodes of the groin [44–46]. Although preserving the fascia lata represents a good compromise for groin lymphadenectomy in genital tumors, where dissection is also aimed for staging purpose, for skin cancers, it does not allow complete clearance of all the potentially metastatic lymph nodes of the groin, and, for this reason, it is never performed for this indication.

Table 14.2 Reconstructive options after IL

Skin grafts
Gracilis or sartorius muscle transposition
Omental flap
Tensor fascia lata flap
Anterolateral thigh flap
Rectus abdominis flap
Rectus femoris flap

Flap Procedures

In an attempt to reduce wound-related morbidities, several reconstructive surgical procedures are used after dissection, mainly when the risk of skin edge necrosis is judged to be very high or a primary closure of the wound is impossible (Table 14.2). Primary reconstructive procedures are generally considered in the presence of bulky disease with suspicious areas of skin infiltration, skin ulceration, previous irradiation of the groin, or systemic chemotherapy. The primary advantage of flap closure of the groin is the protection of the femoral vessels in case of dehiscence. Exposure of the femoral vessels represents a surgical emergency that should be avoided whenever possible. In the presence of extensive skin infiltration, these techniques allow a wound closure without tension. Moreover, the flap procedures allow the covering of the dead space in the femoral triangle preventing seroma. The use of flap procedures should always be tailored to each patient and clinical situation (skin infiltration, previous radiotherapy, and other patient-related risk factors), preferring the simplest technique over more complex techniques.

Skin grafting is sometimes necessary when a large portion of the skin should be resected. The cutaneous edges are sutured to the deep layers, and the residual wound defect is covered by split-thickness skin graft. Sartorius muscle transposition (SMT) was first introduced in 1960 to protect the femoral vessels in case of wound dehiscence [47]. The technique is relatively simple; the muscle is detached from its proximal insertion, rotated medially over the femoral artery and vein, and fixed to the inguinal ligament and adductor muscle with interrupted sutures. In the presence of wound breakdown, the muscle gives reliable protection for the underlying vessels avoiding additional surgery in most cases. The role of SMT in preventing wound-related complications (infections, seroma) itself is less clear. A small randomized controlled trial did not show any benefit of SMT in preventing wound-related morbidity [48]. No statistically significant differences were observed in the incidence of wound cellulitis, wound breakdown, lymphedema, or rehospitalization. Paradoxically, the incidence of seroma was increased in the SMT group. Data coming out from the American College of Surgeons National Surgical Quality Improvement Program (ACS NSQIP) database showed that SMT is used in 16.5% of patients undergoing lymphadenectomy for melanoma, more frequently after inguinoiliac dissection with respect to superficial (inguinal only) [49]. In this study, SMT is associated with a significantly longer operative time: 228 versus 168 min in inguinoiliac dissection and 181 versus 120 min for inguinal dissection. A similar wound complication rate has been detected between patients treated with SMT and those who were not, but when accounting for operative time, SMT group showed fewer complications compared with the non-SMT group. If we consider operative time as a surrogate of disease extension, these data suggest that SMT is potentially able to reduce morbidity in high-risk patients. For these reasons, SMT after IL should be used selectively, mainly for patients with established risk factors for wound complications, first and foremost obesity. SMT can be omitted in all the cases where the skin incision does not directly overlie the vessels, as in the case of a completion lymphadenectomy after sentinel lymph node biopsy. The application of sartorius tendon transposition has recently been proposed as a variant of SMT [50]. Transposition of the tendon obviates the need to cut and skeletonize the muscle, avoiding damage to the lateral femoral nerve (cause of persistent sensory disturbances) and lowering the surgical trauma in the inguinal area. A case-controlled study comparing classic SMT with sartorius tendon transposition showed that tendon preservation is associated with a lower incidence of wound dehiscence and lymphedema as well as a better quality of life. This technical variant, although promising, should be further evaluated within a prospective randomized trial.

Omentum flap is another technique aimed at limiting postoperative morbidity, mainly lymphedema by emptying the dead space in the femoral triangle, covering the femoral vessels after IL [51, 52]. Omentum is mobilized beneath the inguinal ligament, using a double incision, inguinal and abdominal, allowing a passage throughout the femoral canal. The flap, once in the groin, is fixed to the myofascial edges with interrupted sutures. Although preliminary results are encouraging, mainly for lymphedema, omentoplasty is not widely adopted by surgeons performing IL, and the current available literature is limited to a few pilot studies.

In more complex cases, pedicled myocutaneous flaps can be considered [53]. This kind of surgery should be planned in advance, after a thorough evaluation of costs and benefits, patient consent, and in centers where good skills of advanced plastic surgery techniques are available. Pedicled flaps within single-stage procedures are generally preferred over free flaps for their relative simplicity; they provide a good functional and cosmetic result and resistance to postoperative radiotherapy. The donor site morbidity rate is acceptable, but sometimes flap necrosis or infection may occur. A balanced comparison in terms of aesthetic and functional outcome between different flaps needs additional investigation.

Postoperative Care

A careful postoperative management assumes importance comparable to that of surgical techniques used to decrease the risk of complications. It is important to perform a thorough and daily monitoring of the surgical wound, avoiding fecal contamination and removing the bladder catheter only once the patient is able to mobilize from the bed autonomously. Epidermal vacuum dressing showed encouraging results in preventing wound complication after IL, but the efficacy and cost-effectiveness should be better investigated in the near future [16]. Some groups advocate bed rest to reduce surgical morbidity. In one retrospective study, the incidence of wound necrosis is inversely correlated with bed rest and with flexion of the hip and knee [3]. However, in another study, an early mobilization after 5 days IL does not increase the risk of complications [4]. No clear evidence is available as to when it is safe to mobilize the patient after groin dissection. Generally, it is preferred to extend the bed rest when myofascial flaps have been used to allow their engraftment without problems of dehiscence. During the postoperative period, a proper management of the antithrombotic prophylaxis is necessary and must be based on the use of low molecular weight heparin and antithrombus elastic stockings.

At the end of IL, most surgeons place one or more suction drains in the wound. This policy allows blood, reactive fluids and lymph to be drained away, maintaining the dead space of the groin empty. No evidence is available that indicates whether the positioning of a suction drain after IL consistently prevents postoperative complications. The risk of occurrence of an infection is paradoxically increased, as is clearly shown as drainage facilitates the entry of bacteria into the wound, in proportion to the time in which it is held in place. An early removal of drainage, which is typically volume directed (when the output is 30–50 mL in 24 h) or time directed (1 week irrespective of the drain output), is always advisable. The evidence on how to handle drains after IL for malignant disease remains minimal, and no clear guidelines on management can be supported [54]. A prospective trial would be useful to evaluate the real impact of drains use after groin lymphadenectomy. Some RCTs in pelvic lymphadenectomies suggest that the use of drainage after pelvic surgery is not required. Meanwhile, it would be appropriate that each center assess in a critical sense the appropriateness of using drainage in the groin, avoiding their use after combined pelvic lymphadenectomy.

In the early postoperative course, a proactive prevention program for lymphedema should be planned for each patient. In case of persistent and untreated postoperative leg edema, fibrotic process may occur secondary to stagnation of lymph, which worsens lymphedema and makes any further interventions useless. Programs for preventing postoperative lymphedema include an early use of elastic wraps, slow ambulation, and strict leg elevation when the patient is not ambulating. Upon the first sign of lymphedema, the patient should be immediately referred for massage therapy and compression stockings. Leg measurements should be routinely performed by the specialist engaged in follow-up of patients, as the first onset of lymphedema can be delayed, even after 12–24 months.

References

1. Chang SB, Askew RL, Xing Y, Weaver S, Gershenwald JE, Lee JE, et al. Prospective assessment of postoperative complications and associated costs following inguinal lymph node dissection (ILND) in melanoma patients. Ann Surg Oncol. 2010;17(10):2764–72.
2. Söderman M, Thomsen JB, Sørensen JA. Complications following inguinal and ilioinguinal lymphadenectomies: a meta-analysis. J Plast Surg Hand Surg. 2016;5:1–6.

3. Poos HP, Kruijff S, Bastiaannet E, van Ginkel RJ, Hoekstra HJ. Therapeutic groin dissection for melanoma: risk factors for short term morbidity. Eur J Surg Oncol. 2009;35(8):877–83.
4. Wevers KP, Poos HP, van Ginkel RJ, van Etten B, Hoekstra HJ. Early mobilization after ilio-inguinal lymph node dissection for melanoma does not increase the wound complication rate. Eur J Surg Oncol. 2013;39(2):185–90.
5. Glarner CE, Greenblatt DY, Rettammel RJ, Neuman HB, Weber SM. Wound complications after inguinal lymph node dissection for melanoma: is ACS NSQIP adequate? Ann Surg Oncol. 2013;20(6):2049–55.
6. Hyngstrom JR, Chiang YJ, Cromwell KD, Ross MI, Xing Y, Mungovan KS, et al. Prospective assessment of lymphedema incidence and lymphedema-associated symptoms following lymph node surgery for melanoma. Melanoma Res. 2013;23(4):290–7.
7. Mehrara BJ, Greene AK. Lymphedema and obesity: is there a link? Plast Reconstr Surg. 2014;134(1):154e–60e.
8. Sabel MS, Griffith KA, Arora A, Shargorodsky J, Blazer DG 3rd, Rees R, et al. Inguinal node dissection for melanoma in the era of sentinel lymph node biopsy. Surgery. 2007;141(6):728–35.
9. Beitsch P, Balch C. Operative morbidity and risk factor assessment in melanoma patients undergoing inguinal lymph node dissection. Am J Surg. 1992;164(5):462–5.
10. Gopman JM, Djajadiningrat RS, Baumgarten AS, Espiritu PN, Horenblas S, Zhu Y, et al. Predicting postoperative complications of inguinal lymph node dissection for penile cancer in an international multicentre cohort. BJU Int. 2015;116(2):196–201.
11. Faries MB, Thompson JF, Cochran A, Elashoff R, Glass EC, Mozzillo N, et al. The impact on morbidity and length of stay of early versus delayed complete lymphadenectomy in melanoma: results of the multicenter selective lymphadenectomy trial (I). Ann Surg Oncol. 2010;17(12):3324–9.
12. van der Ploeg AP, van Akkooi AC, Schmitz PI, van Geel AN, de Wilt JH, Eggermont AM, Verhoef C. Therapeutic surgical management of palpable melanoma groin metastases: superficial or combined superficial and deep groin lymph node dissection. Ann Surg Oncol. 2011;18(12):3300–8.
13. Coit DG, Peters M, Brennan MF. A prospective randomized trial of perioperative cefazolin treatment in axillary and groin dissection. Arch Surg. 1991;126(11):1366–71. discussion 1371–2
14. Cormier JN, Askew RL, Mungovan KS, Xing Y, Ross MI, Armer JM. Lymphedema beyond breast cancer: a systematic review and meta-analysis of cancer-related secondary lymphedema. Cancer. 2010;116(22):5138–49.
15. Dickson JK, Davies A, Rahman S, Sethu C, Smith JR, Orlando A, et al. Dissections of regional lymph nodes for treatment of skin cancer: predicting annual caseloads that will optimise outcomes. Ann R Coll Surg Engl. 2015;97(1):52–5.
16. Tauber R, Schmid S, Horn T, Thalgott M, Heck M, Haller B, et al. Inguinal lymph node dissection: epidermal vacuum therapy for prevention of wound complications. J Plast Reconstr Aesthet Surg. 2013;66(3):390–6.
17. Sommariva A, Pasquali S, Rossi CR. Video endoscopic inguinal lymphadenectomy for lymph node metastasis from solid tumors. Eur J Surg Oncol. 2015;41(3):274–81.
18. Tremblay C, Grabs D, Bourgouin D, Bronchti G. Cutaneous vascularization of the femoral triangle in respect to groin incisions. J Vasc Surg. 2015;22:pii:S0741-5214(15).
19. Karakousis CP. Therapeutic node dissections in malignant melanoma. Semin Surg Oncol. 1998;14(4):291–301.
20. Kean J, Hough M, Stevenson JH. Skin excision and groin lymphadenectomy: techniques and outcomes. Lymphology. 2006;39(3):141–6.
21. Furukawa H, Hayashi T, Oyama A, Funayama E, Murao N, Yamao T, et al. Effectiveness of intraoperative indocyanine-green fluorescence angiography during inguinal lymph node dissection for skin cancer to prevent postoperative wound dehiscence. Surg Today. 2015;45(8):973–8.
22. Spratt J. Groin dissection. J Surg Oncol. 2000;73:243–62.
23. Manci N, Marchetti C, Esposito F, De Falco C, Bellati F, Giorgini M, et al. Inguinofemoral lymphadenectomy: randomized trial comparing inguinal skin access above or below the inguinal ligament. Ann Surg Oncol. 2009;16(3):721–8.

24. Spillane AJ, Tucker M, Pasquali S. A pilot study reporting outcomes for melanoma patients of a minimal access ilio-inguinal dissection technique based on two incisions. Ann Surg Oncol. 2011;18(4):970–6.
25. La-Touche S, Ayres B, Lam W, Alnajjar HM, Perry M, Watkin N. Trial of ligation versus coagulation of lymphatics in dynamic inguinal sentinel lymph node biopsy for staging of squamous cell carcinoma of the penis. Ann R Coll Surg Engl. 2012;94(5):344–6.
26. Lumachi F, Burelli P, Basso SM, Iacobone M, Ermani M. Usefulness of ultrasound scissors in reducing serous drainage after axillary dissection for breast cancer: a prospective randomized clinical study. Am Surg. 2004;70(1):80–4.
27. Iovino F, Auriemma PP, Ferraraccio F, Antoniol G, Barbarisi A. Preventing seroma formation after axillary dissection for breast cancer: a randomized clinical trial. Am J Surg. 2012;203(6):708–14.
28. Matthey-Gié ML, Gié O, Deretti S, Demartines N, Matter M. Prospective randomized study to compare lymphocele and lymphorrhea control following inguinal and axillary therapeutic lymph node dissection with or without the use of an ultrasonic scalpel. Ann Surg Oncol. 2016;23(5):1716–20.
29. White I, Mills JK, Diggs B, Fortino Hima J, Ellis MC, et al. Sentinel lymph node biopsy for melanoma: comparison of lymphocele rates by surgical technique. Am Surg. 2013;79(4):388–92.
30. Weldrick C, Bashar K, O'Sullivan TA, Gillis E, Clarke Moloney M, et al. A comparison of fibrin sealant versus standard closure in the reduction of postoperative morbidity after groin dissection: a systematic review and meta-analysis. Eur J Surg Oncol. 2014;40(11):1391–8.
31. Carless PA, Henry DA. Systematic review and meta-analysis of the use of fibrin sealant to prevent seroma formation after breast cancer surgery. Br J Surg. 2006;93(7):810–9.
32. Boccardo F, De Cian F, Campisi CC, Molinari L, Spinaci S, Dessalvi S, et al. Surgical prevention and treatment of lymphedema after lymph node dissection in patients with cutaneous melanoma. Lymphology. 2013;46(1):20–6.
33. Catalona WJ. Modified inguinal lymphadenectomy for carcinoma of the penis with preservation of saphenous veins: technique and preliminary results. J Urol. 1988;140(2):306–10.
34. Pearlman NW, Robinson WA, Dreiling LK, McIntyre RC Jr, Gonzales R. Modified ilioinguinal node dissection for metastatic melanoma. Am J Surg. 1995;170(6):647–9. discussion 649–50
35. Zhang SH, Sood AK, Sorosky JI, Anderson B, Buller RE. Preservation of the saphenous vein during inguinal lymphadenectomy decreases morbidity in patients with carcinoma of the vulva. Cancer. 2000;89(7):1520–5.
36. Zhang X, Sheng X, Niu J, Li H, Li D, Tang L, et al. Sparing of saphenous vein during inguinal lymphadenectomy for vulval malignancies. Gynecol Oncol. 2007;105(3):722–6.
37. Dardarian TS, Gray HJ, Morgan MA, Rubin SC, Randall TC. Saphenous vein sparing during inguinal lymphadenectomy to reduce morbidity in patients with vulvar carcinoma. Gynecol Oncol. 2006;101(1):140–2.
38. Kehoe S, Luesley D, Chan KK. A pilot study on early post-operative morbidity and technique of inguinal node dissection in vulval carcinoma. Eur J Gynaecol Oncol. 1998;19(4):374–6.
39. Abbas S, Seitz M. Systematic review and meta-analysis of the used surgical techniques to reduce leg lymphedema following radical inguinal nodes dissection. Surg Oncol. 2011;20(2):88–96.
40. AbuRahma AF, Woodruff BA, Lucente FC. Edema after femoropopliteal bypass surgery: lymphatic and venous theories of causation. J Vasc Surg. 1990;11(3):461–7.
41. Suzuki M, Unno N, Yamamoto N, Nishiyama M, Sagara D, Tanaka H, et al. Impaired lymphatic function recovered after great saphenous vein stripping in patients with varicose vein: venodynamic and lymphodynamic results. J Vasc Surg. 2009;50(5):1085–91.
42. Lawton G, Rasque H, Ariyan S. Preservation of muscle fascia to decrease lymphedema after complete axillary and ilioinguinofemoral lymphadenectomy for melanoma. J Am Coll Surg. 2002;195(3):339–51.

43. Micheletti L, Borgno G, Barbero M, Preti M, Cavanna L, Nicolaci P, et al. Deep femoral lymphadenectomy with preservation of the fascia lata. Preliminary report on 42 invasive vulvar carcinomas. J Reprod Med. 1990;35(12):1130–4.
44. Rouzier R, Haddad B, Dubernard G, Dubois P, Paniel BJ. Inguinofemoral dissection for carcinoma of the vulva: effect of modifications of extent and technique on morbidity and survival. J Am Coll Surg. 2003;196(3):442–50.
45. Bouchot O, Rigaud J, Maillet F, Hetet JF, Karam G. Morbidity of inguinal lymphadenectomy for invasive penile carcinoma. Eur Urol. 2004;45(6):761–5.
46. Hacker NF, Eifel PJ, van der Velden J. Cancer of the vulva. Int J Gynaecol Obstet. 2012;119(Suppl 2):S90–6.
47. Way S. Carcinoma of the vulva. Am J Obstet Gynecol. 1960;79:692–7.
48. Judson PL, Jonson AL, Paley PJ, Bliss RL, Murray KP, Downs LS Jr, et al. A prospective, randomized study analyzing sartorius transposition following inguinal-femoral lymphadenectomy. Gynecol Oncol. 2004;95(1):226–30.
49. Bartlett EK, Meise C, Bansal N, Fischer JP, Low DW, Czerniecki BJ, et al. Sartorius transposition during inguinal lymphadenectomy for melanoma. J Surg Res. 2013;184(1):209–15.
50. Li L, Kou X, Feng X, Liu F, Chao H, Wang L. Clinical application of sartorius tendon transposition during radical vulvectomy: a case control study of 58 cases at a single institution. J Gynecol Oncol. 2015;26(4):320–6.
51. Benoit L, Boichot C, Cheynel N, Arnould L, Chauffert B, Cuisenier J, et al. Preventing lymphedema and morbidity with an omentum flap after ilioinguinal lymph node dissection. Ann Surg Oncol. 2005;12(10):793–9.
52. Logmans A, Kruyt RH, de Bruin HG, Cox PH, Pillay M, Trimbos JB. Lymphedema and lymphocysts following lymphadenectomy may be prevented by omentoplasty: A pilot study. Gynecol Oncol. 1999;75(3):323–7.
53. Murthy V, Gopinath KS. Reconstruction of groin defects following radical inguinal lymphadenectomy: an evidence based review. Indian J Surg Oncol. 2012;3(2):130–8.
54. Thomson DR, Sadideen H, Furniss D. Wound drainage following groin dissection for malignant disease in adults. Cochrane Database Syst Rev. 2014;11(11):CD010933.

Chapter 15
Surgical and Anatomic Considerations of Malignancies Affecting the Groin: Reconstructive Approaches to the Groin

Seyed Amirhossein Razavi, Karan A. Desai, Albert Liao, and Albert Losken

There has been an increasing demand for reconstructive surgery in acute or chronic groin wounds resulting from burn, trauma patients with oncologic defects following groin lymphadenectomy due to urogenital and lower limb malignancies and complications of infrainguinal vascular bypass surgeries [1–6]. The reconstructive surgeon has to select the optimal soft tissue coverage after considering the patient's comorbidities, postoperative quality of life, and functional outcome. In this chapter, we will review the reconstructive options regarding complex wounds in the inguinal region, a brief description of the reconstructive technique, reported outcomes, and comparison between options including use of grafts and flaps.

Primary Closure

Primary closure is always preferred when possible due to lower morbidity. Using the reconstructive ladder as a guideline, primary reconstruction is the most direct form of reconstruction and can be used in a variety of traumatic wounds and skin cancer excisions [7, 8]. Primary closure however is often not possible in wounds that are wide and have been previously irradiated.

S.A. Razavi, M.D. • K.A. Desai, M.D. • A. Liao, M.D.
A. Losken, M.D., F.A.C.S. (✉)
Division of Plastic Surgery, Department of Surgery,
Emory University Hospital Midtown, Emory University School of Medicine,
Medical Office Tower, 550 Peachtree Street, SE Suite 9000, Atlanta, GA 30308, USA
e-mail: srazavi@emory.edu; karan.desai@emory.edu; albert.liao@emory.edu; alosken@emory.edu

K.A. Delman, V.A. Master (eds.), *Malignancies of the Groin*,
https://doi.org/10.1007/978-3-319-60858-7_15

Skin Graft

Skin graft is the transfer of skin from a donor site to a recipient site. When primary closure is not possible secondary to a large defect, a skin graft may be used. In contrast to flap reconstruction, skin grafts develop their blood supply by neovascularization during the first 48–72 h after transfer. Two types of skin grafts are split-thickness and full-thickness skin grafts. Full-thickness skin grafts require more time for revascularization but have less wound contracture profile. In general, full-thickness skin grafts do not have a role in groin reconstruction secondary to the overall large size of the wounds requiring coverage. Groin wounds are most often covered by split-thickness skin grafts from the thigh, buttocks, or trunk and will subsequently heal by epithelialization [8–10]. While split-thickness skin grafts are versatile, their use is restricted in the presence of wounds with exposed vital structures such as bones, nerves, or femoral vessels. Furthermore, reconstruction of a radiated groin wound requires coverage with a well-vascularized tissue to prevent wound-healing complications. In these situations, the clinician should seek more complex methods in the reconstructive ladder such as flap reconstruction [2, 11, 12]. Skin grafts are often used if a wound has been slow to heal and otherwise has a healthy granulation bed or is a superficial defect with a well-vascularized layer subcutaneous tissue above the vessels.

Flap Reconstruction

Flap reconstruction is the transfer of skin and underlying tissue, when it is lifted from a donor site and moved to a recipient site with its blood supply. Pedicled flaps have their vascular supply connected anatomically throughout the flap transfer. Free flaps on the other hand are when vascular supply is disconnected from its donor supply during transfer and reconnected microsurgically to a new vessel at or near the recipient site. The free flaps are often used when no local or pedicle flap is available.

Flap closures are particularly useful for a wound bed with compromised tissue such as in infected or radiated wounds with exposed structures. They provide protection over exposed bones, nerves, or vessels in the wound; increase vascularization to deficient areas; decrease scar formation; and result in a tension-free closure [2, 13].

The main complication associated with flap reconstruction is the risk of vascular compromise. When feasible, pedicled flaps are always preferred over free flaps. Frequent postoperative flap checks including monitoring flap color, temperature, capillary refill, appearance, and use of tissue oximetry and external and implantable Dopplers are necessary for early diagnosis of flap vascular compromise [14–16]. Other potential complications of flap reconstruction include hematoma and seroma formation, surgical site infections, and donor site complications.

There are many different local flap options for groin reconstruction, including gracilis muscle flap (medial femoral circumflex system), sartorius muscle flap

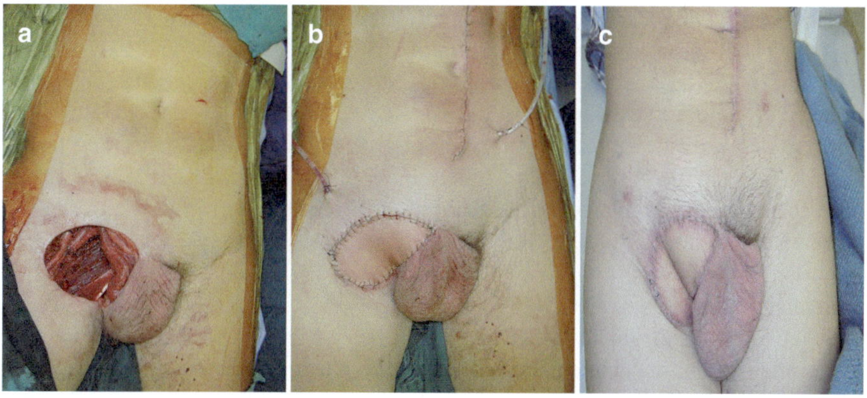

Fig. 15.1 Rectus myocutaneous flap reconstruction of right groin oncologic defect with exposed vessels. (**a**) Pre-reconstruction, (**b**) post-reconstruction, (**c**) first follow-up in the plastic surgery clinic

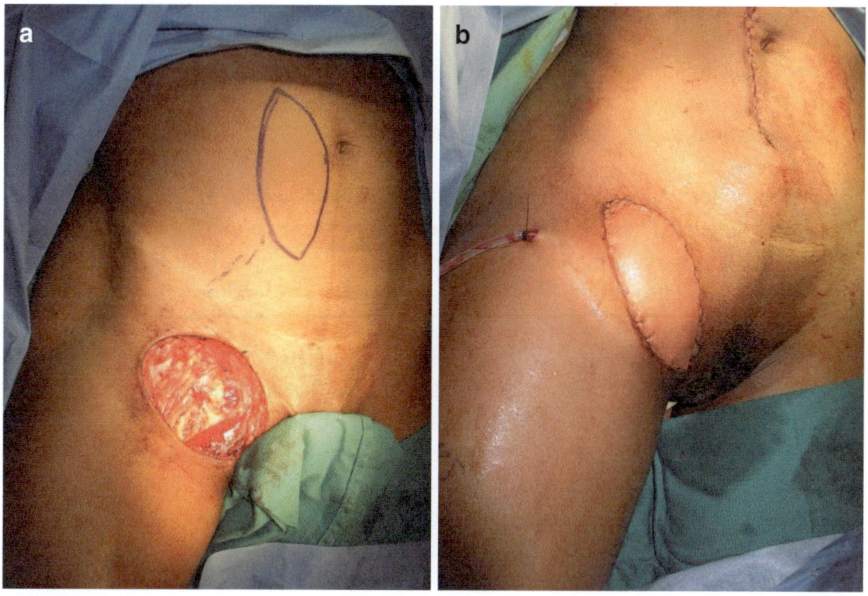

Fig. 15.2 Another example of rectus myocutaneous flap reconstruction. (**a**) Pre-reconstruction, (**b**) post-reconstruction

(superficial circumflex iliac artery from the superficial femoral artery), omental flap (right omental artery from the right gastroepiploic artery), tensor fascia lata flap (ascending branch of lateral femoral circumflex artery), anterolateral thigh flap (descending branch of lateral femoral circumflex artery), rectus abdominis flap (deep inferior epigastric artery), and rectus femoris flap (descending branch of lateral femoral circumflex artery) (Figs. 15.1 and 15.2).

Regional muscle flaps such as the gracilis muscle flap and the sartorius muscle flap are usually useful for reconstruction of small groin defects [17–19]. Sartorius flaps and pedicled gracilis flaps are safe and durable reconstructions for locally infected or exposed vascular grafts and infected groin wound of <10 cm. However, as these flaps consist only of muscle, a skin graft is often needed over the flap for skin coverage in the event of a skin deficit. In their retrospective review of 244 flaps used for complex groin wound reconstruction, Fischer et al. suggested the use of prophylactic sartorius muscle flap for smaller, low-risk wounds, reserving rectus femoris flap for larger, more complex wounds and anterolateral thigh for wounds with larger cutaneous defect [20].

A retrospective review by Ducic and colleagues for the use of extended dissection gracilis flap in high-risk patients with complex groin wounds requiring more coverage than a standard gracilis flap concluded that the extended dissection gracilis flap has greater arc of rotation and no restriction on postoperative ambulation or thigh abduction [21]. First described by Hason et al. [22] in the extended dissection gracilis flap technique, after the identification of the pedicle, the muscle is divided proximally, and pedicle dissection continues deeper to the adductor longus muscle, the perforators, and the vascular network and can be extended all the way to the profunda femoris artery.

There does not appear to be a consensus first choice flap option for large (>10 cm) groin defects; however, anterolateral thigh (ALT) flap (both as a pedicled and free flap) and vertical rectus abdominis myocutaneous flap are considered highly versatile and reliable for this type of defect [3, 7, 23]. In their retrospective review, Lannon et al. determined that the pedicled ALT flap should be the flap of choice for large groin defects. There are several techniques to preserve flap viability, including suprafascial flap harvesting, extended harvesting of fascia, utilization of the fascia to protect the pedicle, coharvest with the vastus lateralis, and preservation of large lateral rectus femoris perforators [23].

A retrospective review by LoGiudice et al. looking at 39 patients who underwent reconstruction with ALT flap and rectus abdominis flap demonstrated shorter time to healing and lower rate of delayed postoperative complications in the ALT group. Incisional hernias were of particular concern in rectus abdominis flap patients [4]. Aslim et al. [3] reported their use of ALT flap and vertical rectus abdominis musculocutaneous (VRAM) flap for large groin defects, both resulting in consistent results with little morbidity. Additionally, both flaps have the benefit of having donor sites that can be closed primarily reducing donor site morbidity.

A review of 50 patients, who underwent VRAM (63%) and extended RAM (37%) flaps by Parrett et al., showed reliable coverage of irradiated thigh and groin oncologic wounds. Extended RAM was used for contralateral and more distal defects. Parret reported significantly improved results when performing immediate compared to delayed reconstruction [5].

Comparing vertical and oblique RAM flaps in 71 patients, Combs and colleagues reported similar complication rates between the two reconstruction techniques. They concluded that an oblique RAM flap is a safe alternative to VRAM with advantages including greater arc of rotation, thinner skin paddle, less bulk, and limited fascial harvest [24].

Alkon et al. [25] reviewed their experience with rectus femoris muscle flaps as an effective and reliable mean for complex groin wound reconstruction. This flap is harvested through a mid-anterior incision extending over the distal two-thirds of the thigh, with muscle elevated on its pedicle and transposed into the groin wound defect. They reported no flap losses and no donor site complications in the 37 rectus femoris flaps performed between 1999 and 2003. Reoperation was required in four patients, one for flap readvancement and three for prosthetic vascular graft removal.

A cost analysis study by Chatterjee and colleagues comparing sartorius ($n = 234$) versus rectus femoris ($n = 62$) flap in the treatment of the infected vascular groin graft wounds notes that the rectus femoris flap is the more cost-effective option with less major complication profile [26].

In conclusion, evaluation for complex groin wound reconstruction starts with considering patients' comorbidities, postoperative quality of life, and functional outcome and going through the reconstructive ladder. The reconstructive surgeon should always consider primary reconstruction and skin grafting as first options and then move to more complex reconstruction options such as pedicled and free flaps.

The gracilis and sartorius flaps are useful options for reconstruction of smaller, low-risk groin wounds (<10 cm), while anterolateral thigh flap and vertical rectus abdominis myocutaneous flap are reliable options for reconstruction of larger groin defects.

References

1. Horch R, Stark GB. Prosthetic vascular graft infection--defect covering with delayed vertical rectus abdominis muscular flap (VRAM) and rectus femoris flap. Vasa. 1994;23:52–6.
2. Murthy V, Gopinath KS. Reconstruction of groin defects following radical inguinal lymphadenectomy: an evidence based review. Indian J Surg Oncol. 2012;3:130–8.
3. Aslim EJ, Rasheed MZ, Lin F, Ong YS, Tan BK. Use of the anterolateral thigh and vertical rectus abdominis musculocutaneous flaps as utility flaps in reconstructing large groin defects. Arch Plast Surg. 2014;41:556–61.
4. LoGiudice JA, Haberman K, Sanger JR. The anterolateral thigh flap for groin and lower abdominal defects: a better alternative to the rectus abdominis flap. Plast Reconstr Surg. 2014;133:162–8.
5. Parrett BM, Winograd JM, Garfein ES, Lee WP, Hornicek FJ, et al. The vertical and extended rectus abdominis myocutaneous flap for irradiated thigh and groin defects. Plast Reconstr Surg. 2008;122:171–7.
6. Perez-Burkhardt JL, Gonzalez-Fajardo JA, Carpintero LA, Mateo AM. Sartorius myoplasty for the treatment of infected groins with vascular grafts. J Cardiovasc Surg (Torino). 1995;36:581–5.
7. Larson JD, Altman AM, Bentz ML, Larson DL. Pressure ulcers and perineal reconstruction. Plast Reconstr Surg. 2014;133:39e–48e.
8. Simman R. Wound closure and the reconstructive ladder in plastic surgery. J Am Col Certif Wound Spec. 2009;1:6–11.
9. Ratner D. Skin grafting. Semin Cutan Med Surg. 2003;22:295–305.
10. Ogawa R, Hyakusoku H, Ono S. Useful tips for successful skin grafting. J Nippon Med Sch. 2007;74:386–92.
11. Fujioka M. Surgical Reconstruction of Radiation Injuries. Adv Wound Care (New Rochelle). 2014;3:25–37.

12. Payne WG, Walusimbi MS, Blue ML, Mosiello G, Wright TE, et al. Radiated groin wounds: pitfalls in reconstruction. Am Surg. 2003;69:994–7.
13. Schoeman BJ. The tensor fascia lata myocutaneous flap in reconstruction of inguinal skin defects after radical lymphadenectomy. S Afr J Surg. 1995;33:175–8.
14. Lin SJ, Nguyen MD, Chen C, Colakoglu S, Curtis MS, et al. Tissue oximetry monitoring in microsurgical breast reconstruction decreases flap loss and improves rate of flap salvage. Plast Reconstr Surg. 2011;127:1080–5.
15. Lohman RF, Langevin CJ, Bozkurt M, Kundu N, Djohan R. A prospective analysis of free flap monitoring techniques: physical examination, external Doppler, implantable Doppler, and tissue oximetry. J Reconstr Microsurg. 2013;29:51–6.
16. Sinna R, Boloorchi A, Mahajan AL, Qassemyar Q, Robbe M. What should define a "perforator flap"? Plast Reconstr Surg. 2010;126:2258–63.
17. Maser B, Vedder N, Rodriguez D, Johansen K. Sartorius myoplasty for infected vascular grafts in the groin. Safe, durable, and effective. Arch Surg. 1997;132:522–5. discussion 525-526
18. Chen SH, Hentz VR, Wei FC, Chen YR. Short gracilis myocutaneous flaps for vulvoperineal and inguinal reconstruction. Plast Reconstr Surg. 1995;95:372–7.
19. Shih PK, Cheng HT, Wu CI, Chang SC, Chen HC, et al. Management of infected groin wounds after vascular surgery. Surg Infect (Larchmt). 2013;14:325–30.
20. Fischer JP, Mirzabeigi MN, Sieber BA, Nelson JA, Wu LC, et al. Outcome analysis of 244 consecutive flaps for managing complex groin wounds. J Plast Reconstr Aesthet Surg. 2013;66:1396–404.
21. Ducic I, Dayan JH, Attinger CE, Curry P. Complex perineal and groin wound reconstruction using the extended dissection technique of the gracilis flap. Plast Reconstr Surg. 2008;122:472–8.
22. Hasen KV, Gallegos ML, Dumanian GA. Extended approach to the vascular pedicle of the gracilis muscle flap: anatomical and clinical study. Plast Reconstr Surg. 2003;111:2203–8.
23. Lannon DA, Ross GL, Addison PD, Novak CB, Lipa JE, et al. Versatility of the proximally pedicled anterolateral thigh flap and its use in complex abdominal and pelvic reconstruction. Plast Reconstr Surg. 2011;127:677–88.
24. Combs PD, Sousa JD, Louie O, Said HK, Neligan PC, et al. Comparison of vertical and oblique rectus abdominis myocutaneous flaps for pelvic, perineal, and groin reconstruction. Plast Reconstr Surg. 2014;134:315–23.
25. Alkon JD, Smith A, Losee JE, Illig KA, Green RM, et al. Management of complex groin wounds: preferred use of the rectus femoris muscle flap. Plast Reconstr Surg. 2005;115:776–83. discussion 784–75
26. Chatterjee A, Kosowski T, Pyfer B, Fisher CS, Tchou JC, et al. A cost-utility analysis comparing the sartorius versus the rectus femoris flap in the treatment of the infected vascular groin graft wound. Plast Reconstr Surg. 2015;135:1707–14.

Chapter 16
Quality of Life and Outcomes in Patients with Malignancies of the Groin

Andrew J. Lambour and Sandra L. Wong

Introduction

Malignancies of the groin can be generally categorized as primary tumors or as regional spread of disease to the inguinal lymph nodes. Prognosis and overall survival are generally related to the natural history of the underlying tumor type as well as available treatment options. Because of the specific anatomic considerations of the groin, it is important to note that malignancies of the groin have a serious impact on the quality of life (QoL) of patients. Oftentimes, this impact occurs as the result of the uncontrolled or progressive disease process or as the direct result of complications of treatment for either primary or secondary disease. As such, QoL considerations span the entire continuum of cancer care, including preemptive or primary operations to longer-term posttreatment effects.

This chapter will provide an overview of QoL, why QoL is important, and how QoL is measured, as well as touch upon the specifics of how certain disease states impact QoL. The four diseases to be covered in this chapter include melanoma, penile cancer, soft tissue sarcoma, and vulvar cancer. Related QoL considerations can include mild symptoms of lymphedema to painful, bulky disease. Generally speaking, there has been little research on the topic despite the severe implications that it has on a patient's overall health.

A.J. Lambour, M.D. • S.L. Wong, M.D., M.S. (✉)
Department of Surgery, Dartmouth-Hitchcock Medical Center, Geisel School of Medicine at Dartmouth, Lebanon, NH, USA
e-mail: Andrew.J.Lambour@hitchcock.org; Sandra.L.Wong@hitchcock.org

K.A. Delman, V.A. Master (eds.), *Malignancies of the Groin*,
https://doi.org/10.1007/978-3-319-60858-7_16

What Is "Quality of Life"?

Aristotle's *Nicomachean Ethics* provides one of the earliest definitions of quality of life. In his discourse on happiness, Aristotle states, "what constitutes happiness is a matter of dispute… some say one thing and some say another, indeed very often the same man says different things at different times: when he falls sick he thinks health is happiness, when he is poor wealth." This in fact informs present-day discussions, when the World Health Organization (WHO) defines *health* as "a state of complete physical, mental, and social well-being, and not merely the absence of disease" [1].

There is actually no consensus as to what constitutes QoL. Much of this stems from the fact that QoL is inherently subjective and personal. It is because of this subjectivity that there exists a deep, indescribable sense of what "quality of life" is. It is ineffable; one may know it when he or she sees it, and physicians may understand it when a patient describes it. QoL is a complex construct, which can be applied across a variety of contexts. Healthcare-related quality of life (HRQoL) is the focus herein. Reasonable proposals for evaluating QoL capture the following areas: general health, physical function, physical symptoms and toxicity, emotional function, cognitive function, role function, social function, sexual function, and existentialism. A possibly more parsimonious list includes categories of physical, emotional, and social aspects of one's life. Any complete definition or measurement of QoL would ideally incorporate these dimensions.

Quality of life attributions are subjective—one patient's perspective on QoL may be very different from that of another patient. A patient's quality of life is often affected by a variety of issues related to their disease. Age, performance status, disease burden, and tolerance of treatment may influence a patient's perspective on QoL. In addition to this, there is a temporal aspect to QoL that may change as external factors or disease processes change. Quantifying QoL is truly unlike many of the objective measurements that are obtained in healthcare, such as quantified lab values.

Instruments used to measure QoL attempt to standardize and validate through an objective and scientific manner. By quantifying QoL, one is able to make comparisons and track changes over the course of care. Measurement of QoL must be patient driven though scales used and reported by clinicians should be distinguished from patient-reported outcomes (PROs)/patient-reported outcome measures (PROMs). The Food and Drug Administration defines "patient-reported outcomes" as any report of the status of a patient's health condition that comes directly from the patient, without interpretation by a clinician or anyone else. As would be expected, there has been great interest in PROs/PROMs, but they have yet to be used widely in clinical practice. As such, there are many methodologic challenges, and currently, there is not enough known about their reliability and how to aggregate the data for longitudinal measurement of performance.

Why Is Assessment of Quality of Life Important?

From its measurement in clinical trials to its impact on treatment decisions, QoL plays a tremendous role in decision-making around care that is provided to patients. Ultimately, QoL is one of the major drivers when taking overarching goals of care into account.

One of the most frequent uses for QoL measurement is to balance treatment effects with effects of treatment. In comparing different treatments rendered with curative intent, regimens that have better "objective" or hard outcomes are often felt to carry the "cost" of increased toxicity. For example, a study by Coates et al. demonstrated that continuous as opposed to intermittent chemotherapy for advanced breast cancer resulted in prolonged survival as well as improved QoL [2]. Similarly, QoL measures are also used to compare treatments for palliative intent, which have classically been felt to trade longer survival for improved symptom management. This long-held belief was disproved when Temel et al. showed that in patients with newly diagnosed advanced non-small cell lung cancer, up-front/early palliative treatment combined with standard oncologic care resulted in improved QoL and survival as compared to standard treatment alone [3]. Increasingly, it appears that improvement in QoL is a predictor of those who have better overall outcomes.

Without the addition of formalized QoL assessments, the so-called primary outcomes such as overall survival or progression-free survival would be incomplete. QoL assessments are complementary to the traditionally used objective measurements and provide a reliable, objective, patient-centric data which may otherwise have gone unnoticed by a clinician focused on survival alone. Importantly, assessments of QoL can be independent endpoints and serve as a valuable tool for informed communications with patients about treatment options.

Specific to malignancies of the groin, treatment considerations revolve around local and regional disease control. The anatomic confines of the groin, whether this is defined by the deep pelvis, upper thigh, or inguinal nodal basin, make this a potentially difficult area to treat. Nearby critical structures such as major nerves, vessels, and muscles must be preserved to maximize function and minimize complications.

How Do You Measure Quality of Life?

Given its complexity, how does one even begin to measure QoL? To begin with, the data that is used to measure QoL must be obtained from the patient. This data is usually obtained through the form of questionnaires that the patient themselves fills out or by having the healthcare provider ask a series of questions. Failure to obtain data in such a fashion impinges upon the unique and subjective nature of QoL. Several studies have shown that healthcare providers or relatives of the patient provide differing responses from that of the patient. These estimates can either overestimate or underestimate a patient's QoL. Observers frequently focus on physical signs and symptoms as a surrogate for QoL. As discussed above, this is only one dimension involved with QoL and disregards the impact that other factors have on a patient. Additionally, when healthcare professionals observe expected physical signs and symptoms, such as nausea or emesis following chemotherapy, they may inadvertently miscalculate their actual impact on a patient's QoL.

The specifics are beyond the scope of this chapter, but in general, the following algorithm is used to develop a QoL assessment. Data obtained from the patient is used to develop numeric score. This numeric score is standardized to a scale score (i.e., 0–100). This process is termed standard scoring method. Newer QoL instruments use T-scores to develop a scale through a process known as norm-based scoring.

There are several different QoL instruments for a variety of purposes (Table 16.1). Some QoL tools focus on a single dimension, such as physical function. Others incorporate several different facets of QoL. Although it is impossible to include every dimension, most researchers agree that tools that incorporate several aspects of QoL are considered more comprehensive and superior to the single-dimension approach. From a historical perspective, early assessments of QoL were primarily objective and based on clinical assessment of physical factors. A patient's pain level or ability to perform daily activities was equated to a patient's QoL. In 1947, the Karnofsky Performance Scale was one of the first tools that moved beyond clinical assessment. In the 1970s–1980s, several tools such as the Nottingham Health Profile incorporated additional dimensions in the score. This multidimensional assessment continues to this day with tools such as the SF-36.

Table 16.1 Commonly used instruments to measure quality of life

Name	Target measurement	Logistics of administration	Comments and use
Generic quality of life assessment tools			
Sickness Impact Profile (SIP)	• Measures perceived health status based on impact on behavior	• Interviewer- or self-administered • 136 items over 12 areas of dysfunction • 20–30 min to complete	• Designed for new treatments and health levels in populations
EuroQol (EQ-5D)	• Measures physical, mental, and social function	• Interviewer- or self-administered • 5 items over five areas • 2 min to complete	• Designed as a general-purpose tool • Useful for cost utility evaluation • Extreme simplicity • Best used with other tools
36-Item Short Form (SF-36)	• Measures physical, social, and emotional function	• Interviewer- or self-administered • 36 items over eight health areas	• Generic health concepts with diverse applications • Most widely used
Patient Generated Index (PGI) and Schedule for Evaluation of Individual Quality of Life (SEIQoL)	• Measures five areas as chosen by the patient	• Interviewer- or self-administered • Scales and scoring are slightly different between PGI and SEIQoL • PGI is more simple	• Can identify issues important to an individual patient • Less helpful for clinical trials and comparing groups

Table 16.1 (continued)

Name	Target measurement	Logistics of administration	Comments and use
Disease-specific assessment tools			
European Organization for Research and Treatment of Cancer (EORTC) QLQ-C30	• Cancer-specific different modules for different diseases • Focus on clinical symptoms and ability to perform	• Interviewer- or self-administered • 30 items • Applicable across multiple cultural settings	• Used for clinical trials • Very sensitive tool
Functional Assessment of Cancer Therapy-General (FACT-G)	• Cancer-specific different modules for different diseases • Focus on feelings and concerns	• Interviewer- or self-administered • 27 items • Applicable across multiple cultural settings	• Similar to EORTC QLQ-C30 difference is the focus
Rotterdam Symptom Checklist (RSCL)	• Cancer specific • Focus on symptoms and side effects	• Interviewer- or self-administered • 30 items • Question style differs from EORTC QLQ-C30 and FACT-G	• Used for clinical trials
Dimension-specific assessment tools			
McGill Pain Questionnaire (MPQ)	• Measures pain levels	• Interviewer- or self-administered • 20 items • 5–15 min to complete	• A shorter SF-MPQ exists
Barthel Index of Disability (BI)	• Measures disability, assesses ADLs	• Usually interviewer-administered • 10 items • 1 min to complete	• Used for measuring rehab outcomes
Multidimensional Fatigue Inventory (MFI)	• Measures fatigue	• Usually self-administered • 20 items	

The SF-36

In the early 1990s, Ware et al. conducted a multicenter trial entitled the Medical Outcomes Study (MOS) [4]. The aim of this study was to produce a health-related quality of life assessment tool that would serve as a "happy medium" between lengthy, all-inclusive tools and the shorter, crude instruments. The result of the MOS was the Short Form 36 (SF-36). The SF-36 is a by-product of previous assessment tools. It incorporates and focuses on the dimensions that are most frequently affected by treatments and disease, as well as those most often measured in healthcare surveys. Because of its efficiency and comprehensive nature, the SF-36 remains one of the most widely used assessments of health-related quality of life. Over the years, various versions and updates have been produced. The most current version is the SF-36v2. Other shorter versions of the SF-36 have been developed: SF-12, SF-8, and SF-6D.

As its name would suggest, the SF-36 consists of a total of 36 questions. The questions span a total of eight health dimensions and are ultimately broken down into an evaluation of physical health and mental health. Within the physical health domain are items measuring physical functioning (ten items), role functioning (four items), bodily pain (two items), and general health (five items). Within the physical health domain are items measuring vitality (four items), social functioning (two items), role emotional (three items), and mental health (five items). Designers of the SF-36 have tested and validated the above questions so that they can be scored against a standard scoring method. Logistically, the questionnaire can either be self-assessed or administered by a trained interviewer.

Malignancies of the Groin

There are four main oncologic diseases that involve the inguinal and groin region: melanoma, penile cancer, soft tissue sarcoma, and vulvar cancer. Soft tissue sarcoma would present as a primary site of disease in the groin. Otherwise inguinal lymph nodes serve as a site of regional metastasis for primary diseases of the pelvis, trunk, and lower extremities. Metastatic disease can be identified at the time of initial presentation or as a site of recurrence.

The mainstay of treatment for malignancies of the groin is surgical resection. Resection can take many forms including extirpation of primary or recurrent disease, sentinel lymph node biopsy, and completion lymph node dissection. Indications and technical aspects of these procedures, including the nuances of critical anatomic structures, are discussed elsewhere. There is often a direct relationship between increased complexity of resection and increased morbidity, with an ensuing negative impact on QoL. The use of radiotherapy is infrequently substituted for resection, but is not uncommonly considered as an adjunct to resection. In such situations, the impact on appearance and function can be severe and must be weighed when considering risks versus benefit of treatment.

As discussed above, quality of life is a complex construct. With regard to malignancies of the groin, this complexity makes no exceptions. One can imagine a multitude of factors that impact a patient's QoL. Often, we experience QoL in the context of a negative impact after treatment, but it should be noted that sometimes patients suffering from oncologic diseases experience a positive effect. The term "response shift" has been used to capture the shift in internal standards, values, and conceptualization of QoL. Receiving a cancer diagnosis gives some patients a different outlook on life and improves QoL measurement in response to treatment. It is difficult, if not nearly impossible, to study all the effects of a disease on QoL.

Over time, several disease-specific assessment tools have been developed. For example, there are QoL instruments that have been validated and tested for vulvar cancer and melanoma—the FACT-V and FACT-M, respectively. Generic instruments, such as the EORTC QLQ-C30 and SF-36, are extremely good for an overall, broad QoL assessment. These tools are also advantageous for comparisons between disease states. As expected, the more refined, disease-specific instruments are better able to detect and compare disease-specific areas of impact. For example, the FACT-M will better detect the psychological and social impact that avoiding sun exposure has on melanoma patients.

The intent of the next several sections is to provide an overview of QoL with specific consideration to disparate disease types, noting currently clinical evidence when it is available. Risks associated with progressive disease and its impact on survival outcomes as well as overall function and QoL are the major considerations when deciding to proceed with an operation. However, because complications are relatively common with resection and/or ILND, risks and benefits must be fully considered.

Quality of Life and Melanoma (with Inguinal Nodal Basin Involvement)

At one time, melanoma was a rare form of cancer. Since the mid-1950s, however, its incidence is increasing faster than any other form of cancer. For some, melanoma is a chronic disease that portends a lifetime of risk factor mitigation and careful surveillance. For intermediate and thick tumors, the likelihood of lymph node metastasis is much more frequent, and when the disease is located on the lower extremities or trunk, regional spread to the inguinal nodes must be evaluated.

In the past, inguinal lymph node dissections (ILND) were performed in either an elective setting (ILND done preemptively to diagnose and treat) or a therapeutic setting (ILND done in the setting of diagnosed (palpable) disease). The presumed trade-offs of elective ILND were early removal of disease at a microscopic level versus unnecessary removal of nodes that never would have developed disease. With the advent of sentinel lymph node biopsy (SLNB), the rates of ILND decreased, since such a procedure was reserved for those with known disease, presumably found at a much earlier stage. Current controversies around the need for completion ILND revolve around the question of whether there is any therapeutic effect of ILND following SLNB.

While many studies considered surgical complications, data on the dimensions of QoL were lacking. The surgical management of metastatic disease to the groin is fraught with morbidity and complications. Overall, wound complication rates following inguinal lymph node dissections are reported in up to 71% of patients [5]. Frequent complications include seroma (17%), wound infection (9%), wound necrosis (3%), and edema (20%) [6]. Fortunately, complications appeared to decline with more minimally invasive procedures. The complication rate for a SLNB is between 4.6 and 10.1% [7]. Several studies, including the Sunbelt Melanoma Trial, have demonstrated that the morbidity with SLNB is less than compared to SLNB plus CLND. Further, surgical techniques that utilize a minimally invasive approach have shown some promise to reduce complications, particularly wound-related ones [8, 9]. Additional longitudinal QoL data will be important as approaches to inguinal nodal disease are continuously evaluated.

Bulky adenopathy of the iliac and inguinal nodes can result in lymphedema, vascular compromise, neuropathy/pain, or locoregional wound complications (either associated with involvement of nearby structures or overlying skin). Similarly, feared complications resulting from inguinal node dissections include lymphedema, ranging from relatively asymptomatic edema based on serial leg measurements to massive lymphedema which limits physical functioning; vascular compromise, including venous thrombosis; neuropathy from direct damage to nerves; or nonhealing wounds/infectious complications. Several techniques accompanying ILND include preservation of the muscle fascia, pedicled omentoplasty, sartorius transposition, and saphenous vein sparing and have been developed in an attempt to mitigate complications, without much improvement. Despite all of the above data around morbidity of groin dissections, studies have shown that the QoL of survivors is essentially comparable to the general population [7, 10]. The exact reason for this lack of impact is unknown though possible explanations include concepts such as response shift, survival bias, as well as a selection bias.

Because the majority of patients have a good functional outcome, many believe that the major impact on QoL is psychological. A systematic review of QoL in melanoma patients supports this claim. Approximately a third of all patients will experience a significant level of distress. This effect peaks around the time of diagnosis and shortly after treatment but then decreases with time. Poor preoperative health status and psychological illness are predictors of postoperative QoL impairment [11]. Even worse, studies have shown an association between QoL impairment, psychological factors, and personality structure that may affect survival [11, 12]. The cause for such psychological distress is largely unknown. The finding that approximately 85% of patients did not receive adequate treatment or assessment for psychological distress during surveillance, however, is equally concerning. Like all of the diseases reviewed here, this highlights the need and importance of continued work with QoL in patients. It also shows how important preoperative and continued postoperative psychosocial care is to QoL.

Quality of Life and Penile Cancer

Penile cancer is rare in industrialized countries with a reported annual incidence of 1/100,000 men. The few patients that are affected by penile cancer experience significant impact on their QoL. Overall, penile cancer treatment negatively impacts a patient's well-being 40% of the time [13]. Penile cancer can significantly impact a patient's sexuality and intimacy, body image, urinary function, mental health, and physical function. Two major areas that have been studied in the literature are the physical impact and psychosocial and sexual impact.

Penile cancer often presents at an advanced stage. Much of this is due to patient factors such as embarrassment, confusion, and ignorance, as well as a delay in diagnosis by healthcare professionals [14, 15]. Increased stage of disease often results in increased complexity and involvement of surgical management. This affects primary disease treatment options, as well as the need and type of groin dissection.

Primary treatment options for penile cancer range from penectomy to partial penectomy to penile sparing procedures (e.g., laser ablation or Mohs micrographic surgery) [16, 17]. It should be mentioned that there are nonsurgical treatment options (e.g., radiotherapy and chemotherapy); however, these are restricted to cases of very limited disease, as well as palliative or adjuvant treatments [18, 19]. Several studies support the claim that more radical surgery negatively impacts QoL [16, 20, 21]. A study by Keiffer et al. found that compared to penile sparing procedures, partial penectomy resulted in more problems with orgasm, body image, life interference, and urination [20]. Because of this negative impact on QoL, there has been a concerted effort on the part of clinicians to utilize organ-sparing procedures [16].

Overall, primary treatment of penile cancer is usually technically uncomplicated and frequently associated with short hospital stays with rapid recovery in physical health [15, 22]. Management of inguinal metastatic disease is more complicated, especially with regard to ILND, either in an up-front manner for clinically apparent disease or completion ILND, performed in the setting of a positive SLNB. In a recent large-scale retrospective review of ILND, postoperative complications occurred in 55.4% of cases [23]. Specific postsurgical complications include wound infection (18%), skin edge necrosis (61%), seroma (5%), and lymphedema (25%) [24]. Not surprisingly, with increased depth and involvement of inguinal nodes, the morbidity and physical impact also increases [25]. Discussions on the impact on QoL that penile cancer treatment, and in particular ILND, will have are important for informing clinical decision-making.

Another dimension of QoL that has been studied in the context of penile cancer is the psychosocial and sexual impact on patients. Though not always related to groin dissections, it is an important component to the overall care of this patient population. Penile cancer is predominantly a disease of older men; however, up to 22% of patients are younger than the age of 40 and a significant portion of all patients report being sexually active [15]. A large-scale meta-analysis found that up to 60% of penile cancer patients have a negative impact on sexual function [13]. The data regarding the psychosocial and sexual impact are somewhat mixed, but a large portion of the studies indicate that there is a profound impact on sexual function and psychosocial issues, such as body image and masculinity [18, 20]. This interplay is complex and touches upon a variety of factors such as surgery type, cultural context, preoperative education, as well as postoperative support and rehabilitation. For example, it has been shown that the more radial the treatment procedure, the more severely impacted one's sexual function [18]. It has also been demonstrated that many patients rush into treatment with little knowledge or understanding of the impact of treatment. This negatively influences their postoperative course and severely limits rehabilitation efforts [15].

Quality of Life and Vulvar Cancer

Vulvar cancer accounts for 4% of all gynecologic malignancies. Over the past decade, however, the incidence is on the rise, especially within the younger population. Similar to penile cancer, the effects of vulvar cancer on QoL are complex and not insignificant given the anatomic confines. Studies have shown an impact on physical, emotional, social, sexual, and body image aspects of QoL [26]. To help better evaluate these changes, a vulvar cancer disease-specific QoL assessment tool has been developed entitled the FACT-V [26, 27]. This instrument is based on the previously mentioned generic FACT assessment. Two major areas that have been studied in the literature are the physical impact and psychosocial and sexual impact.

ILND is an important component of vulvar cancer treatment. Even in early stage disease, up to 35% of cases will have lymph node involvement. Though there is certainly morbidity associated with primary disease treatment, studies have demonstrated that the morbidity of vulvar cancer treatment is largely related to groin dissections [28]. Postoperative complications from groin dissections are mostly related to wound healing and lower extremity lymphedema. In a study by Janda et al., lymphedema was the most common postoperative complaint [26]. Several studies have shown that lymphedema, in and of itself, has a significant negative impact on QoL. This impact spans multiple dimensions including physical, cognitive, emotional, social, fatigue, pain, sleep, and financial [28]. Procedures that limit the extent of groin dissection when no regional disease is suspected (i.e., SLNB) decrease the morbidity and impact on QoL [29]. With regard to primary disease treatment, there has also been a correlation between less radical treatments and decreased impact on morbidity and QoL [30, 31]. From a historical perspective, there has been a shift in the standard treatment from vulvectomy and ILND to wide local excision and SLNB. A focus on QoL has ultimately fostered improved physician patient discussions and leads to changes in individual treatment plans.

Similar to penile cancer, the psychosocial and sexual impact of vulvar cancer treatment has been well established. This involves issues such as body image, depression, fear, lower self-esteem, and loss of self-confidence [31]. Lymphadenectomy has been shown to be an independent risk factor for decreased sexual function [32]. Because of the chronic nature of oncologic diseases as well as the lifelong disfiguring impact that treatments have, this psychological impact is not short lived. Furthermore, many of these issues are more common in younger patients, and with a decreased age of diagnosis, this issue is becoming increasingly more important [26]. Further programs for preoperative assessment and education, as well as postoperative support and rehab, need to be developed to help combat this problem.

Quality of Life and Soft Tissue Sarcoma

Soft tissue sarcomas are a rare group of malignancy that occurs in approximately 1% of the adult population and up to 15% of the pediatric population. Although rare, sarcomas can be a devastating disease, noting poorer prognoses with recurrences and widespread metastasis. There has been a significant amount of QoL research as it relates to sarcoma as a whole. Much of this data relates to the functional and psychosocial impact. Groin involvement is largely attributable to primary disease in the inguinal region because metastatic disease to this area is rare. Because of this, there are few reports of lymphadenectomy in sarcoma and no QoL data concerning inguinal lymph node involvement associated with sarcomas. Approximately 46% of all cases involve the thigh, buttock, and inguinal region (including the spermatic cord). There are little to no groin-specific QoL data related to sarcomas.

Surgical treatment for primary sarcomas varies from wide local excision to radical extirpation. Resections of primary groin sarcomas can be a technically difficult operation. This complexity is often the result of local invasion into important nearby structures (i.e., nerves or major vessels) or the wide margins that are needed to prevent recurrence (i.e., resections involving the involved musculature or the testis). Overall, complications occur in up to 68% of cases [33]. These complications range from local wound infections and poor healing to arterial insufficiency and decreased neurological function.

With the radical nature of some sarcoma surgery for pelvic or upper thigh tumors, patients are often left with severe functionally debilitating outcomes and have resultant problems with daily activities. It has been shown that sarcoma survivors who are unemployed are more likely to experience psychological stress [34]. When treatment involves amputation of an extremity, there is a serious impact on the psychological and emotional aspects of life though QoL reports are sparse. Furthermore, much of the research in this area has been performed in the pediatric population, which magnifies the gravity of the situation. Surprisingly, most of the data directly on the emotional and psychological dimensions is mixed. Several studies have shown a strong but incomplete link with anxiety, depression, and PTSD [35, 36]. Given the potential impact on a young population afflicted with a chronic disease, further research is needed to better define this relationship and lend support to this delicate situation.

Conclusions

Either primary or secondary malignancies of the groin can have a serious impact on the quality of life (QoL) of patients as the result of uncontrolled or progressive disease or complications of treatment. As such, QoL considerations span the entire continuum of care. The importance of QoL considerations must be balanced with complexity of QoL measurement. There is a need for a better understanding of the impact that malignancies of the groin have on a patient's QoL as treatment options are evaluated and discussed with patients.

References

1. Fayers PM, Machin D. Quality of life: the assessment, analysis and reporting of patient-reported outcomes. 3rd ed. West Sussex: John Wiley & Sons; 2016.
2. Coates A, Gebski V, Bishop JF, Jeal PN, Woods RL, Snyder R, et al. Improving the quality of life during chemotherapy for advanced breast cancer. N Engl J Med. 1987;317:1490–149.

3. Temel JS, Greer JA, Muzikansky A, Gallagher ER, Admane S, Jackson VA, et al. Early palliative care for patients with metastatic non-small-cell lung cancer. N Engl J Med. 2010;363:733–42.
4. Ware JE, Sherbourne CD. The MOS 36-item short-form health survey (SF-36): 1. conceptual framework and item selection. Med Care. 1992;30(6):473–83.
5. Serpell JW, Carne PW, Bailey M. Radical lymph node dissection for melanoma. ANZ J Surg. 2003;73(5):294–9.
6. Baas PC, Schraffordt Koops H, Hoekstra HJ, van Bruggen JJ, van der Weele LT, Oldhoff J. Groin dissection in the treatment of lower-extremity melanoma. Short-term and long-term morbidity. Arch Surg. 1992;17(3):281–6.
7. De Vries M, Hoekstra HJ, Hoekstra-Weebers JEHM. Quality of life after axillary or groin sentinel lymph node biopsy, with or without completion lymph node dissection, in patients with cutaneous melanoma. Ann Surg Oncol. 2009;16(10):2840–7.
8. Jakub JW, Terando AM, Sarnaik A, Ariyan CE, Faries MB, Zani S, et al. Safety and feasibility of minimally invasive inguinal lymph node dissection in patients with melanoma (SAFE-MILND): report of a prospective multi-institutional trial. Ann Surg. 2017;265(1):192–6.
9. Abbott AM, Grotz TE, Rueth NM, Nernandez Irizarry RC, Tuttle TM, Jakub JW. Minimally invasive inguinal lymph node dissection (MILND) for melanoma: experience from two academic centers. Ann Surg Oncol. 2013;20(1):340–5.
10. Noorda EM, van Kreij RH, Vrouenraets BC, Nieweg OE, Muller M, Kroon BB, et al. The health-related quality of life of long-term survivors of melanoma treated with isolated limb perfusion. Eur J Surg Oncol. 2007;33(6):776–82.
11. Cornish D, Holterhues C, van de Poll-Franse LV, Coebergh JW, Nijsten T. A systematic review of health-related quality of life in cutaneous melanoma. Ann Oncol. 2009;20(6):51–8.
12. Lehto US, Ojanen M, Dyba T, Arpo A, Pirkko KL. Baseline psychosocial predictors of survival in localized melanoma. J Psychosom Res. 2007;63(1):9–15.
13. Maddineni SB, Lau MM, Sanger VK. Identifying the needs of penile cancer sufferers: a systematic review of the quality of life, psychosexual and psychosocial literature in penile cancer. BMC Urol. 2009;9:8.
14. Meyer CP, Hanske J, Sammon JD. Prediction and predicament: complications after inguinal lymph node dissection for penile cancer. BJU Int. 2015;116(2):165.
15. Bullen K, Edwards S, Marke V, Matthews S. Looking past the obvious: experiences of altered masculinity in penile cancer. Psychooncology. 2010;19(9):933–40.
16. Pietrzak P, Corbishley C, Watkin N. Organ-sparing surgery for invasive penile cancer: early follow-up data. BJU Int. 2004;94(9):1253–7.
17. Hegarty PK, Eardley I, Heidenreich A, McDougal WS, Minhas S, Spiess PE, et al. Penile cancer: organ-sparing techniques. BJU Int. 2014;114:799–805.
18. Opjordsmoen S, Waehre H, Aass N, Fossa SD. Sexuality in patients treated for penile cancer: patients' experience and doctors' judgement. BJU Int. 1994;73(5):554–60.
19. Witty K, Branney P, Evans J, Bullen K, White A, Eardley I. The impact of surgical treatment for penile cancer – patients' perspectives. Eur J Oncol Nurs. 2013;17(5):661–7.
20. Kieffer JM, Djajadiningrat RS, van Muilekom EA, Graafland NM, Horenblas S, Aaronson NK. Quality of life for patients treated for penile cancer. J Urol. 2014;192(4):1105–10.
21. Davis JW, Schellhammer PF, Schlossberg SM. Conservative surgical therapy for penile and urethral carcinoma. Urology. 1999;53(2):386–92.
22. Siow WY, Cheng C. Penile cancer: current challenges. Can J Urol. 2005;12(1):18–23.
23. Gopman JM, Djajadiningrat RS, Baumgarten AS, Espiritu PN, Horenbias S, Zhu Y, et al. Predicting postoperative complication inguinal lymph node dissection for penile cancer in an international multicentre cohort. BJU Int. 2015;116(2):196–201.
24. Ravi R. Morbidity following groin dissection for penile cancer. Br J Urol. 1993;72(6):941–5.
25. Kroon BK, Lont AP, Valdes Olmos RA, Nieweg OE, Horenblas S. Morbidity of dynamic sentinel node biopsy in penile cancer. J Urol. 2005;173(3):813–5.
26. Janda M, Obermair A, Cella D, Crandon AJ, Trimmel M. Vulvar cancer patients' quality of life: a qualitative assessment. Int J Gynecol Cancer. 2004;14(5):875–81.

27. Janda M, Obermair A, Cella D, Perrin LC, Nicklin JL, Ward BG, et al. The functional assessment of cancer-vulvar: reliability and validity. Gynecol Oncol. 2005;97(2):568–75.
28. de Melo Ferreira AP, de Figuiredo EM, Lima RA, Candido EB, de Castro Monteiro MV, de Figueiredo Franco TMR, et al. Quality of life in women with vulvar cancer submitted to surgical treatment: a comparative study. J Obstet Gynecol. 2012;165(1):91–5.
29. Oonk MH, van Os MA, de Bock GH, de Hullu JA, Ansink AC, van der Zee AG. A comparison of quality of life between vulvar cancer patients after sentinel lymph node procedure only and inguinofemoral lymphadenectomy. Gynecol Oncol. 2009;113(3):301–5.
30. Van der Velden K, Ansink A. Primary groin irradiation vs primary groin surgery for early vulvar cancer. Cochrane Database Syst Rev. 2001;4:CD002224.
31. Barton DP. The prevention and management of treatment related morbidity in vulval cancer. Best Pract Res Clin Obstet Gynaecol. 2003;17:683–701.
32. Forner DM, Dakhil R, Lampe B. Quality of life and sexual function after surgery in early stage vulvar cancer. Eur J Sug Oncol. 2014;41(1):40–5.
33. Emori M, Hamado K, Omori S, Joyama S, Tomita Y, Hashimoto N, et al. Surgery with vascular reconstruction for soft-tissue sarcomas in the inguinal region: oncologic and functional outcomes. Ann Vasc Surg. 2012;26(5):693–9.
34. Wiener L, Battles H, Bernstein D, Long L, Derdak J, Mackall CL, et al. Persistent psychological distress in long-term survivors of pediatric sarcoma: the experience at a single institution. Psychooncology. 2006;15(10):898–910.
35. Tang MH, Pan DJ, Castle DJ, Choong PF. A systematic review of the recent quality of life studies in adult extremity sarcoma survivors. Sarcoma. 2012;2012:171342.
36. Paredes T, Canavarro MC, Simoes MR. Anxiety and depression in sarcoma patients: emotional adjustment and its determinants in the different phases of disease. Eur J Oncol Nurs. 2011;15(1):73–9.

Index

K.A. Delman, V.A. Master (eds.), *Malignancies of the Groin*,
https://doi.org/10.1007/978-3-319-60858-7

N

O

P

Q

MIX
Papier aus verantwortungsvollen Quellen
Paper from responsible sources
FSC® C105338

FSC
www.fsc.org

If you have any concerns about our products, you can contact us on
ProductSafety@springernature.com

In case Publisher is established outside the EU, the EU authorized representative is:
Springer Nature Customer Service Center GmbH
Europaplatz 3, 69115 Heidelberg, Germany

Printed by Libri Plureos GmbH
in Hamburg, Germany